Innovation, Technology, and Knowledge Management

Series Editor
Elias G. Carayannis

More information about this series at http://www.springer.com/series/8124

Tugrul U. Daim • Jisun Kim • Ibrahim Iskin
Rimal Abu Taha • Kevin C. van Blommestein
Editors

Policies and Programs for Sustainable Energy Innovations

Renewable Energy and Energy Efficiency

Springer

Editors
Tugrul U. Daim
Engineering and Technology Management
Portland State University
Portland, OR, USA

Ibrahim Iskin
Portland State University
Portland, OR, USA

Kevin C. van Blommestein
Portland State University
Portland, OR, USA

Jisun Kim
Portland State University
Portland, OR, USA

Rimal Abu Taha
Portland State University
Beaverton, OR, USA

ISSN 2197-5698 ISSN 2197-5701 (electronic)
Innovation, Technology, and Knowledge Management
ISBN 978-3-319-16032-0 ISBN 978-3-319-16033-7 (eBook)
DOI 10.1007/978-3-319-16033-7

Library of Congress Control Number: 2015933146

Springer Cham Heidelberg New York Dordrecht London

Printed on acid-free paper

Springer International Publishing AG Switzerland is part of Springer Science+Business Media (www.springer.com)

Series Foreword

The Springer book series *Innovation, Technology, and Knowledge Management* was launched in March 2008 as a forum and intellectual, scholarly "podium" for global/local, transdisciplinary, transsectoral, public–private, and leading/"bleeding" edge ideas, theories, and perspectives on these topics.

The book series is accompanied by the Springer *Journal of the Knowledge Economy*, which was launched in 2009 with the same editorial leadership.

The series showcases provocative views that diverge from the current "conventional wisdom" that are properly grounded in theory and practice, and that consider the concepts of ***robust competitiveness***,[1] ***sustainable entrepreneurship***,[2] and ***democratic capitalism***,[3] central to its philosophy and objectives. More specifically, the aim of this series is to highlight emerging research and practice at the dynamic intersection of these fields, where individuals, organizations, industries, regions, and nations are harnessing creativity and invention to achieve and sustain growth.

[1] We define *sustainable entrepreneurship* as the creation of viable, profitable, and scalable firms. Such firms engender the formation of self-replicating and mutually enhancing innovation networks and knowledge clusters (innovation ecosystems), leading toward robust competitiveness (E. G. Carayannis, *International Journal of Innovation and Regional Development* 1(3), 235–254, 2009).

[2] We understand *robust competitiveness* to be a state of economic being and becoming that avails systematic and defensible "unfair advantages" to the entities that are part of the economy. Such competitiveness is built on mutually complementary and reinforcing low-, medium-, and high-technology and public and private sector entities (government agencies, private firms, universities, and nongovernmental organizations) (E. G. Carayannis, *International Journal of Innovation and Regional Development* 1(3), 235–254, 2009).

[3] The concepts of *robust competitiveness and sustainable entrepreneurship* are pillars of a regime that we call "*democratic capitalism*" (as opposed to "popular or casino capitalism"), in which real opportunities for education and economic prosperity are available to all, especially—but not only—younger people. These are the direct derivatives of a collection of topdown policies as well as bottom-up initiatives (including strong research and development policies and funding, but going beyond these to include the development of innovation networks and knowledge clusters across regions and sectors) (E. G. Carayannis and A. Kaloudis, *Japan Economic Currents*, p. 6–10 January 2009).

Books that are part of the series explore the impact of innovation at the "macro" (economies, markets), "meso" (industries, firms), and "micro" levels (teams, individuals), drawing from such related disciplines as finance, organizational psychology, research and development, science policy, information systems, and strategy, with the underlying theme that for innovation to be useful it must involve the sharing and application of knowledge.

Some of the key anchoring concepts of the series are outlined in the figure below and the definitions that follow (all definitions are from E. G. Carayannis and D. F. J. Campbell, *International Journal of Technology Management*, 46, 3–4, 2009).

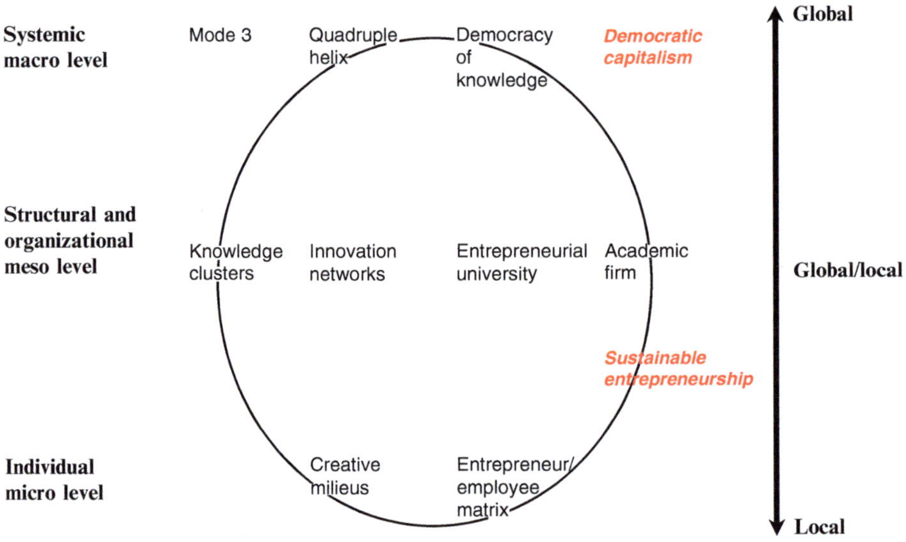

Conceptual profile of the series *Innovation*, *Technology*, and *Knowledge Management*

- The "Mode 3" Systems Approach for Knowledge Creation, Diffusion, and Use: "Mode 3" is a multilateral, multinodal, multimodal, and multilevel systems approach to the conceptualization, design, and management of real and virtual, "knowledge-stock" and "knowledge-flow," modalities that catalyze, accelerate, and support the creation, diffusion, sharing, absorption, and use of cospecialized knowledge assets. "Mode 3" is based on a system-theoretic perspective of socio-economic, political, technological, and cultural trends and conditions that shape the coevolution of knowledge with the "knowledge-based and knowledge-driven, global/local economy and society."
- Quadruple Helix: Quadruple helix, in this context, means to add to the triple helix of government, university, and industry a "fourth helix" that we identify as the "media-based and culture-based public." This fourth helix associates with "media," "creative industries," "culture," "values," "life styles," "art," and perhaps also the notion of the "creative class."

- Innovation Networks: Innovation networks are real and virtual infrastructures and infratechnologies that serve to nurture creativity, trigger invention, and catalyze innovation in a public and/or private domain context (for instance, government–university–industry public–private research and technology development coopetitive partnerships).
- Knowledge Clusters: Knowledge clusters are agglomerations of cospecialized, mutually complementary, and reinforcing knowledge assets in the form of "knowledge stocks" and "knowledge flows" that exhibit self-organizing, learning-driven, dynamically adaptive competences, and trends in the context of an open systems perspective.
- Twenty-First Century Innovation Ecosystem: A twenty-first century innovation ecosystem is a multilevel, multimodal, multinodal, and multiagent system of systems. The constituent systems consist of innovation metanetworks (networks of innovation networks and knowledge clusters) and knowledge metaclusters (clusters of innovation networks and knowledge clusters) as building blocks and organized in a self-referential or chaotic fractal knowledge and innovation architecture,[4] which in tum constitute agglomerations of human, social, intellectual, and financial capital stocks and flows as well as cultural and technological artifacts and modalities, continually coevolving, cospecializing, and cooperating. These innovation networks and knowledge clusters also form, reform, and dissolve within diverse institutional, political, technological, and socioeconomic domains, including government, university, industry, and nongovernmental organizations and involving information and communication technologies, biotechnologies, advanced materials, nanotechnologies, and next-Generation energy technologies.

Who is this book series published for? The book series addresses a diversity of audiences in different settings:

1. *Academic communities:* Academic communities worldwide represent a core group of readers. This follows from the theoretical/conceptual interest of the book series to influence academic discourses in the fields of knowledge, also carried by the claim of a certain saturation of academia with the current concepts and the postulate of a window of opportunity for new or at least additional concepts. Thus, it represents a key challenge for the series to exercise a certain impact on discourses in academia. In principle, all academic communities that are interested in knowledge (knowledge and innovation) could be tackled by the book series. The interdisciplinary (transdisciplinary) nature of the book series underscores that the scope of the book series is not limited a priori to a specific basket of disciplines. From a radical viewpoint, one could create the hypothesis that there is no discipline where knowledge is of no importance.
2. *Decision makers—private/academic entrepreneurs and public* (*governmental, subgovernmental*) actors: Two different groups of decision makers are being

[4] E. G. Carayannis, *Strategic Management of Technological Learning,* CRC Press, 2000.

addressed simultaneously: (1) private entrepreneurs (firms, commercial firms, academic firms) and academic entrepreneurs (universities), interested in optimizing knowledge management and in developing heterogeneously composed knowledge-based research networks; and (2) public (governmental, subgovernmental) actors that are interested in optimizing and further developing their policies and policy strategies that target knowledge and innovation. One purpose of public *knowledge and innovation policy* is to enhance the performance and competitiveness of advanced economies.

3. *Decision makers in general:* Decision makers are systematically being supplied with crucial information, for how to optimize knowledge-referring and knowledge-enhancing decision-making. The nature of this "crucial information" is conceptual as well as empirical (case-study-based). Empirical information highlights practical examples and points toward practical solutions (perhaps remedies), conceptual information offers the advantage of further driving and further-carrying tools of understanding. Different groups of addressed decision makers could be decision makers in private firms and multinational corporations, responsible for the knowledge portfolio of companies; knowledge and knowledge management consultants; globalization experts, focusing on the internationalization of research and development, science and technology, and innovation; experts in university/business research networks; and political scientists, economists, and business professionals.
4. *Interested global readership:* Finally, the Springer book series addresses a whole global readership, composed of members who are generally interested in knowledge and innovation. The global readership could partially coincide with the communities as described above ("academic communities," "decision makers"), but could also refer to other constituencies and groups.

Washington, DC, USA Elias G. Carayannis

Preface

As the world looks for alternative energy resources, renewable energy and energy efficiency shine as the most appropriate alternative solutions. However, there are technical, political, social, economic, and environmental challenges. This volume investigates these challenges.

Chapters 1 through 3 review policies which are likely to increase the adoption of renewable energy technologies. These three chapters provide a good introduction to the different policies for different types of energy sources as well as an approach on how to evaluate these policies for different cases.

Chapters 4 through 9 present cases around the evaluation of renewable energy. Chapter 4 presents a wind energy case from Pakistan. This case presents a technical approach to assess the wind potential. Chapter 5 presents a technology assessment framework evaluating conversion of solid waste to energy. Chapters 6 and 7 provide two different perspectives on biofuels. They review national and local cases. Chapter 8 presents a case on residential solar electric systems. The chapter provides a framework on evaluating the adoption phenomenon. Finally, Chap. 9 provides a case on portfolio optimization in the electricity market and thus provides us with a financial perspective.

Chapters 10 and 11 review energy efficiency technologies and programs as well as approaches to evaluate them. These two chapters provide a solid introduction to the energy efficiency concept through a local case and a comprehensive review of the literature.

Chapters 12 through 15 present different cases in the energy efficiency area. Chapters 12 and 13 present two different approaches to evaluate clothes dryers which are major consumers of power in the residential market. One of the approaches relies on expert judgment quantification while the other one on intelligence through patent analysis and modeling based on such intelligence. Chapter 14 presents a case on furnace fan motors, while Chap. 15 on insulation material for home construction.

We hope that this volume will provide some guidance to those who are just planning to explore renewable energy and energy efficiency as alternatives.

Portland, OR Tugrul U. Daim
Jisun Kim
Ibrahim Iskin
Beaverton, OR Rimal Abu Taha
Portland, OR Kevin C. van Blommestein

Contents

Chapter 1
Review of Policies Toward the Acceleration of the Adoption of Renewable Energy Technologies

Rimal Abu Taha and Tugrul U. Daim

Abstract This chapter reviews energy policy supporting diffusion of renewable energy (RE) and describes different types of available RE. The increased level of carbon dioxide is the main cause of the "Global Warming Effect." One suggested solution to global warming is to replace the current energy technologies with alternatives that have similar or even better performance, but do not emit greenhouse gases (GHGs). Beside environmental concerns, energy availability concerns and political pressure have prompted governments to look for alternative energy resources that can minimize the undesirable effects for current energy systems. Shifting away from the conventional fuel resources and increasing the percentage of generated electricity from renewable resources is an opportunity to guarantee lower (CO_2) emissions and to create better economic opportunities for the United States. RE resources offer a good alternative for the current fossil fuel system with its minimal impact on the environment and unlimited availability. Even with the fact that a diversity of renewable energy resources available in the United States and the development of the technologies themselves are more mature, the use of such resources is still very limited in the United States, but as the fossil fuel system is deteriorating with price increase and supply scarcity the transition to a new era of renewable energy is inevitable (Energy Policy 31:353–367, 2003). Policy can play an important role in promoting the penetration of renewable energies (Energy Policy 39:4726–4741, 2011). This chapter discusses the available policies that can promote RE adoption and deployment as well as the available technologies and literature assessing that adoption.

R. Abu Taha (✉)
Portland State University, 15606 SW Snowy Owl Lane, Beaverton, OR 97007, USA
e-mail: rimalat@yahoo.com

T.U. Daim
Engineering and Technology Management, Portland State University,
1900 SW 4th Avenue 50-02, Portland, OR 97007, USA

T.U. Daim et al. (eds.), *Policies and Programs for Sustainable Energy Innovations*, Innovation, Technology, and Knowledge Management,
DOI 10.1007/978-3-319-16033-7_1

With the foreseeable need for renewable energy (RE) adoption in the power generation sector, many literatures have emerged discussing the need for wide-ranging policy regime to facilitate this adoption. Three main directions were found in the literature: (1) literature explored the main players in policy formulation and their roles (government, private investors, and public), (2) literature explaining the characteristics and impact of different energy policies on one or more RE, (3) studies that demonstrated barriers and benefits of RE technology adoption and analyzed policy targets to maximize deployment and development.

For RE policy to be successful, understanding the economic, legal, and institutional aspects for a technology is needed for increasing diffusion in power sector. Norberg-Bohm discussed an in-depth history of technological development for four electric power technologies: wind turbines, solar photovoltaic, gas turbines, and atmospheric fluidized bed combustion [3]. The paper explained the role of government in technology commercialization and what policies were implemented in the United States (supply-push and demand-pull approaches). Federal policies mainly take the form of R&D funding or financial incentives. Kobos et al. argued that without institutional support, emerging energy technologies are limited by their costs from adoption and reaching consumer market [4]. Their analysis explored the relationship between RD investments, energy cost reduction, and market penetration. The methodology used here is to combine two theoretical frameworks: the estimate of energy cost as a function of cumulative installed capacity (a learning by doing factor) and cumulative RD&D expenditures (a learning by searching factor). The study concluded that institutional policy instruments play an important role for renewable energy technologies to reach sufficient cost reductions and further market adoption [4]. Anderson provided a review of policies and noted that investment followed by R&D activities and environmental policies are key factors for RE development [5]. Jaffe et al. provided a background of environmental policy in the United States and emphasized on a need for a public policy that encourage emission reductions and environmental aspects of a technology [6].

The European Union has emerged in the literature as a leader in RE adoption and formulating energy policy with a target of promoting RE resources in its member countries through establishing an aggressive energy policy [7, 8]. Different studies presented a comparison between two policy mechanisms that are under consideration in the EU to promote RE adoption: tradable green certificates (TGC) and feed-in tariffs (FITs). It was found that the TGC system is likely to be less effective and less efficient than the FIT and that FIT deliver larger and faster penetration of RE than TGC, at lower cost [7, 9, 10]. Söderholm and Pettersson presented an analysis for the effect of policy system in Sweden on increasing offshore wind power deployment [11]. Toke also analyzed offshore wind energy in the UK and the policy effectiveness [12]. In both papers, it was found that there is a kind of policy for promoting such technology as the financial incentives but the public support and acceptance for the slightly increased electricity price is more likely to help achieve targeted capacity goals. This supported the findings in Zografakis et al. of the high attention policy makers should give in planning and implementing renewable energy projects for how much more customers are willing to pay for renewable energy more than for fossil fuel energy [13]. For a summary of renewable energy policies and methods used for assessment, please refer to Table 1.1.

Table 1.1 Summary of energy policy and corresponding literature

Policy	Literature	Methodology
R&D funding	[14]	Technology S-curves to analyze RE performance and R&D investments
	[4]	Experience curves for energy cost as a function of cumulative capacity and R&D investments
	[15]	Comparison between R&D funding between different countries and its effect on wind adoption
	[16]	Patent analysis to investigate the effect of new knowledge on energy investment decisions
Tax credits, grants, and incentives	[17]	Empirical study
	[18]	Case study
	[17]	Quantitative cash flow analysis
Cap and trade	[19]	Case study
	[20]	Scenario analysis
RPS	[18, 21]	Case study
	[22]	Empirical research with incentives as indicator of magnitude and capacity
	[23]	Fixed-effect model to evaluate the effectiveness of RPS and percentage of RE generation
	[24]	Case study
	[25]	Scenario analysis, numerical simulation
	[26]	Linear regression
REC	[27]	Comparison between RPS requirement of different states and the effect of integrating REC
	[24]	Case study
Feed in tariff	[8]	Comparative study for different energy policies adopted in European countries
	[28]	Case study
	[29]	Case study for different model to structure FIT
Mandatory green power option	[30]	Fixed-effect model
	[26]	Linear regression

Other studies have analyzed the status of diffusion of renewable energies and its implications. Cantono and Silverberg developed a network model of new technology diffusion to analyze the relationship between the diffusion of a new technology, learning economies, and financial support [31]. Rao and Kishore took another path where they utilized diffusion models to understand the development of RET and the barriers affecting that adoption [32]. To shift to a new energy system involves changes of the whole ecosystem nature regarding structural, organizational, economic, and social entities. Tsoutsos and Stamboulis argued that the basic of a successful renewable energy policy is to consider that renewable technologies are a different system from conventional resources and have different barriers and stimulants [33]. With this recognition of system complexity, there is a need for a tool that can integrate a large number of variables in the energy policy system and have results and analysis presented in a usable manner. Bassi and Shilling presented a computer simulation model (Threshold 21) as a tool for comprehensive national policy planning that can analyze the long-term implications of policies and strategies including their positive and negative impacts [34]. Running the simulation model for different scenarios for multiple elements in the United States (society, economy, environment) revealed that the direction in policy making should be directed to enforcing new government regulations to manage energy consumption, focus on developing new clean technologies, and improve energy efficiency and conservation while the current policy trends in the United States will only increase the use of foreign fuels which will lead to more price fluctuations.

1.1 Renewable Energy Policy in the United States

In spite of the federal government efforts in pursuing solutions to deploy RE through several policy regimes [35], it looks like those efforts lack coordination between states themselves and a comprehensive long-term planning which led state and local governments to step in with several policy approaches and distinguish themselves as renewable energy policy establishers [23, 36]. Surveying the literature revealed three main directions or focus for policy intended to increase RE adoption:

- Mandated regulations:
  Those kinds of policies are mandatory regulations that power generators or other stakeholders like customers should comply with. This direction of policies is targeted to maintain a certain level of renewable energy sources in the power generation energy mix or keep GHG emissions under a certain value.
- Market-based policy:
  Renewable energies are supposed to establish themselves in an already mature fossil fuel market with low prices, resource availability, and already accessible technologies. These policies are designed to facilitate communication between generator and consumer as well as establish a competitive price for renewables.
- Financial based:
  Policies designed to overcome the financial obstacles facing renewable energies to increase development and deployment, both consumers and developers benefit from such policies.

1.1.1 Federal Policy

It is necessary to provide incentives for the development and diffusion of the renewable energy in different fields. The federal Energy Policy Act of 2005 set up a tax deduction for energy-efficient commercial buildings in the United States. A tax deduction is given to owners of new or existing buildings who install (1) interior lighting; (2) building envelope, or (3) heating, cooling, ventilation, or hot water systems that reduce the building's total energy and power cost by 50 % or more [37]. This policy act had also impacted R&D efforts for different types of renewable energy which had increased the installed capacity of several resources [38].

- Financial incentives:
  Production Tax Credit (PTC) and Federal Investment Tax Credit (ITC) are federal policies in the form of financial incentives that provides an amount (cents/kWh) to private investors and investor-owned utilities that deploy RE resources in their production. PTC has been found that it had a great effect on wind energy deployment and development in the United States [2, 18, 26]. Under The American Recovery and Reinvestment Act of 2009 ("ARRA 2009") projects that are eligible to receive the PTC can instead choose the ITC, which is a credit of 30 % of the cost of development depending on the source of energy used [17]. There are other financial incentives in the form of loans and grants to fund projects and buy equipment. Clean Renewable Energy Bonds (CREBs) are another financial incentive that can finance renewable energy projects. Eligible technologies are generally the same as that used for the federal renewable energy PTC.

- Carbon tax or carbon cap and trade:
  This is a federal environmental policy that is designed to gradually reduce CO_2 and other GHG emissions in a cost-effective manner. Under this policy, a limited amount called "cap" is allocated on each large-scale generators for the emissions produced and issues permits as a share of that cap, the holders for these permits can trade them to other ammeters which create a market for green energy. The caps will become lower over time which leads to less and less GHG emissions until the desired reduction goal are met [39]. A cap and trade program has been proposed by the Obama administration but hasn't passed the congress yet. There are similar programs endorsed by the Clean Air Act of 1990 [40], which target other emissions and it had met their intended emission levels. Examples of these policies are nationwide Acid Rain Program and the regional NOx Budget Trading Program in the Northeast and the Clean Air Interstate Rule (CAIR).

1.1.2 State and Local Government Policy

Several states have adopted different energy policies to promote the adoption of RE [41]. Roach presented a survey of current energy policy in the United States and it noted that since each state is different, they should be allowed to tailor their specific policy according to their specific needs [19]. State government actions can take different forms like financial incentives, direct regulations, and regulatory changes. Menz and Vachon presented an empirical analysis on the effect of RE policy to determine which policy has more effect on wind energy capacity installed [26]. The paper analyzed the effect of different state policies, namely, renewable portfolio standards (RPS), fuel generation disclosure rules, mandatory green power options, and public benefits fund on wind power development. The results of this study showed that RPS is an effective policy for promoting RE deployment. Carley builds on Menz and Vachon 2006 efforts and tests directly the relation between RE generation percentage across states and state RPS and electricity-based RE policy incentives with an empirical investigation using casual effect models (fixed effects vector decomposition model, FEVD) [23]. In a report by National Renewable Energy Laboratory, the effect of various RPS and cap-and-trade policy options on the US. electricity sector was examined. The analysis uses the simulation model ReEDS to estimate the least-cost expansion of electricity generation capacity and examine the impact of an emissions cap as well as scenario analysis. The report also examines the effects of merging RPS policy with the emissions caps [20].

- Renewable portfolio standards (RPS):
  Renewable portfolio standard is a state policy that requires a predetermined percentage of the electricity produced or sold in a state to be from qualifying renewable energy resources. Twenty-nine states plus the District of Columbia have already established RPS [42] but the regulation is different from one state to another. Eight other states have set voluntary renewable goals where it only applies to private investors and it is nonbinding [43]. Electricity providers are given the chance to satisfy the RPS requirements by purchasing renewable energy credits (REC) from other qualifying producers. Kydes analyzed the impact of imposing a federal RPS of 20 % nonhydropower renewable generation levels on the US energy markets by 2020 [44]. The analysis was conducted by using the December 2001 version of the National Energy Modeling System (NEMS) of the Energy Information Administration (EIA) and the assumptions and results of the Annual Energy Outlook 2002 (AEO2002) reference case. The conclusion of the paper was that this policy seems to be effective in encouraging the adoption of renewable energy technologies. With this agreement on the benefits of RPS, surprisingly Sovacool and Cooper disagree. They argue that state RPS has several deficiencies and the transmission to a federal mandatory level would provide more clarity and unity in objectives and initiate a national commitment to RE generation [45].

- Financial incentives:
  State governments offer financial incentives to encourage the adoption of RE and their main target is to overcome the financial obstacles that make RE investment unattractive to investors. These financial incentives can have several types; it includes tax deductions and credits, subsidies (grants, loans, etc.).

- Feed in tariff:
  A FIT is an energy policy dedicated to support the development of new renewable power generation. Under this policy in the United States, generators are paid a cost-based price for the renewable electricity they produce where utilities are required to buy electricity from eligible renewable energy generators. Fifteen states and three other municipal utilities have considered FIT legislation [28].

- Green certificate market:
  This is a market based policy that is designed to open markets for power generated from renewable resources. This policy defines an obligation from consumers to utilities to purchase a certain amount of electricity generated from renewable energy sources. This market assurance is an incentive to supply green certificates since they can increase their revenue [7, 10, 46].
- Mandatory green power:
  It is a state policy that obligates utilities to offer their customers the option to get their electricity from renewable resources. Kneifel estimated the effect of different state policies in all 50 states using state fixed-effect model. It was found from the analysis that RPS, funding, and green power option have a great impact on increasing RE capacity in each state [30].

1.2 Alternative Energy Technologies

Alternative energy is a general term that refers to any source of energy that is going to supplement or replace current energy sources. The use of this term has changed along the history where different types of energy resources had replaced others like coal replacing wood as a source for heat and energy. Alternative energy now refers to any source of energy that replaces the current fossil fuel system without having the consequences on the environment or energy security as fossil fuels do [47]. Different factors are considered when searching for sustainable alternative energy sources such as availability, cost, and environmental impact. The combustion of fossil fuels is accompanied by emissions of large quantities of carbon dioxide into the atmosphere which enhances the greenhouse effect and consequently global warming and climate change. There is a pressing need to develop a highly efficient energy deployment processes to substitute current energy sources and control the current CO_2 matter [48]. Alternative energies can be utilized to substitute fossil fuels in different capacities such as power generation or transportation. Although some new technologies are not renewable such as new energy efficiency technologies, they can still be considered alternatives as it minimizes the effects of fossils.

1.2.1 Energy Efficiency Technologies

Many different stakeholders can be involved in the evaluation and preparation of energy policy. Government can play a key role in formulating the policy for any new energy technology adoption and commercialization path [49]. Chai and Zhang explored the technological and policy issues in China that facilitates the adoption of a more sustainable energy system, energy efficiency technologies, and renewable energy technologies [50]. It was found that for the transition to a sustainable energy system additional actions can be done such as R&D support of new innovative sustainable energy technologies, improving manufacturing capacity of these energy technologies, and offering more economic incentives for research and development. On the other hand, Luiten et al. investigated and analyzed the effect of government intervention as R&D support for four different energy efficiency industrial processes and its innovation [51]. The analysis of four industrial processes revealed that although government support for R&D is an important policy in motivating energy efficiency technologies, the effect of this policy is dependent on the stakeholders in the industry itself. Noailly and Batrakova investigate the relation between technological innovation for energy efficiency technologies and energy policy in the Dutch building sector [52]. By developing a patent analysis method, the authors aimed to explain the innovation in energy efficiency technologies in buildings and presented a complete historical overview of the energy policy instruments in Netherlands as a case study. A review of the diverse Dutch policy initiatives shows that environmental policy has greater effect on adopting energy efficiency technologies rather than financial incentives.

1.2.2 Alternative Transportation Fuels

Although the transportation system in the United States is still highly dependent on petroleum, energy security and environmental issues have been pushing to consider alternative transportation fuels. Hydrogen is considered one of the possible "good fuels" to replace current transportation fuel system [53]. Alternative fuels can be ethanol, hydrogen, biodiesel, and others, see Fig. 1.1. Ethanol is still the most

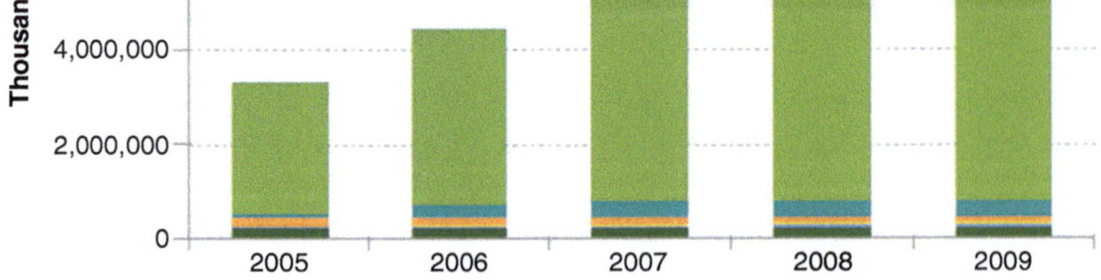

Fig. 1.1 Consumption of alternative fuel in the United States [54]

produced fuel in the United States. However, biodiesel production in the United States has seen an enormous increase in the last decade.

Meyer and Winebrake evaluated the diffusion of hydrogen technology in the automobile sector in the united States and analyzed barriers affecting this diffusion as well as related policy [55]. The analysis utilized system dynamic modeling and scenario analysis to evaluate the diffusion of hydrogen technology (vehicles and refueling) as complementary goods. Hydrogen refueling infrastructures are considered complementary goods, and complementary goods are goods that must be used together since they operate as a system (DVD players and DVDs). The results prevailed that both hydrogen vehicles and needed infrastructure should get attention by policy makers and given more incentives to achieve market penetration.

1.2.3 *Renewable Energy*

Even though wind and solar technologies have exhibited large growth rates [56], nonhydropower renewable in total (wind, solar, geothermal, biomass) still count for a small percentage of total US power consumption [57], (see Fig. 1.2 and Table 1.2).

There is a need for an energy system that helps to overcome barriers for RE adoption and guarantee the maximum benefits of such adoption [58]. Holmes and Papay analyzed the US energy system and noted that national targets of 10 % by 2020 and 20 % by 2035 of the US power supply to be from nonhydro renewable can be achieved given the coordination between policy regimes, technology development, and capital allocation [59]. Kajikawa et al. opted to understand the research structure of renewable energy in two studies [60, 61]. The first explored what emerging technologies are in the field of RE by using citation network analysis. Their

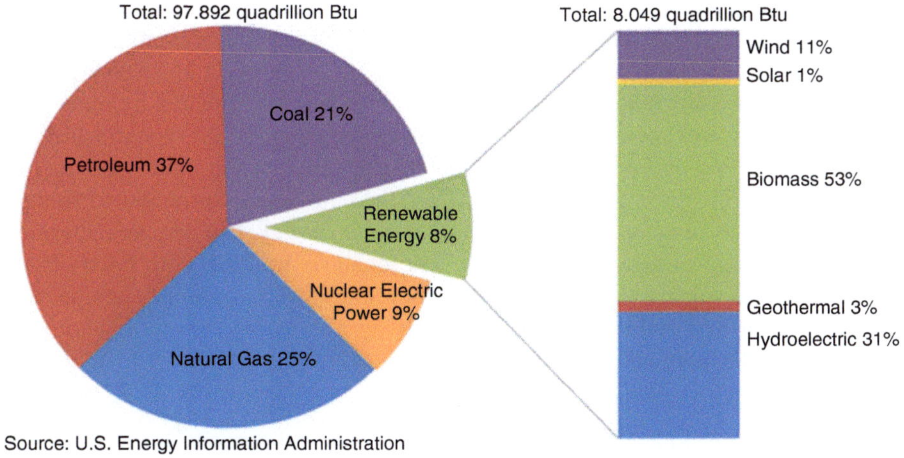

Fig. 1.2 Renewable energy consumption in the US energy supply, 2010 [56]

Table 1.2 Renewable energy sources

Energy	Description
Solar energy	Solar energy technology is a variety of technologies that have been developed to harness the solar energy from the sun and convert it to electricity or heat. It includes different technologies: concentrating solar power systems, photovoltaic systems, solar hot water, and others. Solar power can be used in either large-scale applications or small residential application
Wind energy	It is the technology to harness the power of wind by wind turbines. Wind turbines can be used as stand-alone applications for water pumping or as part of utility power grid to generate electricity. Wind energy is the fastest growing renewable energy technology worldwide
Hydroelectric	The technology to benefit from the running water power. It can be harnessed by building large dams on natural reservoirs and used mainly for electricity generation
Biomass	Biomass energy is the energy from plants and plant-derived materials. Wood is still the largest biomass energy resource today, but other sources of biomass can also be used: food crops, residues from agriculture or forestry, and industrial wastes. Biomass can be converted directly into liquid fuels (biofuels) or used to generate electricity
Geothermal	Geothermal energy is taking advantage of heat from the earth. This heat can be drawn from several sources: hot water or steam reservoirs deep in the earth. It can be used on both large and small scales, either to drive generators and produce electricity or to provide heating and cooling in homes and other buildings
Ocean	The oceans cover over 75 % of our planet. The ocean energy can be harnessed from the tidal waves and ocean temperature difference to generate electricity

analysis found out that the fuel cell and solar energy are rapidly growing domains in energy research. The second paper looked more in details at the biomass as a sustainable and renewable energy growing rapidly. In this paper, they performed a citation network analysis of scientific publications to unfold the current structure of biomass research. Their work revealed the taxonomic structure of biomass research especially focusing on biofuel and bio-energy research. Shen et al. examined how different policy goals (energy, environmental, economic) is satisfied by renewable energy resources in Taiwan. The analysis utilized AHP model and scenario analysis and concluded that hydropower, wind, and solar energy are the three technologies that could meet the three policy goals [62]. On the other hand, Shrimali and Kniefel analyzed the US state policy to determine which policy had led to increasing installed renewable energy capacity [2]. The analysis utilized state fixed-effects model and case studies for four types of renewables (wind, biomass, geothermal, and solar). For a summary of literature studying alternative energy technologies adoption and the methods utilized to assess this adoption, please refer to Table 1.3.

- Wind energy:
  China is the world leader in cumulative installed wind capacity followed by the United States. Wind energy is the fastest growing renewable energy in the United States with Texas leading the states in wind installation in 2010 [54].

Table 1.3 Summary of technologies studied and methodologies used for evaluating their adoption

Technology	Literature	Methodology
Alternative fuels	[55]	System dynamic
	[67]	Case study
Energy efficiency technologies	[52]	Patent analysis
	[51]	Case studies
	[69]	AHP/DEA
Wind (offshore, wind farms)	[70]	MCDM
	[15]	Empirical study as a comparison between wind installed capacity in United States, Japan, and Europe in terms of R&D funding and policy measure
	[71, 18]	Case study
	[26]	Linear regression
	[72]	Bass diffusion model installed capacity
	[73]	Case study
Solar (PV, CTP)	[74]	Experience curves
	[75–77]	MCDM
	[29]	Case study, levelized cost method
	[78]	Bass diffusion model
Other renewables (biomass, geothermal)	[67, 79]	
	[80]	Case study

Bird et al. analyzed several drivers and policies that contribute to wind energy development in the United States and noted that it is not feasible to determine one single driver for wind power development but several drivers function as a package and influence one another's effectiveness such as renewable portfolio standard in combination with financial incentives as well as developing the market for green power [18]. Although considerable investments are necessary for initial stages of RE technologies used in power generation, wind turbines are already economically competitive under favorable conditions, and wind power subsidies have resulted in a rapid growth in the number of wind turbines installed. Niji discussed the possibility for cost reductions in renewable energy technologies on the long run. Experience curves were utilized to analyze the prospects for diffusion and adoption of renewable energy technologies, with more stress on wind turbines and photovoltaic (PV) modules [63]. Loiter and Norberg-Bohm also presented a study on the development of wind power in the United States [64]. The primary conclusion is that demand-side policies are needed to encourage diffusion of wind energy and the innovation in the technology.

- Solar energy:
  Solar energy utilization for electricity generation has grown consistently by about 20 % yearly over the past 20 years. This increase has happened as a result of manufacturing technology improvements, increasing efficiency of solar modules, and economies of scale which lead to decrease in cost. European countries and Japan that have more compelling solar policies lead the world in solar PV

deployment. In the same way, the states with more solar incentives have more cumulative and annual capacity installations in 2010 (California, New Jersey, Colorado, Arizona, and Nevada) [54].

Sawyer had examined the characteristics of the first homeowners who installed solar energy systems and integrated the information in a diffusion model to anticipate future solar market penetration patterns [65]. Sawyer looked at their socioeconomic characteristics, purchase motivations, and satisfaction levels and noticed that these individuals conform to the "early adopter" type identified in innovation diffusion research and not the "innovator" type that would be expected at this early stage of commercialization. In conclusion, he suggested that rapid market penetration is possible if the high capital costs characteristic of solar energy systems are effectively addressed.

- Hydropower:
  Hydropower remains the largest and oldest source of renewable energies. Although all the states utilize hydropower for power generation, the Pacific Northwest accounts for about 60 % of this production [54].

- Geothermal:
  Although the United States is a world leader in installed geothermal electricity capacity, this capacity has almost remained constant for the last decade. Geothermal is distinct from wind and solar that it has no intermittency and can provide reliable and constant power as well as direct heat [54].

- Biomass:
  Biomass energy is the energy from plants and plant-derived materials mainly agricultural residues. Biomass can be used to generate power or converted directly into liquid fuels (biofuels) for transportation. Biopower generation currently is 53 % of all renewable energy consumed in the United States [54].

  Biofuels use is of showing interest in different usage for its diversity and economy. Hoekman analyzed policy issues and adoption drivers for biofuels in the United States, describes usage trends, and emphasized on the role of R&D efforts to promote development of biofuel technologies [66]. Charles et al. addressed the issue of adopting renewable energy sources from a comprehensive public policy viewpoint [23]. The article explained the driving forces of current biofuel promotion policies, environmental and socioeconomic benefits, and problems and policy implications. Consequently, the authors concluded that the different policy instruments currently used or proposed by governments in developed nations to promote biofuels appear to be doubtful and open to discussion and debate [67].

- Ocean:
  Despite the huge potential available in ocean energy [68], this technology had not matured yet and not many projects are deployed worldwide. The interest of this energy had begun to grow recently in the United States with many prototype projects in testing stage (Table 1.3).

1.3 Conclusions

This chapter explored studies of renewable energy policy. Although there is an increasing number of literature exploring renewable energy policies [2, 22, 23, 26], the link between degree of adoption of renewable energies in the power generation sector and relative effectiveness of these policies is not yet well established. Many of the present research involves case studies that explain drivers and enablers for RE adoption [1].

Renewable energy technologies are becoming more and more important portion of the electricity supply mix, but they still face some challenges involving the large-scale deployment and commercialization. Economic consideration, availability of resources and their intermittent nature, social effects, and technology maturity impose a pressure on policy makers, scientists, and private investor to outline the way for a renewable future. Deploying renewable electricity and making it available on a scale that would make a major contribution to US electricity generation would require overcoming several obstacles like the need for huge investment and investors, development of technologies to be cost competitive, and human resources. Even with the federal government efforts and state regulations and mandates to encourage the adoption of renewable energies, this will only be successful with the help of private sector investment and support of innovation [19, 73]. The Office and Technology Assessment noted in the report (renewing our energy future) that energy policy had focused in the beginning at commercialization and operation of renewable with public support, but lately it had focused more on tax and subsidies policy that opened the way for more private investors [81].

RE adoption has so far been supported by environmental and socioeconomic dynamics and political systems. As seen from the literature, wind, solar, and biomass energy are now more deployed into the energy portfolio. However, most RE technologies (except wind and hydropower) are still at a very early stage of the diffusion. This delay of diffusion can be referred to the still undeveloped markets and ineffective policies. There is an imperative need to find a way that RE can penetrate the market in an already established energy market with cheap, available, and already mature fuel market. The ability to overcome market barriers and integrate renewable power in this market is a key factor to successful deployment of renewable electricity.

References

1. D.L. Klass, A critical assessment of renewable energy usage in the USA. Energy Policy **31**, 353–367 (2003)
2. G. Shrimali, J. Kniefel, Are government policies effective in promoting deployment of renewable electricity resources? Energy Policy **39**, 4726–4741 (2011)
3. V. Norberg-Bohm, Creating incentives for environmentally enhancing technological change: lessons from 30 years of U.S. Energy technology policy. Technol. Forecast. Soc. Change **65**, 125–148 (2000)

4. P.H. Kobos, J.D. Erickson, T.E. Drennen, Technological learning and renewable energy costs: implications for US renewable energy policy. Energy Policy **34**, 1645–1658 (2006)
5. D. Anderson, Renewable energy technology and policy for development. Annu. Rev. Energy Environ. **22**, 187–215 (2003)
6. A.B. Jaffe, R.G. Newell, R.N. Stavins, A tale of two market failures: technology and environmental policy. Ecol. Econ. **54**, 164–174 (2005)
7. A. Bergek, S. Jacobsson, Are tradable green certificates a cost-efficient policy driving technical change or a rent-generating machine? Lessons from Sweden 2003–2008. Energy Policy **38**, 1255–1271 (2010)
8. N.I. Meyer, Learning from wind energy policy in the EU: lessons from Denmark, Sweden and Spain. Eur. Environ. **17**, 347–362 (2007)
9. D. Fouquet, T.B. Johansson, European renewable energy policy at crossroads—focus on electricity support mechanisms. Energy Policy **36**(11), 4079–4092 (2008)
10. T. Unger, E.O. Ahlgren, Impacts of a common green certificate market on electricity and CO_2-emission markets in the Nordic countries. Energy Policy **33**, 2152–2163 (2005)
11. P. Söderholm, M. Pettersson, Offshore wind power policy and planning in Sweden. Energy Policy **39**, 518–525 (2011)
12. T. David, The UK offshore wind power programme: a sea-change in UK energy policy? Energy Policy **39**, 526–534 (2011)
13. N. Zografakis, E. Sifaki, M. Pagalou, G. Nikitaki, V. Psarakis, K.P. Tsagarakis, Assessment of public acceptance and willingness to pay for renewable energy sources in Crete. Renew. Sust. Energy Rev. **14**, 1088–1095 (2010)
14. M.A. Schilling, M. Esmundo, Technology S-curves in renewable energy alternatives: analysis and implications for industry and government. Energy Policy **37**, 1767–1781 (2009)
15. P. Harborne, C. Hendry, Pathways to commercial wind power in the US, Europe and Japan: the role of demonstration projects and field trials in the innovation process. Energy Policy **37**, 3580–3595 (2009)
16. D. Popp, I. Hascic, N. Medhi, Technology and the diffusion of renewable energy. Energy Econ. **33**, 648–662 (2011)
17. M. Bolinger, R. Wiser, K. Cory, T. James, PTC, ITC, or cash grant? An analysis of the choice facing renewable power projects in the United States, National Renewable Energy Laboratory, 2009
18. L. Bird, M. Bolinger, T. Gagliano, R. Wiser, M. Brown, B. Parsons, Policies and market factors driving wind power development in the United States. Energy Policy **33**(11), 1397–1407 (2005)
19. R. Craig, America's energy and climate change policy. Electr. J. **24**, 16–26 (2011)
20. L. Bird, C. Chapman, J. Logan, J. Sumner, W. Short, Evaluating Renewable Portfolio Standards and Carbon Cap Scenarios in the U.S. Electric Sector, National Renewable Energy Laboratory, 2010
21. M.-Y. Huang, J.R.R. Alavalapati, D.R. Carter, M.H. Langholtz, Is the choice of renewable portfolio standards random? Energy Policy **35**(11), 5571–5575 (2007)
22. H. Yin, N. Powers, Do state renewable portfolio standards promote in-state renewable generation? Energy Policy **38**, 1140–1149 (2010)
23. S. Carley, State renewable energy electricity policies: an empirical evaluation of effectiveness. Energy Policy **37**, 3071–3081 (2009)
24. K.S. Cory, B.G. Swezey, Renewable portfolio standards in the states: balancing goals and rules. Electr. J. **20**, 21–32 (2007)
25. K. Palmer, D. Burtraw, Cost-effectiveness of renewable electricity policies. Energy Econ. **27**, 873–894 (2005)
26. F.C. Menz, S. Vachon, The effectiveness of different policy regimes for promoting wind power: experiences from the states. Energy Policy **34**, 1786–1796 (2006)
27. P. Mozumder, A. Marathe, Gains from an integrated market for tradable renewable energy credits. Ecol. Econ. **49**, 259–272 (2004)
28. K. Cory, T. Couture, C. Kreycik, Feed-in Tariff Policy: Design, Implementation, and RPS Policy Interactions. National Renewable Energy Laboratory, 2009

29. G.R. Timilsina, L. Kurdgelashvili, P.A. Narbel, Solar energy: markets, economics and policies. Renew. Sust. Energy Rev. **16**, 449–465 (2012)
30. J. Kneifel, Effects of State Government Policies on Electricity Capacity from Non-hydropower Renewable Sources, Unpublished Manuscript, Jun, 2008
31. S. Cantono, G. Silverberg, A percolation model of eco-innovation diffusion: the relationship between diffusion, learning economies and subsidies. Technol. Forecast. Soc. Change **76**, 487–496 (2009)
32. K.U. Rao, V.V.N. Kishore, A review of technology diffusion models with special reference to renewable energy technologies. Renew. Sust. Energy Rev. **14**, 1070–1078 (2010)
33. T.D. Tsoutsos, Y.A. Stamboulis, The sustainable diffusion of renewable energy technologies as an example of an innovation-focused policy. Technovation **25**(7), 753–761 (2005)
34. A.M. Bassi, J.D. Shilling, Informing the US energy policy debate with threshold 21. Technol. Forecast. Soc. Change **77**, 396–410 (2010)
35. G. Birgisson, E. Petersen, Renewable energy development incentives: strengths, weaknesses and the interplay. Electr. J. **19**(3), 40–51 (2006)
36. K.H. Engel, B.Y. Orbach, Micro-motives and state and local climate change initiatives. Harvard Law Policy Rev. **2**, 119 (2008)
37. Federal Energy-Efficient Commercial Buildings Tax Deduction (2012), http://www.dsireusa.org/incentives/incentive.cfm?Incentive_Code=US40F&re=1&ee=1
38. K. Malmedal, B. Kroposki, P.K. Sen, Energy Policy Act of 2005 and Its Impact on Renewable Energy Applications in USA, in *Power Engineering Society General Meeting, 2007* (IEEE, 2007), pp. 1–8
39. Cap and Trade | US EPA (2012), http://www.epa.gov/captrade/
40. Cap and Trade 101 (2012), http://www.americanprogress.org/issues/2008/01/capandtrade101.html
41. F.C. Menz, Green electricity policies in the United States: case study. Energy Policy **33**, 2398–2410 (2005)
42. EERE State Activities and Partnerships: States with Renewable Portfolio Standards (2012), http://apps1.eere.energy.gov/states/maps/renewable_portfolio_states.cfm
43. DSIRE (2011), Database of State Incentives for Renewable Energy, http://www.dsireusa.org/
44. A.S. Kydes, Impacts of a renewable portfolio generation standard on US energy markets. Energy Policy **35**(2), 809–814 (2007)
45. B.K. Sovacool, C. Cooper, Big is beautiful: the case for federal leadership on a national renewable portfolio standard. Electr. J. **20**, 48–61 (2007)
46. S. Patrik, The political economy of international green certificate markets. Energy Policy **36**, 2051–2062 (2008)
47. Alternative Energy—Wind, Solar, Hydro and Other Alt Energy Sources for Home Power (2012), http://www.altenergy.org/
48. R. Shinnar, F. Citro, Decarbonization: achieving near-total energy independence and near-total elimination of greenhouse emissions with available technologies. Technol. Soc. **30**(1), 1–16 (2008)
49. U. Katrin, Who formulates renewable-energy policy? A Swedish example. Energy Policy **38**, 6674–6683 (2010)
50. Q. Chai, X. Zhang, Technologies and policies for the transition to a sustainable energy system in china. Energy **35**, 3995–4002 (2010)
51. E. Luiten, H. van Lente, K. Blok, Slow technologies and government intervention: energy efficiency in industrial process technologies. Technovation **26**, 1029–1044 (2006)
52. J. Noailly, S. Batrakova, Stimulating energy-efficient innovations in the Dutch building sector: empirical evidence from patent counts and policy lessons. Energy Policy **38**, 7803–7817 (2010)
53. W.W. Clark Ii, J. Rifkin, A green hydrogen economy. Energy Policy **34**, 2630–2639 (2006)
54. 2010 Renewable Energy Data Book, U.S. Department of Energy, 2012
55. P.E. Meyer, J.J. Winebrake, Modeling technology diffusion of complementary goods: the case of hydrogen vehicles and refueling infrastructure. Technovation **29**, 77–91 (2009)

56. Annual Energy Outlook 2010, U.S. Energy Information Administration, 2010
57. V. Fthenakis, J.E. Mason, K. Zweibel, The technical, geographical, and economic feasibility for solar energy to supply the energy needs of the US. Energy Policy **37**, 387–399 (2009)
58. E. Martinot, F. Beck, Renewable Energy Policies and Barriers. Renewable Energy Policy Project, 2004
59. K.J. Holmes, L. Papay, Prospects for electricity from renewable resources in the United States. J. Renew. Sust. Energy **3**, 042701–042714 (2011)
60. Y. Kajikawa, Y. Takeda, Structure of research on biomass and bio-fuels: a citation-based approach. Technol. Forecast. Soc. Change **75**(9), 1349–1359 (2008)
61. Y. Kajikawa, J. Yoshikawa, Y. Takeda, K. Matsushima, Tracking emerging technologies in energy research: toward a roadmap for sustainable energy. Technol. Forecast. Soc. Change **75**(6), 771–782 (2008)
62. Y.C. Shen, C.J. Chou, G.T.R. Lin, The portfolio of renewable energy sources for achieving the three E policy goals. Energy **36**(5), 2589–2598 (2011)
63. L. Neij, Use of experience curves to analyse the prospects for diffusion and adoption of renewable energy technology. Energy Policy **25**(13), 1099–1107 (1997)
64. J.M. Loiter, V. Norberg-Bohm, Technology policy and renewable energy: public roles in the development of new energy technologies. Energy Policy **27**(2), 85–97 (1999)
65. S.W. Sawyer, Leaders in change: solar energy owners and the implications for future adoption rates. Technol. Forecast. Soc. Change **21**(3), 201–211 (1982)
66. S.K. Hoekman, Biofuels in the U.S.—challenges and opportunities. Renew. Energy **34**, 14–22 (2009)
67. M.B. Charles, R. Ryan, N. Ryan, R. Oloruntoba, Public policy and biofuels: the way forward? Energy Policy **35**(11), 5737–5746 (2007)
68. Renewable Energy Ocean Wave Energy Development (2012), http://www.oregon.gov/ENERGY/RENEW/Hydro/Ocean_Wave.shtml
69. S.K. Lee, G. Mogi, S.C. Shin, J.W. Kim, in *An AHP/DEA Hybrid Model for Measuring the Relative Efficiency of Energy Efficiency Technologies*, pp. 55–59
70. F. Cavallaro, L. Ciraolo, A multicriteria approach to evaluate wind energy plants on an Italian island. Energy Policy **33**, 235–244 (2005)
71. E.W.E. Association, Delivering Offshore Wind: Policy Recommendations for Large-Scale Deployment of Offshore Wind Power in Europe by 2020, 2007
72. K. Usha Rao, V.V.N. Kishore, Wind power technology diffusion analysis in selected states of India. Renew. Energy **34**, 983–988 (2009)
73. V. Dinica, Initiating a sustained diffusion of wind power: the role of public–private partnerships in Spain. Energy Policy **36**(9), 3562–3571 (2008)
74. D. Poponi, Analysis of diffusion paths for photovoltaic technology based on experience curves. Sol. Energy **74**, 331–340 (2003)
75. F. Cavallaro, Multi-criteria decision aid to assess concentrated solar thermal technologies. Renew. Energy **34**, 1678–1685 (2009)
76. F. Cavallaro, A comparative assessment of thin-film photovoltaic production processes using the ELECTRE III method. Energy Policy **38**, 463–474 (2010)
77. F. Cavallaro, Fuzzy TOPSIS approach for assessing thermal-energy storage in concentrated solar power (CSP) systems. Appl. Energy **87**, 496–503 (2010)
78. M. Guidolin, C. Mortarino, Cross-country diffusion of photovoltaic systems: modelling choices and forecasts for national adoption patterns. Technol. Forecast. Soc. Change **77**, 279–296 (2010)
79. C. Sherrington, J. Bartley, D. Moran, Farm-level constraints on the domestic supply of perennial energy crops in the UK. Energy Policy **36**(7), 2504–2512 (2008)
80. A. Jäger-Waldau, H. Ossenbrink, Progress of electricity from biomass, wind and photovoltaics in the European Union. Renew. Sust. Energy Rev. **8**, 157–182 (2004)
81. Renewing our Energy Future, U.S. Congress, Office of Technology Assessment, 1995

Chapter 2
Renewable Energy Technology Adoption in the Pacific Northwest: A Technology Policy Review

Rimal Abu Taha and Tugrul U. Daim

Abstract The term Northwest in the United States is used loosely to indicate the area between the states of Alaska, Washington, Oregon, northern California, Idaho, Montana, and Wyoming. In the heart of Northwest are two states (Oregon and Washington) bordering the Pacific Ocean and are geographically and culturally similar, those two cities are referred to as the Pacific Northwest. The State of Oregon has been strategically weighing energy demand, supply, and resources to give Oregonians a more sustainable and convenient energy future. Renewable energy is perceived by the Oregonians as a source of energy independence, rural community development, and cleaner air. After the oil crises in 1973, Governor Tom McCall launched an emergency energy conservation program and in 1975 The Oregon Department of Energy (ODOE) was established to support energy conservation and renewable energy policy planning. Many of these policies and plans are still active until today but with some modifications over time. Another crisis that hit the Pacific Northwest was the low rain levels in the years 2000–2001 which lead to lower hydropower year and increased electricity demand with few power plants being built.

Literature on renewable energy technologies in the Pacific Northwest discussed various subjects and examined different areas concerning this issue. Daim et al. developed a model to create renewable energy portfolio and assess renewable energy technologies in Oregon that would be used to achieve the mandated levels by the new Renewable Portfolio Standards (RPS) [1]. The Pacific Northwest is one of the regions in the United States that has significant wind power potential and wind power projects either installed or planned. Washington currently ranks sixth in the nation in the total capacity of wind power installation while Oregon currently has the seventh

R. Abu Taha (✉)
Portland State University, 15606 SW Snowy Owl Lane, Beaverton, OR 97007, USA
e-mail: rimalat@yahoo.com

T.U. Daim
Engineering and Technology Management, Portland State University,
1900 SW 4th Avenue 50-02, Portland, OR 97007, USA

T.U. Daim et al. (eds.), *Policies and Programs for Sustainable Energy Innovations*, Innovation, Technology, and Knowledge Management,
DOI 10.1007/978-3-319-16033-7_2

most wind power capacity installed of any state with the Biglow project as the eighth largest wind farm in the country [2]. Yin analyzed policies and financial incentives in Oregon and their role in wind development [3]. Sailor et al. investigated several scenarios of climate change effect on wind power generation potential in the Northwest [4, 5]. Solar energy in the United States is still considered expensive and only accounts for a small amount of the overall energy usage. Nevertheless, Oregon solar installed capacity of grid-connected photovoltaic has grown exponentially from 2.8 MW in 2007 to 14 MW in 2009 which demonstrates 400 % growth just in 2 years [6]. Daim et al. developed a model to assess the available renewable energy resources in the State of Oregon that can lead the way to satisfy the RPSs by 2025. It seems that in spite of the common perception that the Northwest climate is cloudy and lacks sunshine, there is still an abundant sun energy in many regions to harvest [7, 8]. On the other hand, other literature assessed the potential of ocean and wave energy resources along the coastal area of the Pacific Northwest [9–11] although those later two are not deployed currently in the region and not likely to be in the near future.

2.1 Why the Shift to Renewables?

A variety of factors have encouraged renewable energy development and deployment in the Pacific Northwest including market conditions, policy enhancement, skilled labor, and the cultural and environmental concerns of consumers. One of famous publication about the Northwest energy paradigm is a book called Transition, A Book on Future Energy: Nuclear or Solar? [12]. In this book, the author described the energy dilemma that the region was facing that time "…as energy prices rose, it became apparent that the energy systems so many had taken for granted were almost entirely outside of our control. In fact, about 95 % of the energy we use in Oregon is imported." This clearly outlined the problem and first warning that a considerable amount of the regions budget was going to export foreign or out of state energy.

Oregon's first commercial-scale wind power project was the 25-MW in 1998. The project was planned after PGE agreed to develop renewable energy generation to nuclear energy. There was a few years delay in wind energy development until the year 2001 where wind energy development picked up to satisfy regional supply shortage caused by the lack of rain and the shortage of supply from California State [13]. BPA then had selected seven projects in Oregon and Washington out of 25 proposals to start operating [14].

The population of the Pacific Northwest will continue to increase which means a greater demand for energy. That growing demand requires the search for new energy resources. The people of the northwest are environmentally oriented and are concerned about health and natural habitat. Since renewable energy resources offer these environment benefits, they have been very welcomed in the region. Utilities in Oregon and Washington offer green pricing options for consumers which had helped to encourage consumer demand for renewable power. Federal and State policies that are aimed to stimulate investing in renewable energy resources have contributed enormously in the development and deployment for renewables from consumer and developer point of view [3, 15].

2.2 Energy Resources

Eight states: Washington, California, Oregon, New York, Idaho, Alabama, Montana, and Texas have provided almost 70 % of the US renewable energy generated in 2006 [16]. The Pacific Northwest region is rich with many forms of renewable energy resources including various types of biofuel, geothermal, hydropower, wind, solar, and marine energy resources. All renewable energy sources can be used for power generation. In addition to that solar, geothermal, and biomass can also be used to for heat generation. Alternative transportation fuels are extracted from biomass. Washington leads the states in hydropower supply where it forms 71 % of its total energy sources [17] (Fig. 2.1). The percentage of non-hydro renewable energy share in the power generation sector in Oregon had went up (2000–2010) from 1.8 to 11.9 and in Washington from 0.9 to 5.2 MW [13]. See Figs. 2.2 and 2.3 for cumulative consumption rates of different renewable energies in the Pacific Northwest (Fig. 2.4).

- Hydroelectric power:
  Hydropower is the main source of power in Oregon's and Washington's electricity portfolio. The Bonneville Power Administration manages and markets power from 31 hydropower facilities in the Northwest, 14 of which are located in Oregon which benefits the consumer-owned utilities with cheaper rates of power from the BPA power system. New growth in the hydropower sector is most likely to occur in three areas: irrigation systems, pumped storage, and the addition of power facilities on existing dams, especially federal dams. Other than providing reliable, cheap source of energy in the Northwest, hydropower is used to balance the variation of wind energy production and balancing the load-supply aspects.

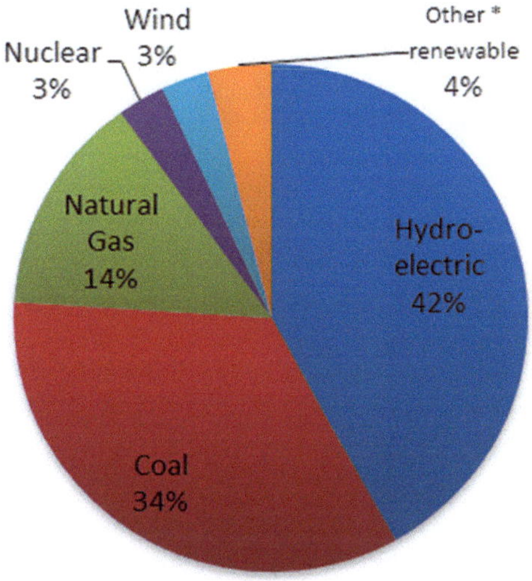

Fig. 2.1 Oregon's power source portfolio [17]

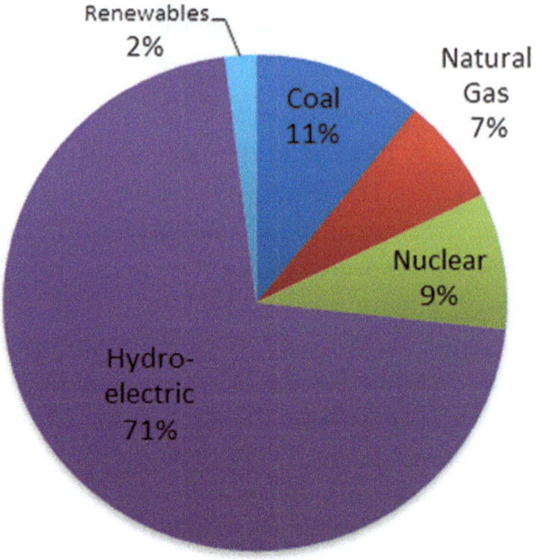

Fig. 2.2 Washington's power source portfolio [18]

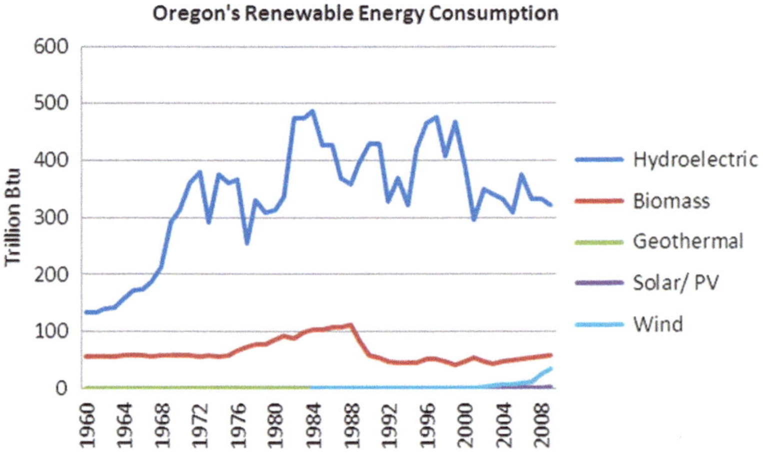

Fig. 2.3 Oregon's renewable energy consumption (trillion Btu) 1960–2008 (*Data source*: U.S. Energy Information Administration [17])

- Wind:
  Wind is the second most deployed renewable energy after hydropower in the Pacific Northwest. The first wind farm in Oregon was installed in 1998 and began operating at a capacity of 25 MW. Oregon currently ranks the seventh in the nation for installed wind power with 2,305 MW currently working and 9,361 MW Wind projects waiting for permits or transmission lines. Washington State is an early leader in the wind industry and currently ranks sixth in the nation with 2,357 MW currently installed and 5,831 MW waiting in queue [18]. It is worth to mention that Oregon is home to the European wind farm operator Iberdrola Renewables and North American headquarters of wind turbine manufacturer Vestas both of which are important players in wind energy development in the

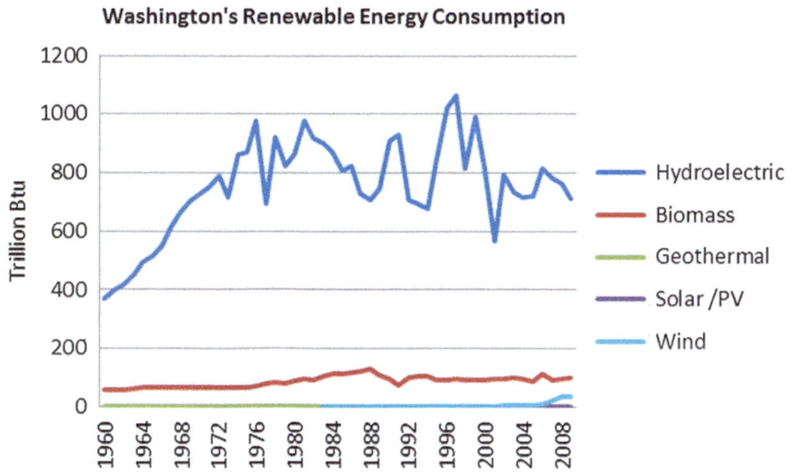

Fig. 2.4 Washington's renewable energy consumption (trillion Btu) 1960–2008 (*Data source*: U.S. Energy Information Administration [17])

Fig. 2.5 Wind energy development in the Pacific Northwest [13]

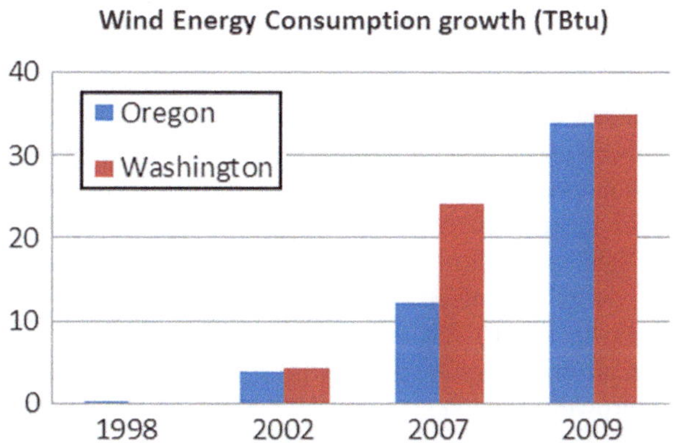

region [19]. California already purchases more than half the wind power generated in the Northwest. And when Shepherds Flat is completed, all of its subsidized output is contracted to go to Southern California Edison (Fig. 2.5).

- Biomass:
  Biofuel is a term that includes liquids, solids, and gas fuels that are produced from biomass. These biofuels can be used for transportation, thermal energy, and power generation. Biomass in the State of Oregon includes agricultural residues, forest slash, and mill residuals. It is used to thermal heat for the forest industry, heat and electricity for homes, schools, and hospitals. The biomass development is a result of collaboration between public-private sectors like the Forest Biomass Working Group and the Forest Cluster Economic Development Team. In Oregon, 2009 legislation authorized a new low-carbon fuel standard designed to reduce the carbon intensity of transportation fuels by 10 % by 2020.

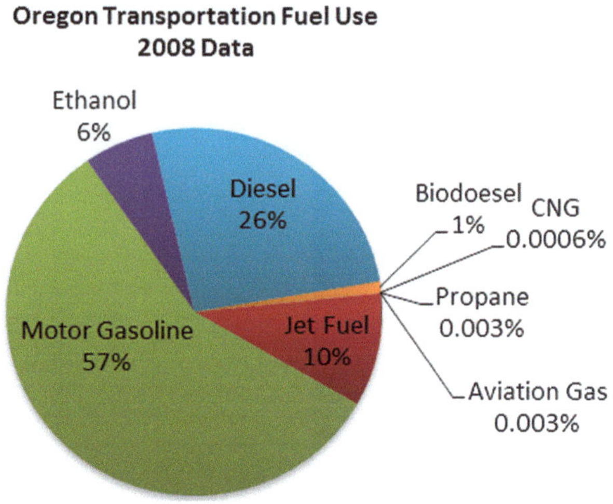

Fig. 2.6 Fuel mix in Oregon [13]

Suppliers can meet this target utilizing different alternative fuels such as ethanol and biodiesel. These alternatives to gasoline and diesel are already being produced and use (Fig. 2.6).

- Solar:
  Solar energy is one of the renewable energies that is relatively mature and has been used for a long time in different applications. Although it doesn't seem so, some parts of eastern and northern Oregon receive annual solar energy as Europe or Florida do. Solar energy can be harvested as direct light into buildings for light and heat, heating water through roof-mounted collectors, and to convert sunlight to electricity using photovoltaic panels (PV) or concentrated Solar Power (CSP). There is still no utility of large-scale solar plants in Oregon as in California, but the residential PV market in Oregon has experienced a significant growth in the year 2010. This increase can be referred to the financial state and federal incentives which helped reducing the cost of PV systems and increase maturity of the technology (Fig. 2.7).
- Geothermal power generation:
  Geothermal energy is the energy extracted the natural heat of the earth which provides a constant base load energy. Geothermal energy in Oregon is not yet used for power generation, rather it is used for agricultural purposes as a heat source, for space heating, and to heat swimming pools at a number of spas and resorts.
- Wave energy:
  Ocean waves energy can be converted into clean, reliable, and cost-effective electricity that has minimal impacts on the environment. In spite of its availability, until now there are only three working projects, and they are all experimental sites with a single device deployed: Makah Bay, Washington; Kaneohe Bay, Hawaii; and off the coast of New Jersey [20]. However, there are several different projects are developing along the Pacific Northwest coast of

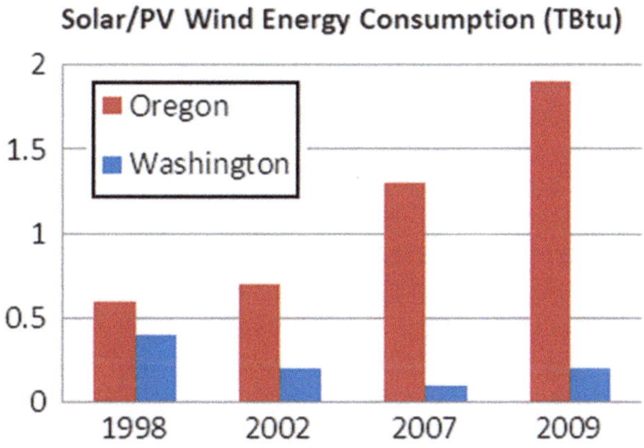

Fig. 2.7 Solar/PV energy development in the Pacific Northwest [13]

which seven located at the coast of Oregon. The state has established itself as the leader in wave energy and become the national center for wave energy research and commercial demonstration [10]. The combination of the potential wave resource and coastline transmission capacity of Oregon and Oregon State University's research facilities has identified Oregon as an ideal location for wave energy conversion as well as a leader in the United States in wave energy development [21].

- Waste (landfill gas):
  It is a mix of gases (methane, CO_2, and water vapor) that is generated by decomposition of organic material and waste at landfill disposal sites. The methane in landfill gas can be burnt to generate electricity or thermal energy. Waste Management Company currently collects waste from Seattle and ships to a massive landfill in north-central Oregon. The methane gas produced as garbage decomposes is collected and burns it to generate electricity which goes back to the city of Seattle [22]. Currently in Oregon, an energy plant in McMinnville has been using landfill gas to make electricity since last June [23].

2.3 Policies

There are a few federal policies that are intended to promote RETs deployment. Since RE resources vary by location and climate, it is typically more efficient to address the deployment issue at the state level. The Pacific Northwest have adopted a number of federal and state policies and incentives to support RE technologies in the form of financial, market, and obligatory forms.

- Renewable Portfolio Standard (RPS):
  Oregon RPS States that by 2025, large electric utilities that serve more than 3 % of Oregon's electric load are required to generate 25 % of Oregon's electric load from eligible renewable energy. For the three largest utilities (Portland General Electric, Pacific Power, and Eugene Water and Electric Board), the targets are

5 % in 2011, 15 % in 2015, 20 % in 2020, and 25 % in 2025. Smaller utilities are excepted from this requirement with targets of 5 % or 10% by 2025, depending on the size of the utility but required not to use coal in new power generation or the targets for the large utilities apply. Eligible renewable resources in the State of Oregon are hydropower, biomass, wind, solar PV, solar thermal, geothermal, wave, tidal, and ocean [24]. Utilities can fulfill this commitment using any way of the following: Build a new eligible facility, buy power from another eligible facility, or buy renewable energy credits. The three largest utilities in Oregon have confirmed that they have achieved the 5 % 2011 goal. Washington passed a renewable energy standard (RES) through ballot initiative in 2006. The RES requires utilities that serve more than 25,000 customers to obtain 15 % of their electricity from renewables by 2020 and invest in energy efficiency.

- Renewable fuel standard (RFS):
  RFS was first established with the enactment of the Energy Policy Act of 2005 with a target to open the transportation fuel market to fuels other than petroleum fuels like biofuels. It is a federal mandate policy that requires a certain amount of biofuels is to be used in the national transportation fuel supply each year [25]. The RFS is managed by the Environmental Protection Agency (EPA), where it has the authority to waive the RFS requirements for a state if they ask for a waiver. In Oregon, all diesel fuel sold or offered for sale must contain a minimum of 5 % by volume biodiesel, creating a B5 biodiesel blend [14]. In the State of Washington, beginning on December 1st, 2008, 2 % of the diesel fuel sold in the State by volume will be biodiesel [26].

- Feed-in-tariff (FIT):
  The traditional FIT policy is to pay a premium for electricity generated by utilities from renewable resources. This type of financial policy has proven to be the world's most effective renewable energy policy [13]. Oregon's model is a little different where utilities pay customers who have solar panels for the power they produce and use. Oregon passed FIT legislation in 2009 where it will be used for compliance with the state's RPS and that only for solar energy. The rate the FIT will be paid is not yet identified and still is determined by utilities and approved by PUC [27]. Washington State is one of the three states with any form of active feed-in tariffs in the United States. Unlike Oregon's, Washington policy requires a full system of feed-in tariffs for all renewable energy technologies. The bill is shaped in line of Germany's successful FIT policy and includes different tariffs not only for solar energy but also for wind [28]. FIT is different than net metering policy as incentive rate is provided for participants for the energy they generate and use themselves, rather than the energy they feed back to the grid. In Oregon, Portland General Electric and Pacific Power began in 2010 their feed-in tariff pilot programs only for solar photovoltaic panels [29].

- Net metering:
  Net metering is a state policy that allows customers to use their own renewable power generation systems to compensate for their consumption feeding the power they generate backward into the grid when it exceeds their demand which

gives them retail credit. Net metering is a low-cost and effective method to encourage private owner to invest in renewable energy technologies. Forty-three states and Washington, D.C. now offer net metering options for their customers [30]. Both Oregon and Washington states have initiated net metering requirements standards for the state's primary investor-owned utilities for projects of 100 kW or less [13], for its municipal utilities and residential systems. Qualifying systems are solar power, wind power, hydropower, fuel cells, or biomass.

- Green pricing (green power option):
  Green pricing is a state policy that requires all electric utilities to offer customers an optional green-power program by paying a premium on their electricity bills to support the incremental cost of the additional renewable energy. Oregon and Washington are the second and third top states selling this offer to customers [31]. A predetermined portion of the electricity sold by a utility as green power must be generated using qualifying renewables, and each utility should declare customers of the sources of the electricity included in its green power program. This policy by creating the market for renewable energy markets provide an additional revenue stream for renewable energy projects and increase consumer knowledge of the benefits of renewable energy. Beginning of 2002, customers served by Oregon's investor-owned utilities were offered a range of service and had access to the several renewable energy options: (1) New Wind Energy—Customers each month can choose to buy certain amount of new wind generation through PGE's Clean Wind program or Pacific Power's Blue Sky program. (2) Renewable Energy Blend—Customers can purchase 100 % of their actual electricity usage from Green Mountain Energy Company generated from wind and geothermal sources. (3) Renewable Energy and Habitat Restoration—Customers can purchase 100 % of their electricity from renewable sources and in the same time help restore native fish habitat. Washington State signed the bill in 2001 requiring the state's electric utilities to offer customers green power option beginning in January 1, 2002. Utilities are required to regularly notify customers of this option of purchasing renewable energy at fixed or variable rates. Qualified energy sources are: wind, solar, geothermal, landfill gas, wastewater treatment gas, wave or tidal action, biomass, and low-impact hydro [32].

- Financial incentives:
  Since renewable energies have relatively higher initial cost, financial policies are needed to encourage investment and deployment of such systems. Federal and state financial incentives vary from several Tax credits, Rebate, Grant or Loan Program. By the American Recovery and Reinvestment Act of 2009), a federal renewable energy cash grant program was created that may be taken instead of the federal business energy investment tax credit (ITC). Only tax-paying entities are eligible for this grant where as federal, state, and local government bodies are not eligible to receive this credit up to 30 % depending on the energy source. Production Tax Credit (PTC) is a federal incentive that provides a tax credit adjusted for electricity produced from renewable energy sources, including wind, biomass, and geothermal. See Table 2.1 for details on some Oregon's state policies and Table 2.2 for Washington's.

Table 2.1 Summary of major Oregon State RE Financial policies [33]

Policy	Summary	Eligible technologies
Biomass producer or collector tax credit	Tax credit for agricultural producers or collectors of biomass	Biomass, biodiesel
Business energy tax credit (BETC)	Tax credit for investments in energy conservation, recycling, and renewable energy resources	Energy efficiency technologies
Tax credit for renewable energy equipment manufacturers	Tax credit equals 50 % of the construction costs of a facility which will manufacture renewable energy systems	Solar, wind, biomass, geothermal heat pumps, hydroelectric, tidal energy, wave energy
Oregon pilot solar volumetric incentive rates and payments program	Rebates for customers who install PV whether they choose net metering or not	Photovoltaic
Energy trust—small wind incentive program	Provides cash incentives for customers of Portland General Electric and Pacific Power that are installing turbines up to 100 kW	Wind
Other rebates, tax credits	Cities offer different rebates and tax credits on energy efficiency equipment, installing wind, PV	Wind, solar
Energy trust of Oregon [34]	This policy requires large utilities to collect a 3 % charge from their customers to support renewable energy and energy efficiency projects through January 1, 2026. These funds are allocated 56.7 % to support energy efficiency programs and 17.1 % as financial incentives to renewables while the remaining funds support low-income housing energy assistance and K-12 school energy-conservation efforts	Solar water heat, solar space heat, solar thermal electric, photovoltaics, wind, biomass, hydroelectric, geothermal electric, fuel cells using renewable fuels, geothermal direct-use, energy efficiency equipment, and technologies

Table 2.2 Summary of major Washington State RE Financial policies [33]

Policy	Summary	Eligible technologies
Federal clean renewable energy bonds (CREBs) and qualified energy conservation bonds (QECBs)	Both bonds (CREBs) may be used by certain entities—primarily in the public sector—to finance renewable energy projects	Solar thermal electric, photovoltaics, landfill gas, wind, biomass, hydroelectric, geothermal electric, municipal solid waste, hydrokinetic power, anaerobic digestion, tidal energy, wave energy, ocean thermal
Renewable energy production incentive (REPI)	Provides incentive payments for electricity generated and sold by new qualifying renewable energy facilities	Solar thermal electric, photovoltaics, landfill gas, wind, biomass, geothermal electric, anaerobic digestion, tidal energy, wave energy, ocean thermal

(continued)

Table 2.2 (continued)

Policy	Summary	Eligible technologies
Federal residential renewable energy tax credit	A taxpayer may claim a credit of 30 % of qualified expenditures for a system that serves a residence located in the United States and used as a residence by the taxpayer	Solar water heat, photovoltaics, wind, fuel cells, geothermal heat pumps, other solar electric technologies, fuel cells using renewable fuels
City of seattle—density bonus for green buildings	An incentive for the construction of green buildings	Whole building
Tax abatement for solar manufacturers	Reduced business tax rate for Washington manufacturers and wholesale marketers of solar-electric systems	Photovoltaics, stirling converters
Washington REPIs	Incentive amount paid to the producer starts at a base rate of $0.15 per kilowatt-hour (kWh) and is adjusted	Solar thermal electric, photovoltaics, wind
Solar express loan program	This program provides rebates and loans to support residential and commercial installations of solar photovoltaics (PV) and solar water heating for residential customers	Solar water heat, photovoltaics
Solar water heater rebate	A rebate to customers who install a solar water heating system	Solar water heat
Fuel mix disclosure	Washington's retail electric suppliers must disclose details regarding the fuel mix of their electric generation to customers	Renewable energy use disclosed in fuel mix data
Green power purchasing	The Bellingham City Council adopted a policy to begin purchasing 100 % green power for all facilities owned by the city—one of the most aggressive such goals in the United States	RE
Mandatory utility green power option	Requires all electric utilities serving more than 25,000 customers to offer customers the option of purchasing renewable energy	Solar thermal electric, photovoltaics, landfill gas, wind, biomass, hydroelectric, geothermal electric, municipal solid waste, tidal energy, wave energy

2.4 Conclusions

With plenty of wind and abundant water resources, the Pacific Northwest has the nation's greatest penetration of renewable energy. The Pacific Northwest region has established a comprehensive renewable energy adoption policy that insured creating a rich, suitable environment for developing and investing in renewable energy technologies which had reflected on decreasing emissions, developing economy, and creating jobs. However, the region is still trying to make those two resources work together. The challenge rises with the intermittency of wind and increase of electricity prices.

As the increase in the percentage of renewables in the energy portfolio to meet the RPS, it is expected that more natural gas-fired power plants will be working to provide base load electricity and avoid intermittent characteristics of renewables.

The federal energy PTC will expire by the end of this year which created uncertainty in the market that's affecting job cuts in Portland since it's affecting orders for renewable-energy producers such as Iberdrola and Vestas Wind Systems. "Without the certainty of that extension, project developers are not doing projects in the U.S., and manufacturers are not getting orders," Portland Mayor Sam Adams said [35].

Although one of the main objectives of RE adoption is reducing the Green House Gases (GHG) emissions, it is shown from Fig. 2.8 that transportation sector is the largest sector responsible for such emissions followed by residential and commercial activities with no noticeable decrease. On the other hand, emissions from the industrial sector have decreased significantly. This can be a result of increasing efficiency in industrial processes and a successful economy.

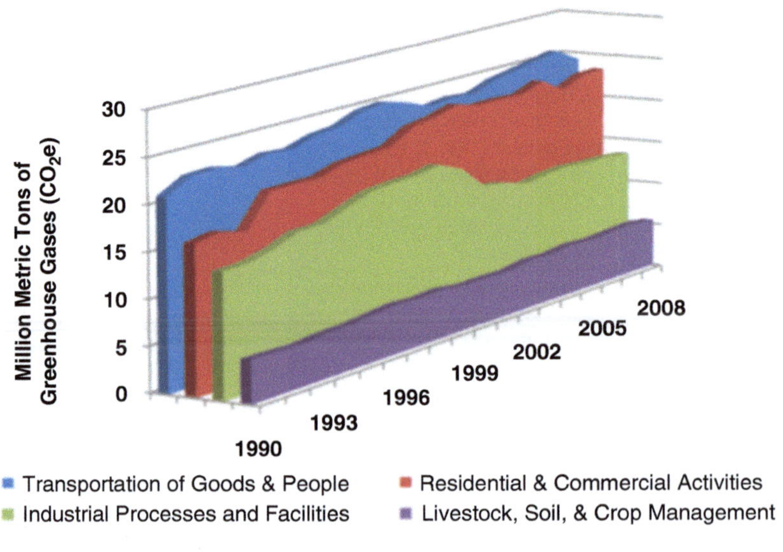

Fig. 2.8 GHG emissions level in Oregon [14]

References

1. T.U. Daim, G. Kayakutlu, K. Cowan, Developing Oregon's renewable energy portfolio using fuzzy goal programming model. Comput. Ind. Eng. **59**, 786–793 (2010)
2. www.AWEA.org
3. Y. Yao, A socio-political analysis of policies and incentives applicable to community wind in Oregon. Energy Policy **42**, 442–449 (2012)
4. D.J. Sailor, M. Smith, M. Hart, Climate change implications for wind power resources in the Northwest United States. Renew. Energy **33**(11), 2393–2406 (2008)
5. R.W. Baker, E.W. Hewson, N.G. Butler, E.J. Warchol, Wind power potential in the Pacific Northwest. J. Appl. Meteorol. **17**, 1814–1825 (1978)
6. Solar-Oregon.org (2012), http://www.solar-oregon.org/
7. Y.A. Zia, Solar energy in the Northwest U.S. Energy **10**, 1159–1164 (1985)
8. C.W. Richard, Solar energy collection in the Pacific Northwest. Sol. Energy **1**, 3–8 (1957)
9. T.L. Kelly-Yong, K.T. Lee, A.R. Mohamed, S. Bhatia, Potential of hydrogen from oil palm biomass as a source of renewable energy worldwide. Energy Policy **35**(11), 5692–5701 (2007)
10. M. Stefanovich, Wave energy and public opinion in the state of Oregon, U.S.A., in *OCEANS 2009, MTS/IEEE Biloxi-Marine Technology for Our Future: Global and Local Challenges*, 2009, pp. 1–10
11. M.A.P. Stefanovich, Does concern for global warming explain support for wave energy development? A case study from Oregon, U.S.A., in *OCEANS, 2011 IEEE-Spain*, 2011, pp. 1–9
12. P. Oregon, Oregon Office of Energy Research and Planning, *Transition: A Book on Future Energy: Nuclear or Solar?* (Prometheus Unbound, Specialty Books, Portland, 1977)
13. State Electricity Profiles 2010 (2012), http://www.eia.gov/electricity/state/pdf/sep2010.pdf
14. State of Oregon Energy Plan, Oregon Department of Energy, 2012
15. Biofuels Impact Study, Oregon Department of Energy, 2010
16. Energy Report-Overview: Renewable Energy (2012), http://www.window.state.tx.us/specialrpt/energy/renewable/
17. EIA-State Energy Data System (2012), http://www.eia.gov/state/seds/hf.jsp?incfile=sep_use/total/use_tot_ORcb.html&mstate=Oregon
18. State Fact Sheets (2012), http://www.awea.org/learnabout/publications/factsheets/factsheets_state.cfm
19. Behind Iberdrola's Portland layoffs: expiring wind-energy tax credits, declining demand, overloaded grid, cheap natural gas | OregonLive.com (2012), http://www.oregonlive.com/business/index.ssf/2012/01/behind_iberdrolas_portland_lay.html
20. Ecological Effects of Wave Energy Development in the Pacific North-west (2012), http://hmsc.oregonstate.edu/waveenergy/
21. Renewable Energy Ocean Wave Energy Development (2012), http://www.oregon.gov/ENERGY/RENEW/Hydro/Ocean_Wave.shtml
22. Landfill Gas—Northwest Renewable News (2012), http://nwrenewablenews.wordpress.com/category/landfill-gas/
23. Pacific Northwest landfills powering 7,500 homes | BrighterEnergy.org (2012), http://www.brighterenergy.org/24138/news/bioenergy/pacific-northwest-landfills-powering-7500-homes/
24. Senate Bill 838 (2007), http://www.oregon.gov/ENERGY/RENEW/docs/sb0838.a.pdf?ga=t
25. R. Schnepf, B. Yacobucci, in *Renewable Fuel Standard (RFS): Overview and Issues*, Congressional Research Service, 2010
26. Renewable Diesel Technology (2012), http://agr.wa.gov/bioenergy/docs/RenewableDieselWhitePaperFINAL.pdf
27. Interstate Renewable Energy Council (2009), 2009 Updates and Trends
28. Washington State Introduces Feed-in Tariff | Renewable Energy News Article (2012), http://www.renewableenergyworld.com/rea/news/article/2009/02/washington-state-introduces-feed-in-tariff-54645

29. Oregon Utilities to Start Feed-in Tariff Programs | Daily Journal of Commerce (2012), http://djcoregon.com/news/2010/06/15/oregon-utilities-to-start-feed-in-tariff-programs/
30. Green Power Network: Net Metering (2012), http://apps3.eere.energy.gov/greenpower/markets/netmetering.shtml
31. J. Heeter, L. Bird, in *Status and Trends in U.S. Compliance and Voluntary Renewable Energy Certificate Markets (2010 Data)*, National Renewable Energy Laboratory, 2011
32. Green Power Network: Mandatory Utility Green Power Option (2012), http://apps3.eere.energy.gov/greenpower/markets/state_policies.shtml#content
33. Energy Central—Member Sign In (2012), http://pro.energycentral.com/professional/news/power/news_article.cfm?id=23171135
34. DSIRE, Database of State Incentives for Renewable Energy (2011), http://www.dsireusa.org/
35. Oregon—Community Renewable Energy (2012), http://nwcommunityenergy.org/resources/financing-incentives/state-incentives/copy_of_oregon

Chapter 3
Evaluating Policies Toward the Adoption of Renewable Energy Technologies

Rimal Abu Taha and Tugrul U. Daim

Abstract This chapter is to use MCDM technique to assess the degree of impact and effectiveness of different renewable energy policies on wind energy deployment in the power generation sector in the State of Oregon. There are a wide array of federal and state policies that are designed to increase the benefits and demolish barriers and hence encourage investors and customers to adopt new resources. This multidimensional issue can be seen as a multi-criteria decision-making problem. Multi-criteria decision analysis (MCDA) provide a flexible tool that is able to handle and connect a wide range of variables and thus offer useful assistance to the decision maker in mapping out the situation.

The demand for renewable energy as a power generation source has increased in the last two decades as a response to major concerns for future projected scarcity in fossil fuel supply as well as climate change issues. Despite the efforts to replace conventional fuels with renewable ones, oil and coal are still the two main sources of energy in the United States. Over the past decade, federal and state governments have initiated policies to accelerate the development and adoption of renewable energy technologies as energy sources for the nation. Designing and selecting comprehensive and coordinated group of policies that focus on energy adoption goals is faced with a number of challenges such as energy demand, fluctuating price, or GHG emissions level. Energy policy design analysis is a multidimensional approach since it involves several elements. One issue is debatable whether renewable energy technologies had reached a mature state and therefore further development and deployment actions can be left for industry and market forces [1]. If not, what form of public policy should be formulated to increase adoption and diffusion?

R. Abu Taha (✉)
Portland State University, 15606 SW Snowy Owl Lane, Beaverton, OR 97007, USA
e-mail: rimalat@yahoo.com

T.U. Daim
Engineering and Technology Management, Portland State University,
1900 SW 4th Avenue 50-02, Portland, OR 97007, USA

T.U. Daim et al. (eds.), *Policies and Programs for Sustainable Energy Innovations*, Innovation, Technology, and Knowledge Management,
DOI 10.1007/978-3-319-16033-7_3

Renewable energy adoption can be formed to target barriers of adoption as well as benefits. The most two barriers to the more RE deployment until now is the higher initial cost which makes it not cost competitive with fossil fuels and that not all technologies are mature enough to overcome the technical obstacles that are associated with the deployment. As known, the returns from such adoption are huge such as having a limitless free primary source of energy, create new jobs, minimize environmental impacts, and nourish the economy.

3.1 Research Design

This chapter applies the Hierarchical Decision Model (HDM) methodology. This methodology allows for breaking down the problem into a hierarchical structure in order to analyze the relationship between a mission, objectives, and alternatives (Fig. 3.1). HDM is used to quantify expert qualitative judgments and convert it to numerical values using pairwise comparison method.

Since each energy source is unique in terms of availability and characteristics and each state has its own cultural and economic values, it is believed that energy policy planning should consider this diversity in resources and loads. The state of Oregon has a diversity of natural resources and is one of the leading states in wind energy utilization.

For the purpose of this study, the focus is only on current energy policy in the state of Oregon and their relative effect on wind energy adoption.

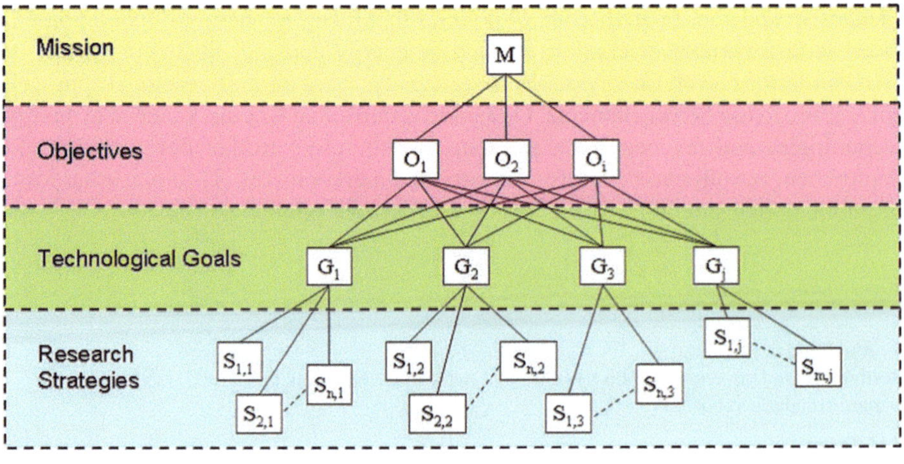

Fig. 3.1 Generic form of HDM with four decision levels [2]

For the design of the model literature, review was carried out to determine the appropriate criteria, subcriteria, and alternatives. In the HDM model, an expert panel should be formed to give judgments that can be translated to quantitative values. For the purpose of this model, experts were chosen from the region. Electricity in Oregon is generated by federal and private-owned Utilities [3]. Expert panel from the region was chosen based on their experience in investor relations, transmission, and technical expertise and knowledge of current policies. The first step was interviews with two of the experts. In the interview the model and criteria were discussed and then the experts were asked if they would accept to fill out the questionnaire. The model was finalized according to their notes and other experts were invited. A total of seven experts responded with their judgments. Five experts were from the region and two from academia. Five of the total experts filled out the whole questionnaire while two of them only put weights for the first two levels.

3.2 Oregon Wind Power Adoption Policy Model

Four-level hierarchical model was constructed for this research (Fig. 3.2). The model is designed to grasp the expert judgment of how does different policies (alternatives) effect wind energy adoption and how does each policy work to satisfy its intended goals. For each level the judgments were collected and converted into weights, the alternative with the maximum weight sum would be the best "fit" to our mission. There is no one perfect solution and it is expected to expand more in the future to include more policies and criteria.

- Mission: As described before, wind energy is picking up fast in Oregon. The objective of this model is to assess the current deployed policies effect on wind energy deployment in Oregon which will lead to develop a policy portfolio that would increase this adoption.
- Criteria: The criteria are the factors that are imbedded in energy policy planning. For the definition of subcriteria, refer to Table 3.1.
  - Economic considerations: how the policy is effective on energy adoption economics to insure that wind energy is or will be cost competitive to other conventional resources; reduce production cost, offer possibilities for future further cost reductions, and lower initial investment.
  - Social effect: what are the social considerations that policy can help overcome and maximize their benefits; create jobs, increase public acceptance for RE projects, and shape a market for RE where customers are willing to pay the price when it sometimes gets higher.
  - Political consideration: it is not "political" in the meaning of the word, rather it means other organizational or institutional support that is considered

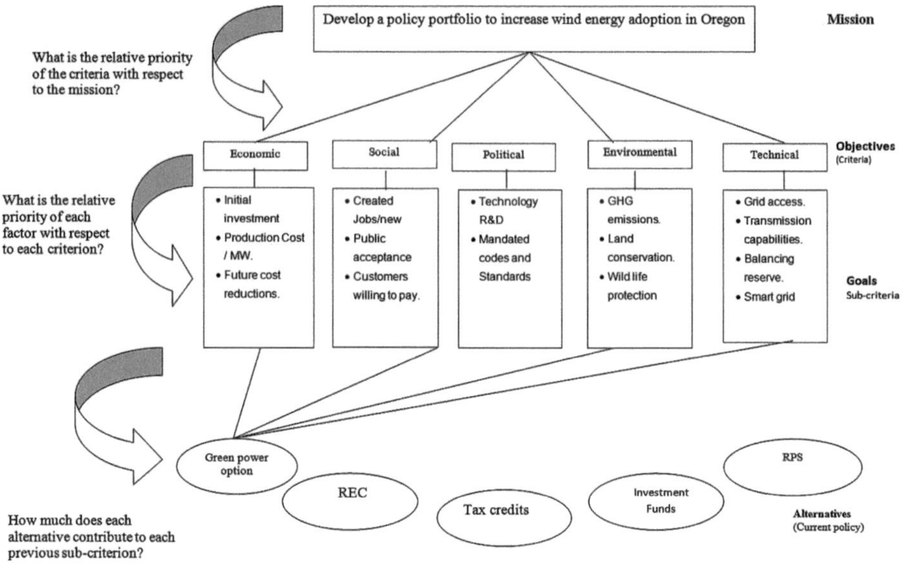

Fig. 3.2 Wind energy policy HDM model

Table 3.1 Criteria and subcriteria in the HDM model

Criteria	Subcriteria	Definition
Economic [4–6]	Initial investment	Startup cost and initial capital of a project. This is usually higher for RE projects. Decreasing this cost or offering loans is favorable for adoption
	Production cost/MW	This cost covers the cost that power is produced, taxes, fees, and any other inherent costs
	Future cost reductions	Guaranteed selling price and market for wind is an assurance for the generator that no hidden cost will be encountered in the future or even further cost reductions using the economies of scale
Social [7–11]	Created jobs/new	With this deteriorating economic situation, starting successful projects (plants, manufacturing equipment) and creating new jobs is favorable
	Public acceptance	Wind farms had faced some opposing point of views with respect to the land area or visual disturbance. Policy should face this issue by creating ways to avoid conflict and increase public satisfaction
	Customers willing to pay	Customers are ready to pay the extra premium for RE usage or for installing equipment. People usually pay when they see a perceived value in the product. Increase in the number of RE customers will lead to the need of more generation

(continued)

Table 3.1 (continued)

Criteria	Subcriteria	Definition
Political	Technology R&D	The maturity of technology is one of the main obstacles to adoption. Increasing R&D to enhance the status of supporting technologies for wind generation (noisy motors, large blades, storage batteries, forecasting,) is crucial for more adoption
	Mandated codes and standards	The equipment and systems used for alternative energy are required to comply with numerous electrical, mechanical and safety codes, and installation requirements. Policy for adopting should insure that new facilities comply with these codes and standards
Environmental [12–14]	GHG emissions	While RE adoption should reduce GHGs, unstructured building for wind farms had put the pressure on gas-fired generators which intern will increase GHGs
	Land conservation	Utility scale RE projects require large area of land. Wind farms are usually remotely positioned but this
	Wildlife protection	Wind farms are creating controversy with respect to their effect on birds and bats. Other RE resources have effect on different natural habitats. RE policy should find ways to protect wild life
Technical [15–18]	Grid access	Intermittent power need to be queued into the grid with guaranteed purchase when generation peak takes place. The aim is to have a better grid access
	Transmission capabilities	Transmission capabilities are limited in the current grid. A policy should enhance transmission capabilities and group generators in geographic areas to facilitate building transmission lines with minimum cost
	Balancing reserve	Schedule and the committed source of energy needed to balance variation of wind energy are important criteria for further wind deployment. Real time response or storage mechanisms is a way to balance this variation
	Smart grid integration	To insure real time response between load and supply and maximum efficiency of reserve balancing

important for wind integration; increase R&D to support wind technology and minimize unwanted outcomes like bigger turbines or forecasting methods. Increase codes and standards unity between states by creating a federal public mandate that will increase wind deployment while managing the risks of no market, for example.

- Environmental effects: environmental considerations are always important for RE policy planning. Although RE are perceived to be environmentally friendly, other aspects of the environment are disturbed by the large wind

Table 3.2 List of alternatives in the HDM model

Alternatives	Description
A1: Green power option	Green power option is a state policy that requires all electric utilities to offer customers an optional green power program by paying a premium on their electricity bills to support the incremental cost of the additional renewable energy
A2: Renewable energy credits (REC)	RECs correspond to the environmental attributes of renewable power generation and are a component of all renewable electricity products. If a certain organization cannot meet its electricity production requirements from renewable energy, they can buy certificate from another supplier that has abundance of renewable or from the market
A3: Business tax credits (BET)	Tax credit for investments in energy conservation, recycling, and renewable energy resources
A4: Investment funding	Federal or state funding for startup renewable energy private-investor owned
A5: Renewable portfolio standards	Oregon RPS States that by 2025, large electric utilities that serve more than 3 % of Oregon's electric load are required to generate 25 % of Oregon's electric load from eligible renewable energy. Smaller utilities are excepted from this requirement with targets of 5 or 10 % by 2025, depending on the size of the utility but required not to use coal in new power generation or the targets for the large utilities apply

turbines; lower GHG emissions are always a target which might be a problem with the need to support wind power with other more constant power, land conservation since some RE technologies require large area of land which creates a debate for residents, wildlife protection and minimize the effect on natural habitat.

- Technological criteria: there is a wide range of technological factors that are important to consider in wind deployment but some are used here for the demonstration of the model; guarantee timely grid access means to systematize power transmission to the grid according to generating peak time, enhance transmission capabilities and group generators in geographic areas to facilitate building transmission lines with minimum cost, and plan balancing of reserves according to the state's excess capacity of other resources, and integrate smart grid options which will help more regulation of loads.

• Alternatives: for the purpose of this model only, alternatives are policies that are implemented in Oregon and benefit wind energy whether it was a marketing, mandates, or financial policy. For the purpose of this model we will consider only these alternatives: Green power option for customers, renewable energy credits (REC) to satisfy requirements, Business tax credits (BET), Investments funds, and renewable portfolio standards (RPS). For information about alternatives, refer to Table 3.2.

3.3 Model Analysis

Experts in the field will be asked to "assign weights" to each element in each level of the model in the form of pairwise comparison out of 100 %; example, economic 36: environmental 64 means that environmental criteria has twice the importance over economic when planning for RE adoption policy, forms of pairwise comparison are provided at the end of this document (see Appendix "PCM software results for pairwise comparison."). After gathering all questionnaire form experts, the values were entered in the PCM software. Normalized weights of each criteria, subcriteria, and alternatives were calculated as well as the inconsistency. Any level of inconsistency <0.1 is acceptable, otherwise we go back to the experts and ask if anything was not understood or if it was just an error (Table 3.3).

Final weights are calculated and aggregated for the alternatives with respect to the mission in the final pay off matrix (see Table 3.4). According to our criteria, policy alternatives rank as A5, A4, A3, A1, and A2 consequently. Renewable portfolio standard is the favorable policy that had contributed to wind energy adoption in Oregon followed by investment funding while the green power option came as the least favorable. Economic aspect of policies and environmental had the highest weights while social aspect was the least important for our experts. For the economic criteria, initial investment is the subcriterion that had the most importance. Socially, all subcriteria were approximately of the same relative priority. For political criteria, it seems that what is needed for wind integration is either enhancing current standards or even change and update them more important that R&D enhancement. Balancing the reserves is the highest priority subcriterion for the technical aspects. A policy that can organize reserves and guarantee maximum efficiency is favorable.

Table 3.3 Weights of criteria and subcriteria as obtained from PCM

Criteria	Subcriteria	Normalized weights of subcriteria	Relative weights
Economic (C1) 0.24	Reduce cost of electricity (S1)	0.4	0.096
	Reduce price for customers (S2)	0.32	0.077
	Increase domestic energy production (S3)	0.28	0.067
Social (C2) 0.14	New jobs created (S4)	0.35	0.049
	Increase of employment (S5)	0.33	0.046
	Public acceptance (S6)	0.32	0.045
Political (C3) 0.2	Increase R&D efforts (S7)	0.39	0.090
	Enforce codes and standards (S8)	0.61	0.140
Environmental (C4) 0.23	Lower GHG emissions (S9)	0.42	0.084
	Land conservation (S10)	0.2	0.040
	Wildlife protection (S11)	0.38	0.076
Technical (C5) 0.19	Guarantee grid access (S12)	0.26	0.049
	Enhance transmission capabilities (S13)	0.27	0.051
	Balancing reserves (S14)	0.3	0.057
	Smart grid integration (S15)	0.17	0.032

Table 3.4 Final pay off matrix and priorities of alternatives

Criteria weights			Subcriteria weights	A1		A2		A3		A4		A5		
0.24	S1	0.4	0.096	0.11	0.0106	0.15	0.0144	0.18	0.0173	0.23	0.0221	0.33	0.0317	
	S2	0.32	0.077	0.13	0.0100	0.16	0.0123	0.17	0.0131	0.26	0.0200	0.28	0.0215	
	S3	0.28	0.067	0.11	0.0074	0.18	0.0121	0.17	0.0114	0.26	0.0175	0.28	0.0188	
0.14	S4	0.35	0.049	0.15	0.0074	0.1	0.0049	0.18	0.0088	0.23	0.0113	0.34	0.0167	
	S5	0.33	0.046	0.29	0.0134	0.14	0.0065	0.16	0.0074	0.14	0.0065	0.27	0.0125	
	S6	0.32	0.045	0.39	0.0175	0.16	0.0072	0.18	0.0081	0.14	0.0063	0.13	0.0058	
0.23	S7	0.39	0.090	0.11	0.0099	0.17	0.0152	0.19	0.0170	0.2	0.0179	0.33	0.0296	
	S8	0.61	0.140	0.25	0.0351	0.15	0.0210	0.17	0.0239	0.21	0.0295	0.22	0.0309	
0.2	S9	0.42	0.084	0.39	0.0328	0.12	0.0101	0.16	0.0134	0.17	0.0143	0.16	0.0134	
	S10	0.2	0.040	0.13	0.0052	0.12	0.0048	0.2	0.0080	0.21	0.0084	0.34	0.0136	
	S11	0.38	0.076	0.17	0.0129	0.13	0.0099	0.18	0.0137	0.19	0.0144	0.33	0.0251	
0.19	S12	0.26	0.049	0.12	0.0059	0.13	0.0064	0.19	0.0094	0.25	0.0124	0.31	0.0153	
	S13	0.27	0.051	0.09	0.0046	0.14	0.0072	0.27	0.0139	0.27	0.0139	0.23	0.0118	
	S14	0.3	0.057	0.1	0.0057	0.12	0.0068	0.26	0.0148	0.29	0.0165	0.23	0.0131	
	S15	0.17	0.032	0.15	0.0048	0.14	0.0045	0.23	0.0074	0.3	0.0097	0.18	0.0058	
1			1		0.1831		0.1433		0.1875		0.2205		0.2656	1
Rank					4		5		3		2		1	

3.4 Conclusion

This chapter demonstrated an assessment model of renewable energy policies to test the methodology. Additional information will be added in the future such as different policies and other criteria, subcriteria to get a comprehensive analysis. This model is missing the differentiation between federal and state policy, taking account of risk apportionment and unintended consequences for the adoption. This analysis was held from the investor's and the operator's point of view which explains the results. Mandatory levels of renewable resources such as the Renewable Portfolio Standards seem to be more effective than voluntary actions (Green Power Option) in this matter. Financial policies as expected had a great effect as a result of how renewable energy plants are capital exhausting and not financially competitive without some kind of support. If we look a little deeper into the model, we can see that economic plus political support (R&D and Standards) are the most two important criteria for policy making. Enhancing mandatory standards and codes for wind technology had more priority than enhancing R&D efforts, this can be referred to that wind energy is already considered a mature technology for investors.

The results of the model suggested that policy makers should focus on creating a policy portfolio in the future that has targets of more aggressive mandates of renewable energy resources while providing the financial incentives for such projects and the technical support.

Appendix

Criteria Comparison

Name: ______________________________

Use pairwise comparison to quantify your judgment. Allocate a total of 100 points to express your judgment about the ratio of one criterion to the other one in pair. Please review the model and criteria before starting.

Example:

If Criterion 1 is 4 times more important than criterion 2, criterion 1 will take 80 and criterion 2 will take 20.

If Criterion 1 is 2 times more important than criterion 2, criterion 1 will take 67 and criterion 2 will take 33.

You can assign any relative importance between the two criteria.

Level 1 (Objectives)

In your expert opinion, what is the relative priority for each criterion over the other when planning for energy policy to promote wind energy adoption?

Criterion	Weight	Weight	Criterion
Economic			Social
Economic			Political
Economic			Environmental
Economic			Technical
Social			Political
Social			Environmental
Social			Technical
Political			Environmental
Political			Technical
Environmental			Technical

Level 2 (Goals)

In your expert opinion, under the economic criteria, what is the relative priority for each goal over the other?

Criterion	Weight	Weight	Criterion
Initial investment			Production cost
Initial investment			Future cost reductions
Production cost			Future cost reductions

In your expert opinion, under the social criteria, what is the relative priority for each goal over the other?

Criterion	Weight	Weight	Criterion
Created jobs/new			Public acceptance
Created jobs/new			Customers willing to pay
Public acceptance			Customers willing to pay

In your expert opinion, under the political criteria, what is the relative priority for each goal over the other?

Criterion	Weight	Weight	Criterion
Technology R&D			Mandated standards

In your expert opinion, under the Environmental criteria, what is the relative priority for each goal over the other?

Criterion	Weight	Weight	Criterion
GHG emissions			Land conservation
GHG emissions			Wildlife protection
Land conservation			Wildlife protection

In your expert opinion, under the Technical criteria, what is the relative priority for each goal over the other?

Criterion	Weight	Weight	Criterion
Grid access			Transmission capabilities
Grid access			Balancing reserves
Grid access			Smart grid
Transmission capabilities			Balancing reserves
Transmission capabilities			Smart grid
Balancing reserves			Smart grid

Level 3 (Alternatives)

In your expert opinion, for the goal of reducing initial investment, what is the relative priority of each policy (alternative) over the other?

Alternative	Weight	Weight	Alternative
Green power option			Renewable energy credits
Green power option			Tax credits
Green power option			Investment funds
Green power option			RPS
Renewable energy credits			Tax credits
Renewable energy credits			Investment funds
Renewable energy credits			RPS
Tax credits			Investment funds
Tax credits			RPS
Investment funds			RPS

In your expert opinion, for the goal of reducing the production cost, what is the relative priority of each policy (alternative) over the other?

Alternative	Weight	Weight	Alternative
Green power option			Renewable energy credits
Green power option			Tax credits
Green power option			Investment funds
Green power option			RPS
Renewable energy credits			Tax credits
Renewable energy credits			Investment funds
Renewable energy credits			RPS
Tax credits			Investment funds
Tax credits			RPS
Investment funds			RPS

In your expert opinion, for the goal of future cost reduction, what is the relative priority of each policy (alternative) over the other?

Alternative	Weight	Weight	Alternative
Green power option			Renewable energy credits
Green power option			Tax credits
Green power option			Investment funds
Green power option			RPS
Renewable energy credits			Tax credits
Renewable energy credits			Investment funds
Renewable energy credits			RPS
Tax credits			Investment funds
Tax credits			RPS
Investment funds			RPS

In your expert opinion, for the goal of creating new jobs, what is the relative priority of each policy (alternative) over the other?

Alternative	Weight	Weight	Alternative
Green power option			Renewable energy credits
Green power option			Tax credits
Green power option			Investment funds
Green power option			RPS
Renewable energy credits			Tax credits
Renewable energy credits			Investment funds
Renewable energy credits			RPS
Tax credits			Investment funds
Tax credits			RPS
Investment funds			RPS

In your expert opinion, for the goal of increasing public acceptance, what is the relative priority of each policy (alternative) over the other?

Alternative	Weight	Weight	Alternative
Green power option			Renewable energy credits
Green power option			Tax credits
Green power option			Investment funds
Green power option			RPS
Renewable energy credits			Tax credits
Renewable energy credits			Investment funds
Renewable energy credits			RPS
Tax credits			Investment funds
Tax credits			RPS
Investment funds			RPS

In your expert opinion, for the goal of increasing customers willing to pay, what is the relative priority of each policy (alternative) over the other?

Alternative	Weight	Weight	Alternative
Green power option			Renewable energy credits
Green power option			Tax credits
Green power option			Investment funds
Green power option			RPS
Renewable energy credits			Tax credits
Renewable energy credits			Investment funds
Renewable energy credits			RPS
Tax credits			Investment funds
Tax credits			RPS
Investment funds			RPS

In your expert opinion, for the goal of reducing GHG emissions, what is the relative priority of each policy (alternative) over the other?

Alternative	Weight	Weight	Alternative
Green power option			Renewable energy credits
Green power option			Tax credits
Green power option			Investment funds
Green power option			RPS
Renewable energy credits			Tax credits
Renewable energy credits			Investment funds
Renewable energy credits			RPS
Tax credits			Investment funds
Tax credits			RPS
Investment funds			RPS

In your expert opinion, for the goal of enhancing land conservation, what is the relative priority of each policy (alternative) over the other?

Alternative	Weight	Weight	Alternative
Green power option			Renewable energy credits
Green power option			Tax credits
Green power option			Investment funds
Green power option			RPS
Renewable energy credits			Tax credits
Renewable energy credits			Investment funds
Renewable energy credits			RPS
Tax credits			Investment funds
Tax credits			RPS
Investment funds			RPS

In your expert opinion, for the goal of protecting wild life, what is the relative priority of each policy (alternative) over the other?

Alternative	Weight	Weight	Alternative
Green power option			Renewable energy credits
Green power option			Tax credits
Green power option			Investment funds
Green power option			RPS
Renewable energy credits			Tax credits
Renewable energy credits			Investment funds
Renewable energy credits			RPS
Tax credits			Investment funds
Tax credits			RPS
Investment funds			RPS

In your expert opinion, for the goal of developing more R&D, what is the relative priority of each policy (alternative) over the other?

Alternative	Weight	Weight	Alternative
Green power option			Renewable energy credits
Green power option			Tax credits
Green power option			Investment funds
Green power option			RPS
Renewable energy credits			Tax credits
Renewable energy credits			Investment funds
Renewable energy credits			RPS
Tax credits			Investment funds
Tax credits			RPS
Investment funds			RPS

In your expert opinion, for the goal of forcing codes and standards, what is the relative priority of each policy (alternative) over the other?

Alternative	Weight	Weight	Alternative
Green power option			Renewable energy credits
Green power option			Tax credits
Green power option			Investment funds
Green power option			RPS
Renewable energy credits			Tax credits
Renewable energy credits			Investment funds
Renewable energy credits			RPS
Tax credits			Investment funds
Tax credits			RPS
Investment funds			RPS

In your expert opinion, for the goal of giving better grid access, what is the relative priority of each policy (alternative) over the other?

Alternative	Weight	Weight	Alternative
Green power option			Renewable energy credits
Green power option			Tax credits
Green power option			Investment funds
Green power option			RPS
Renewable energy credits			Tax credits
Renewable energy credits			Investment funds
Renewable energy credits			RPS
Tax credits			Investment funds
Tax credits			RPS
Investment funds			RPS

In your expert opinion, for the goal of increasing transmission capabilities, what is the relative priority of each policy (alternative) over the other?

Alternative	Weight	Weight	Alternative
Green power option			Renewable energy credits
Green power option			Tax credits
Green power option			Investment funds
Green power option			RPS
Renewable energy credits			Tax credits
Renewable energy credits			Investment funds
Renewable energy credits			RPS
Tax credits			Investment funds
Tax credits			RPS
Investment funds			RPS

In your expert opinion, for the goal of ensuring balancing of reserves, what is the relative priority of each policy (alternative) over the other?

Alternative	Weight	Weight	Alternative
Green power option			Renewable energy credits
Green power option			Tax credits
Green power option			Investment funds
Green power option			RPS
Renewable energy credits			Tax credits
Renewable energy credits			Investment funds
Renewable energy credits			RPS
Tax credits			Investment funds
Tax credits			RPS
Investment funds			RPS

In your expert opinion, for the goal of smart grid integration, what is the relative priority of each policy (alternative) over the other?

Alternative	Weight	Weight	Alternative
Green power option			Renewable energy credits
Green power option			Tax credits
Green power option			Investment funds
Green power option			RPS
Renewable energy credits			Tax credits
Renewable energy credits			Investment funds
Renewable energy credits			RPS
Tax credits			Investment funds
Tax credits			RPS
Investment funds			RPS

PCM Software Results for Pairwise Comparison

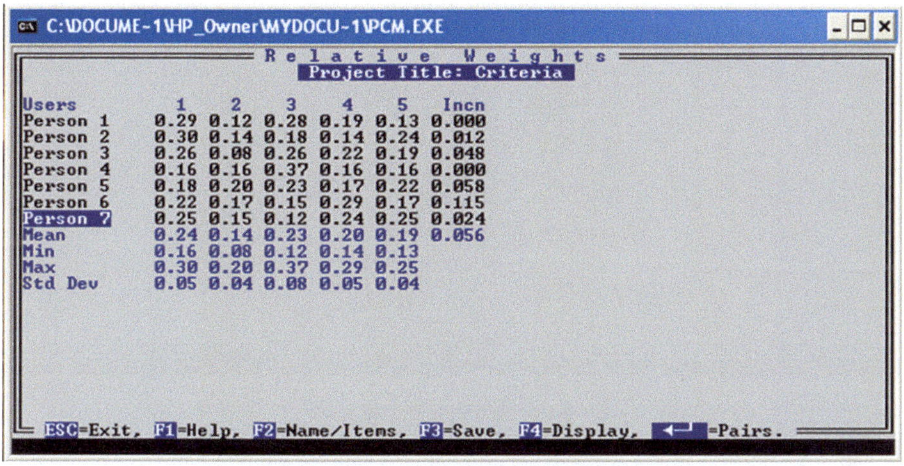

C:\DOCUME~1\HP_Owner\MYDOCU~1\PCM.EXE

Relative Weights

Project Title: Criteria

Users	1	2	3	4	5	Incn
Person 1	0.29	0.12	0.28	0.19	0.13	0.000
Person 2	0.30	0.14	0.18	0.14	0.24	0.012
Person 3	0.26	0.08	0.26	0.22	0.19	0.048
Person 4	0.16	0.16	0.37	0.16	0.16	0.000
Person 5	0.18	0.20	0.23	0.17	0.22	0.058
Person 6	0.22	0.17	0.15	0.29	0.17	0.115
Person 7	0.25	0.15	0.12	0.24	0.25	0.024
Mean	0.24	0.14	0.23	0.20	0.19	0.056
Min	0.16	0.08	0.12	0.14	0.13	
Max	0.30	0.20	0.37	0.29	0.25	
Std Dev	0.05	0.04	0.08	0.05	0.04	

ESC=Exit, F1=Help, F2=Name/Items, F3=Save, F4=Display, ↵=Pairs.

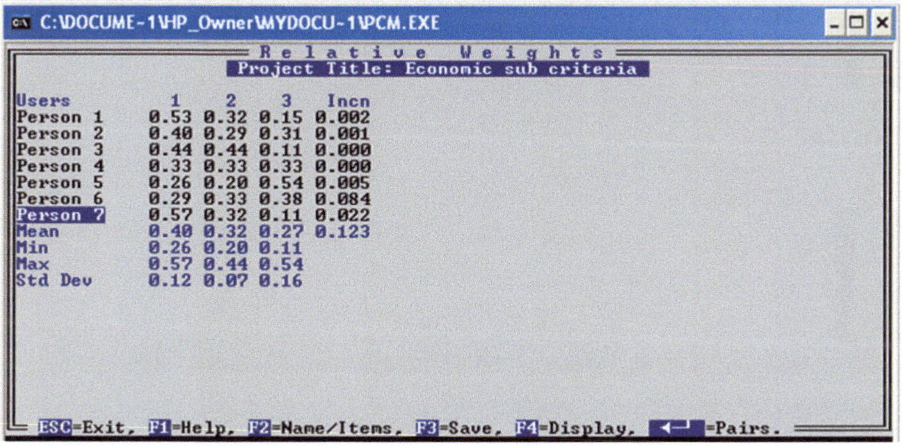

C:\DOCUME~1\HP_Owner\MYDOCU~1\PCM.EXE

Relative Weights

Project Title: Economic sub criteria

Users	1	2	3	Incn
Person 1	0.53	0.32	0.15	0.002
Person 2	0.40	0.29	0.31	0.001
Person 3	0.44	0.44	0.11	0.000
Person 4	0.33	0.33	0.33	0.000
Person 5	0.26	0.20	0.54	0.005
Person 6	0.29	0.33	0.38	0.084
Person 7	0.57	0.32	0.11	0.022
Mean	0.40	0.32	0.27	0.123
Min	0.26	0.20	0.11	
Max	0.57	0.44	0.54	
Std Dev	0.12	0.07	0.16	

ESC=Exit, F1=Help, F2=Name/Items, F3=Save, F4=Display, ↵=Pairs.

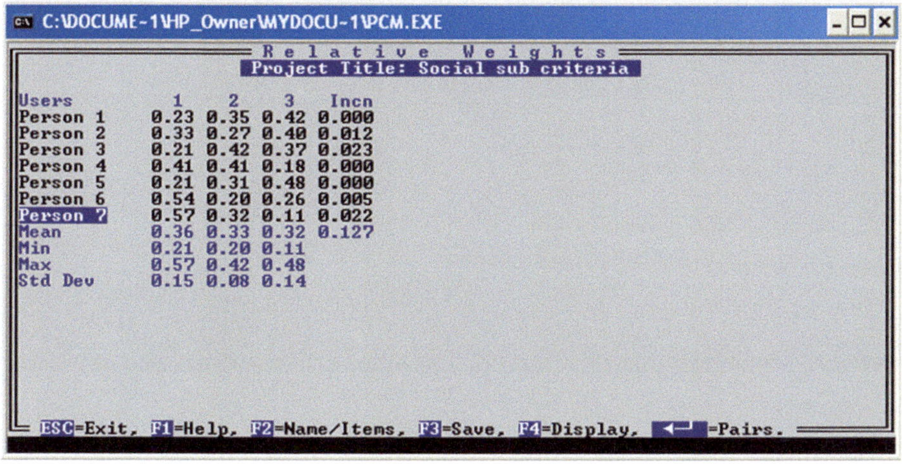

C:\DOCUME~1\HP_Owner\MYDOCU~1\PCM.EXE

Relative Weights

Project Title: Social sub criteria

Users	1	2	3	Incn
Person 1	0.23	0.35	0.42	0.000
Person 2	0.33	0.27	0.40	0.012
Person 3	0.21	0.42	0.37	0.023
Person 4	0.41	0.41	0.18	0.000
Person 5	0.21	0.31	0.48	0.000
Person 6	0.54	0.20	0.26	0.005
Person 7	0.57	0.32	0.11	0.022
Mean	0.36	0.33	0.32	0.127
Min	0.21	0.20	0.11	
Max	0.57	0.42	0.48	
Std Dev	0.15	0.08	0.14	

ESC=Exit, F1=Help, F2=Name/Items, F3=Save, F4=Display, ↵=Pairs.

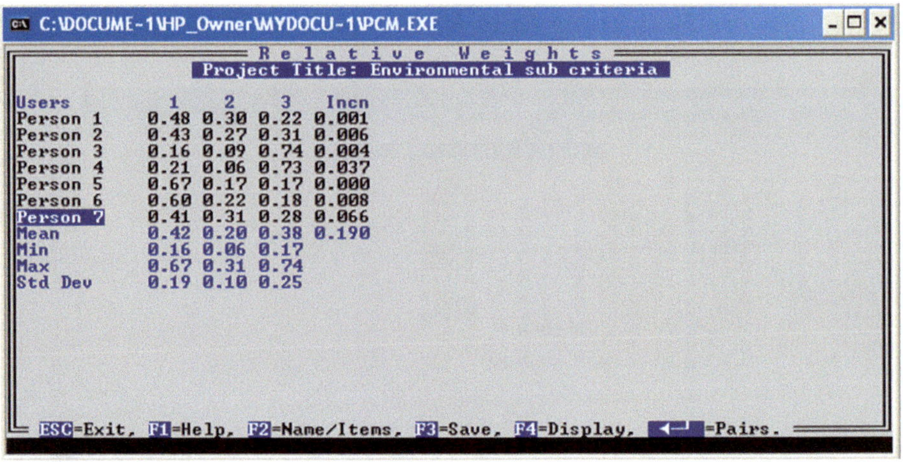

C:\DOCUME~1\HP_Owner\MYDOCU~1\PCM.EXE

Relative Weights

Project Title: Environmental sub criteria

Users	1	2	3	Incn
Person 1	0.48	0.30	0.22	0.001
Person 2	0.43	0.27	0.31	0.006
Person 3	0.16	0.09	0.74	0.004
Person 4	0.21	0.06	0.73	0.037
Person 5	0.67	0.17	0.17	0.000
Person 6	0.60	0.22	0.18	0.008
Person 7	0.41	0.31	0.28	0.066
Mean	0.42	0.20	0.38	0.190
Min	0.16	0.06	0.17	
Max	0.67	0.31	0.74	
Std Dev	0.19	0.10	0.25	

ESC=Exit, F1=Help, F2=Name/Items, F3=Save, F4=Display, ←┘=Pairs.

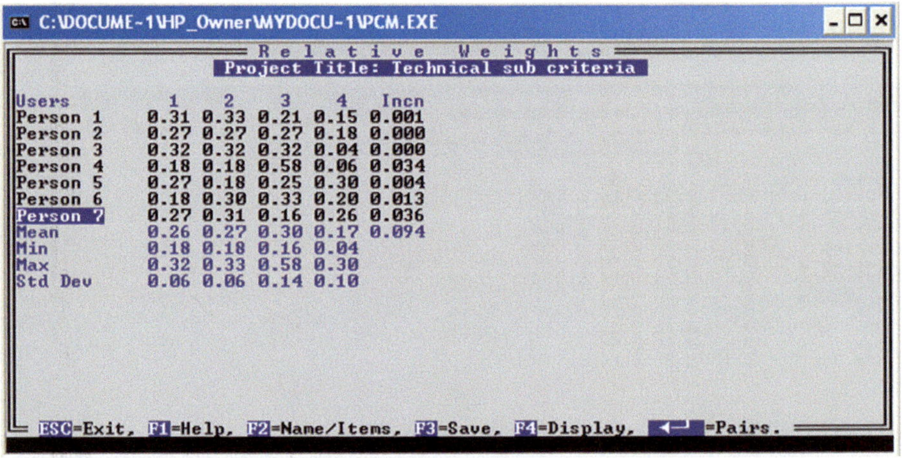

C:\DOCUME~1\HP_Owner\MYDOCU~1\PCM.EXE

Relative Weights

Project Title: Technical sub criteria

Users	1	2	3	4	Incn
Person 1	0.31	0.33	0.21	0.15	0.001
Person 2	0.27	0.27	0.27	0.18	0.000
Person 3	0.32	0.32	0.32	0.04	0.000
Person 4	0.18	0.18	0.58	0.06	0.034
Person 5	0.27	0.18	0.25	0.30	0.004
Person 6	0.18	0.30	0.33	0.20	0.013
Person 7	0.27	0.31	0.16	0.26	0.036
Mean	0.26	0.27	0.30	0.17	0.094
Min	0.18	0.18	0.16	0.04	
Max	0.32	0.33	0.58	0.30	
Std Dev	0.06	0.06	0.14	0.10	

ESC=Exit, F1=Help, F2=Name/Items, F3=Save, F4=Display, ←┘=Pairs.

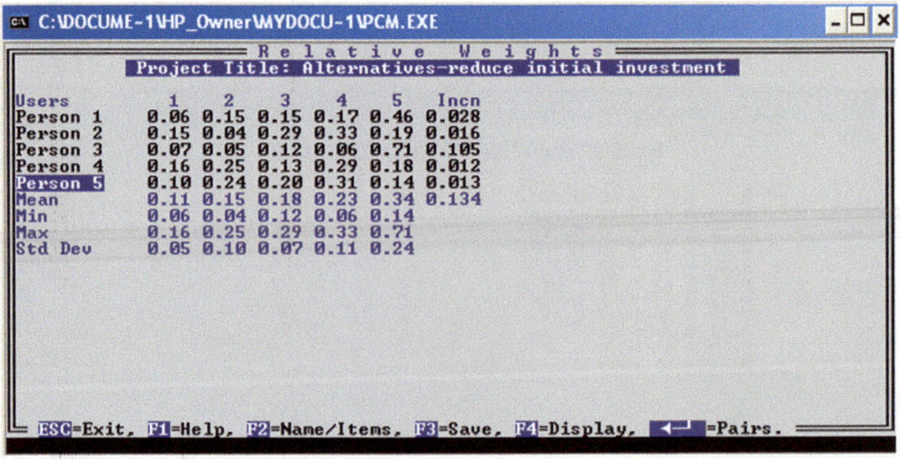

C:\DOCUME~1\HP_Owner\MYDOCU~1\PCM.EXE

Relative Weights

Project Title: Alternatives-reduce initial investment

Users	1	2	3	4	5	Incn
Person 1	0.06	0.15	0.15	0.17	0.46	0.028
Person 2	0.15	0.04	0.29	0.33	0.19	0.016
Person 3	0.07	0.05	0.12	0.06	0.71	0.105
Person 4	0.16	0.25	0.13	0.29	0.18	0.012
Person 5	0.10	0.24	0.20	0.31	0.14	0.013
Mean	0.11	0.15	0.18	0.23	0.34	0.134
Min	0.06	0.04	0.12	0.06	0.14	
Max	0.16	0.25	0.29	0.33	0.71	
Std Dev	0.05	0.10	0.07	0.11	0.24	

ESC=Exit, F1=Help, F2=Name/Items, F3=Save, F4=Display, ←┘=Pairs.

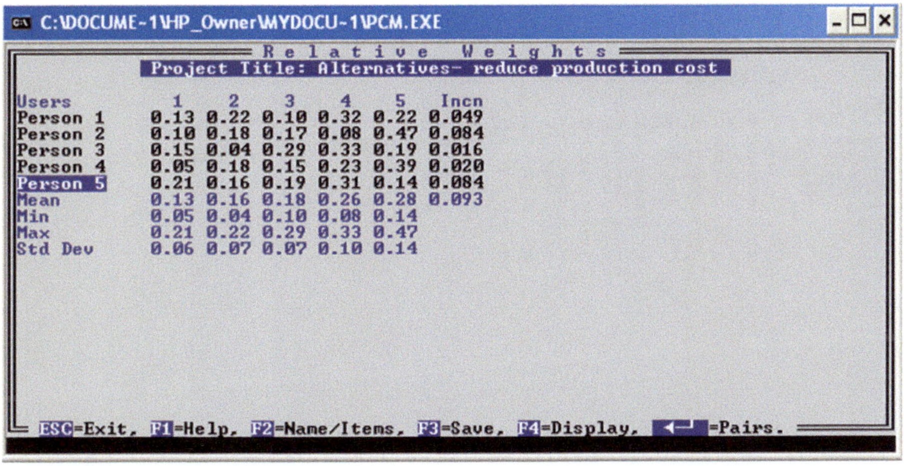

C:\DOCUME~1\HP_Owner\MYDOCU~1\PCM.EXE

Relative Weights

Project Title: Alternatives- reduce production cost

Users	1	2	3	4	5	Incn
Person 1	0.13	0.22	0.10	0.32	0.22	0.049
Person 2	0.10	0.18	0.17	0.08	0.47	0.084
Person 3	0.15	0.04	0.29	0.33	0.19	0.016
Person 4	0.05	0.18	0.15	0.23	0.39	0.020
Person 5	0.21	0.16	0.19	0.31	0.14	0.084
Mean	0.13	0.16	0.18	0.26	0.28	0.093
Min	0.05	0.04	0.10	0.08	0.14	
Max	0.21	0.22	0.29	0.33	0.47	
Std Dev	0.06	0.07	0.07	0.10	0.14	

ESC=Exit, F1=Help, F2=Name/Items, F3=Save, F4=Display, ←┘=Pairs.

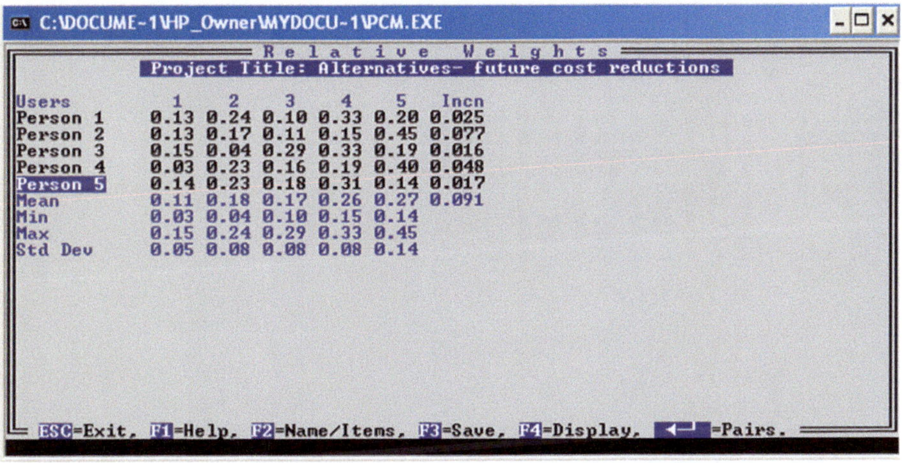

C:\DOCUME~1\HP_Owner\MYDOCU~1\PCM.EXE

Relative Weights

Project Title: Alternatives- future cost reductions

Users	1	2	3	4	5	Incn
Person 1	0.13	0.24	0.10	0.33	0.20	0.025
Person 2	0.13	0.17	0.11	0.15	0.45	0.077
Person 3	0.15	0.04	0.29	0.33	0.19	0.016
Person 4	0.03	0.23	0.16	0.19	0.40	0.048
Person 5	0.14	0.23	0.18	0.31	0.14	0.017
Mean	0.11	0.18	0.17	0.26	0.27	0.091
Min	0.03	0.04	0.10	0.15	0.14	
Max	0.15	0.24	0.29	0.33	0.45	
Std Dev	0.05	0.08	0.08	0.08	0.14	

ESC=Exit, F1=Help, F2=Name/Items, F3=Save, F4=Display, ←┘=Pairs.

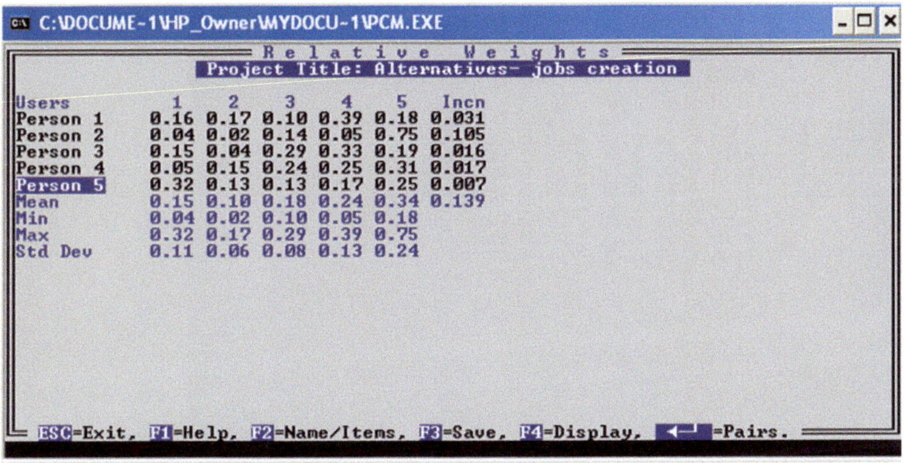

C:\DOCUME~1\HP_Owner\MYDOCU~1\PCM.EXE

Relative Weights

Project Title: Alternatives- jobs creation

Users	1	2	3	4	5	Incn
Person 1	0.16	0.17	0.10	0.39	0.18	0.031
Person 2	0.04	0.02	0.14	0.05	0.75	0.105
Person 3	0.15	0.04	0.29	0.33	0.19	0.016
Person 4	0.05	0.15	0.24	0.25	0.31	0.017
Person 5	0.32	0.13	0.13	0.17	0.25	0.007
Mean	0.15	0.10	0.18	0.24	0.34	0.139
Min	0.04	0.02	0.10	0.05	0.18	
Max	0.32	0.17	0.29	0.39	0.75	
Std Dev	0.11	0.06	0.08	0.13	0.24	

ESC=Exit, F1=Help, F2=Name/Items, F3=Save, F4=Display, ←┘=Pairs.

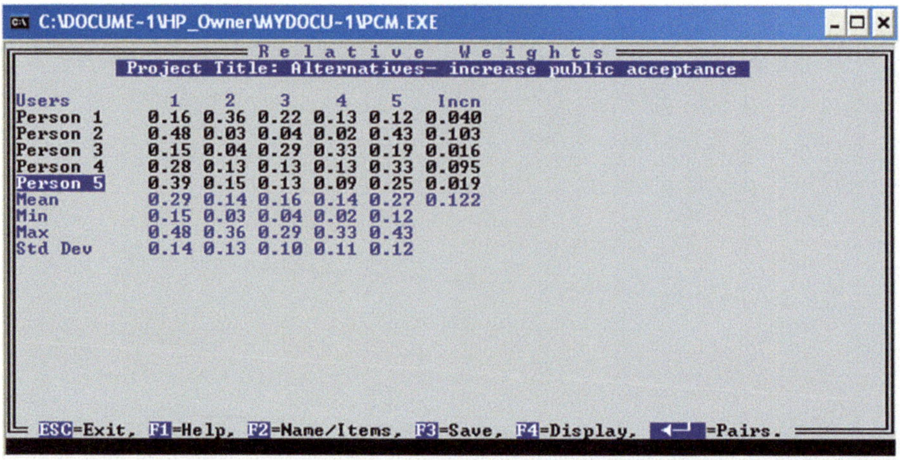

C:\DOCUME~1\HP_Owner\MYDOCU~1\PCM.EXE

Relative Weights

Project Title: Alternatives- increase public acceptance

Users	1	2	3	4	5	Incn
Person 1	0.16	0.36	0.22	0.13	0.12	0.040
Person 2	0.48	0.03	0.04	0.02	0.43	0.103
Person 3	0.15	0.04	0.29	0.33	0.19	0.016
Person 4	0.28	0.13	0.13	0.13	0.33	0.095
Person 5	0.39	0.15	0.13	0.09	0.25	0.019
Mean	0.29	0.14	0.16	0.14	0.27	0.122
Min	0.15	0.03	0.04	0.02	0.12	
Max	0.48	0.36	0.29	0.33	0.43	
Std Dev	0.14	0.13	0.10	0.11	0.12	

ESC=Exit, F1=Help, F2=Name/Items, F3=Save, F4=Display, ←┘=Pairs.

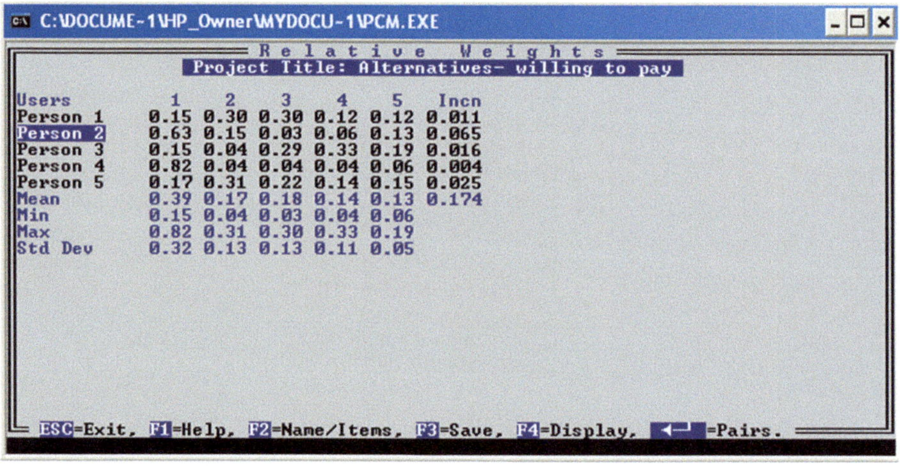

C:\DOCUME~1\HP_Owner\MYDOCU~1\PCM.EXE

Relative Weights

Project Title: Alternatives- willing to pay

Users	1	2	3	4	5	Incn
Person 1	0.15	0.30	0.30	0.12	0.12	0.011
Person 2	0.63	0.15	0.03	0.06	0.13	0.065
Person 3	0.15	0.04	0.29	0.33	0.19	0.016
Person 4	0.82	0.04	0.04	0.04	0.06	0.004
Person 5	0.17	0.31	0.22	0.14	0.15	0.025
Mean	0.39	0.17	0.18	0.14	0.13	0.174
Min	0.15	0.04	0.03	0.04	0.06	
Max	0.82	0.31	0.30	0.33	0.19	
Std Dev	0.32	0.13	0.13	0.11	0.05	

ESC=Exit, F1=Help, F2=Name/Items, F3=Save, F4=Display, ←┘=Pairs.

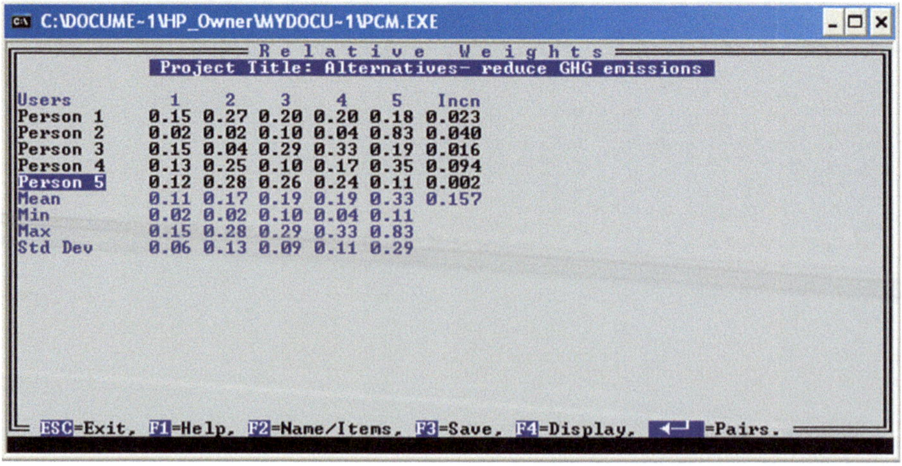

C:\DOCUME~1\HP_Owner\MYDOCU~1\PCM.EXE

Relative Weights

Project Title: Alternatives- reduce GHG emissions

Users	1	2	3	4	5	Incn
Person 1	0.15	0.27	0.20	0.20	0.18	0.023
Person 2	0.02	0.02	0.10	0.04	0.83	0.040
Person 3	0.15	0.04	0.29	0.33	0.19	0.016
Person 4	0.13	0.25	0.10	0.17	0.35	0.094
Person 5	0.12	0.28	0.26	0.24	0.11	0.002
Mean	0.11	0.17	0.19	0.19	0.33	0.157
Min	0.02	0.02	0.10	0.04	0.11	
Max	0.15	0.28	0.29	0.33	0.83	
Std Dev	0.06	0.13	0.09	0.11	0.29	

ESC=Exit, F1=Help, F2=Name/Items, F3=Save, F4=Display, ←┘=Pairs.

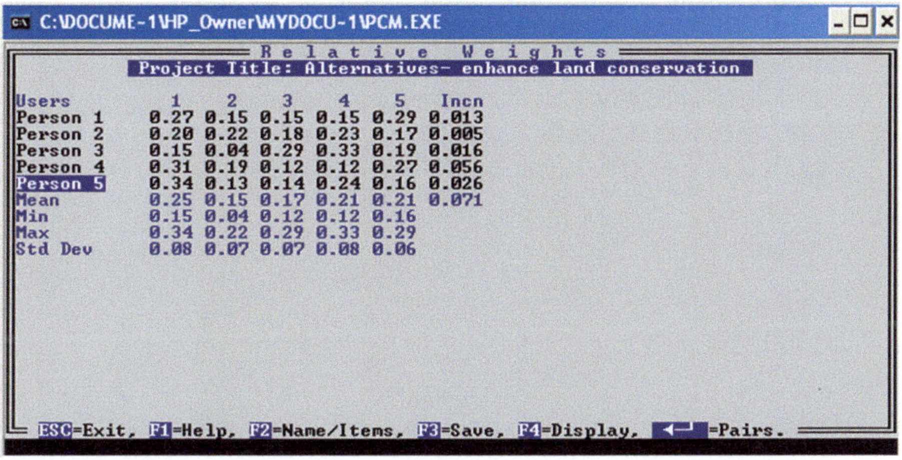
C:\DOCUME~1\HP_Owner\MYDOCU~1\PCM.EXE

Relative Weights

Project Title: Alternatives- enhance land conservation

Users	1	2	3	4	5	Incn
Person 1	0.27	0.15	0.15	0.15	0.29	0.013
Person 2	0.20	0.22	0.18	0.23	0.17	0.005
Person 3	0.15	0.04	0.29	0.33	0.19	0.016
Person 4	0.31	0.19	0.12	0.12	0.27	0.056
Person 5	0.34	0.13	0.14	0.24	0.16	0.026
Mean	0.25	0.15	0.17	0.21	0.21	0.071
Min	0.15	0.04	0.12	0.12	0.16	
Max	0.34	0.22	0.29	0.33	0.29	
Std Dev	0.08	0.07	0.07	0.08	0.06	

ESC=Exit, F1=Help, F2=Name/Items, F3=Save, F4=Display, ←┘ =Pairs.

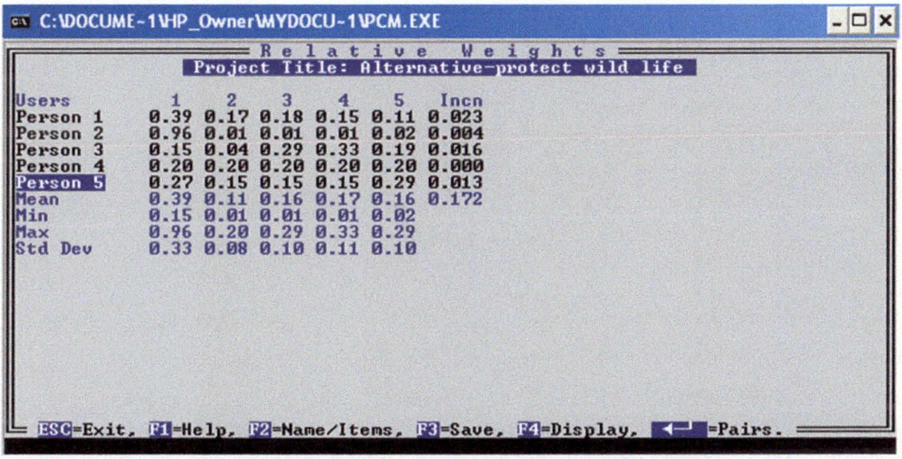
C:\DOCUME~1\HP_Owner\MYDOCU~1\PCM.EXE

Relative Weights

Project Title: Alternative-protect wild life

Users	1	2	3	4	5	Incn
Person 1	0.39	0.17	0.18	0.15	0.11	0.023
Person 2	0.96	0.01	0.01	0.01	0.02	0.004
Person 3	0.15	0.04	0.29	0.33	0.19	0.016
Person 4	0.20	0.20	0.20	0.20	0.20	0.000
Person 5	0.27	0.15	0.15	0.15	0.29	0.013
Mean	0.39	0.11	0.16	0.17	0.16	0.172
Min	0.15	0.01	0.01	0.01	0.02	
Max	0.96	0.20	0.29	0.33	0.29	
Std Dev	0.33	0.08	0.10	0.11	0.10	

ESC=Exit, F1=Help, F2=Name/Items, F3=Save, F4=Display, ←┘ =Pairs.

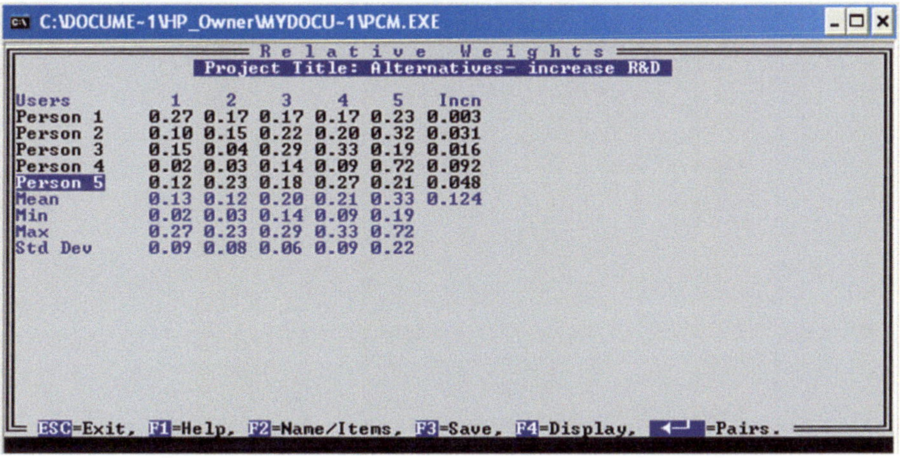
C:\DOCUME~1\HP_Owner\MYDOCU~1\PCM.EXE

Relative Weights

Project Title: Alternatives- increase R&D

Users	1	2	3	4	5	Incn
Person 1	0.27	0.17	0.17	0.17	0.23	0.003
Person 2	0.10	0.15	0.22	0.20	0.32	0.031
Person 3	0.15	0.04	0.29	0.33	0.19	0.016
Person 4	0.02	0.03	0.14	0.09	0.72	0.092
Person 5	0.12	0.23	0.18	0.27	0.21	0.048
Mean	0.13	0.12	0.20	0.21	0.33	0.124
Min	0.02	0.03	0.14	0.09	0.19	
Max	0.27	0.23	0.29	0.33	0.72	
Std Dev	0.09	0.08	0.06	0.09	0.22	

ESC=Exit, F1=Help, F2=Name/Items, F3=Save, F4=Display, ←┘ =Pairs.

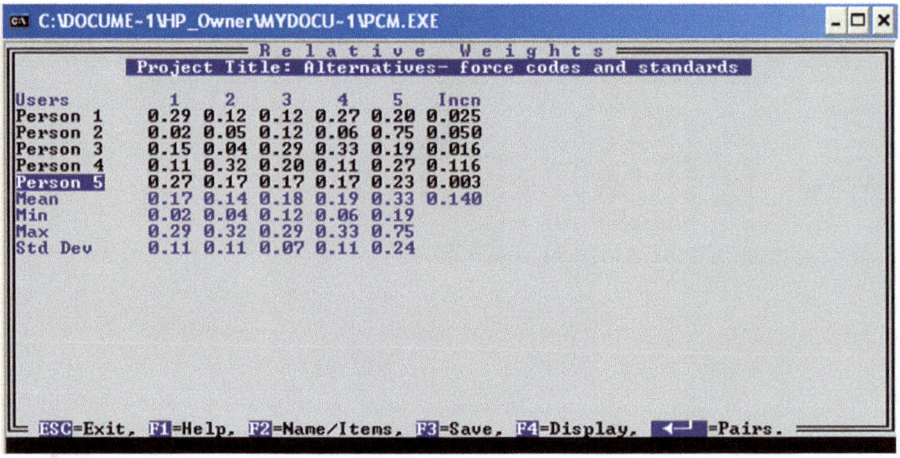

C:\DOCUME~1\HP_Owner\MYDOCU~1\PCM.EXE

Relative Weights

Project Title: Alternatives- force codes and standards

Users	1	2	3	4	5	Incn
Person 1	0.29	0.12	0.12	0.27	0.20	0.025
Person 2	0.02	0.05	0.12	0.06	0.75	0.050
Person 3	0.15	0.04	0.29	0.33	0.19	0.016
Person 4	0.11	0.32	0.20	0.11	0.27	0.116
Person 5	0.27	0.17	0.17	0.17	0.23	0.003
Mean	0.17	0.14	0.18	0.19	0.33	0.140
Min	0.02	0.04	0.12	0.06	0.19	
Max	0.29	0.32	0.29	0.33	0.75	
Std Dev	0.11	0.11	0.07	0.11	0.24	

ESC=Exit, F1=Help, F2=Name/Items, F3=Save, F4=Display, ←┘=Pairs.

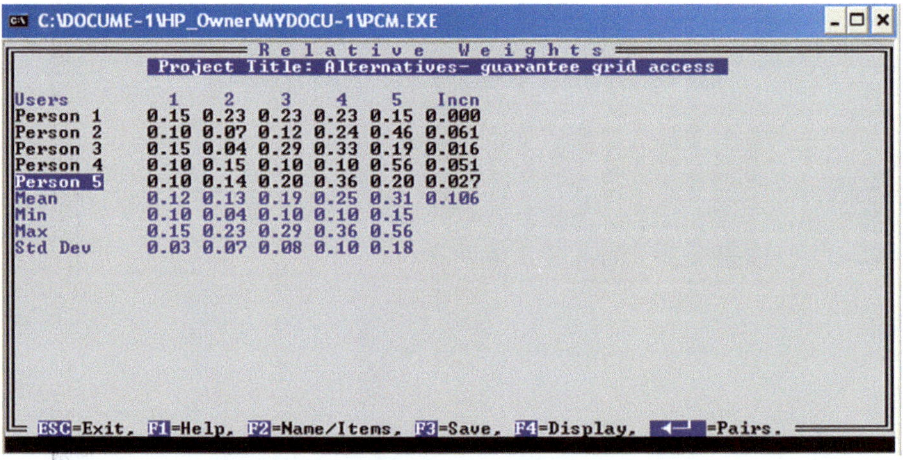

C:\DOCUME~1\HP_Owner\MYDOCU~1\PCM.EXE

Relative Weights

Project Title: Alternatives- guarantee grid access

Users	1	2	3	4	5	Incn
Person 1	0.15	0.23	0.23	0.23	0.15	0.000
Person 2	0.10	0.07	0.12	0.24	0.46	0.061
Person 3	0.15	0.04	0.29	0.33	0.19	0.016
Person 4	0.10	0.15	0.10	0.10	0.56	0.051
Person 5	0.10	0.14	0.20	0.36	0.20	0.027
Mean	0.12	0.13	0.19	0.25	0.31	0.106
Min	0.10	0.04	0.10	0.10	0.15	
Max	0.15	0.23	0.29	0.36	0.56	
Std Dev	0.03	0.07	0.08	0.10	0.18	

ESC=Exit, F1=Help, F2=Name/Items, F3=Save, F4=Display, ←┘=Pairs.

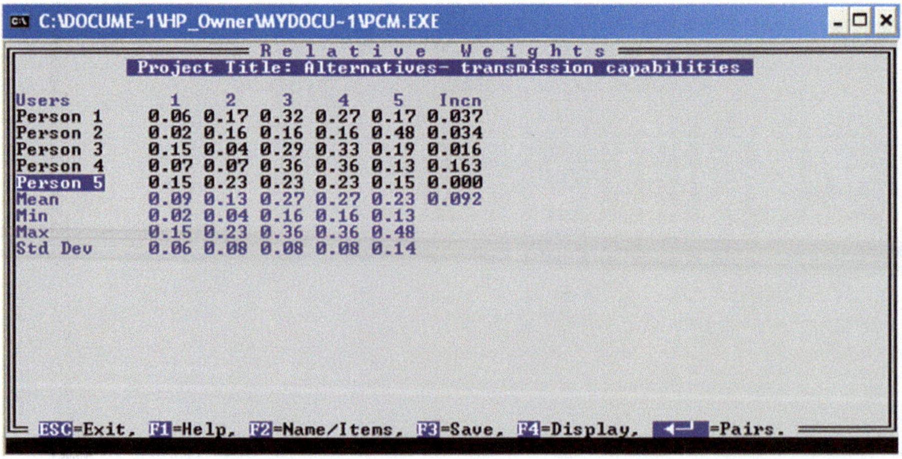

C:\DOCUME~1\HP_Owner\MYDOCU~1\PCM.EXE

Relative Weights

Project Title: Alternatives- transmission capabilities

Users	1	2	3	4	5	Incn
Person 1	0.06	0.17	0.32	0.27	0.17	0.037
Person 2	0.02	0.16	0.16	0.16	0.48	0.034
Person 3	0.15	0.04	0.29	0.33	0.19	0.016
Person 4	0.07	0.07	0.36	0.36	0.13	0.163
Person 5	0.15	0.23	0.23	0.23	0.15	0.000
Mean	0.09	0.13	0.27	0.27	0.23	0.092
Min	0.02	0.04	0.16	0.16	0.13	
Max	0.15	0.23	0.36	0.36	0.48	
Std Dev	0.06	0.08	0.08	0.08	0.14	

ESC=Exit, F1=Help, F2=Name/Items, F3=Save, F4=Display, ←┘=Pairs.

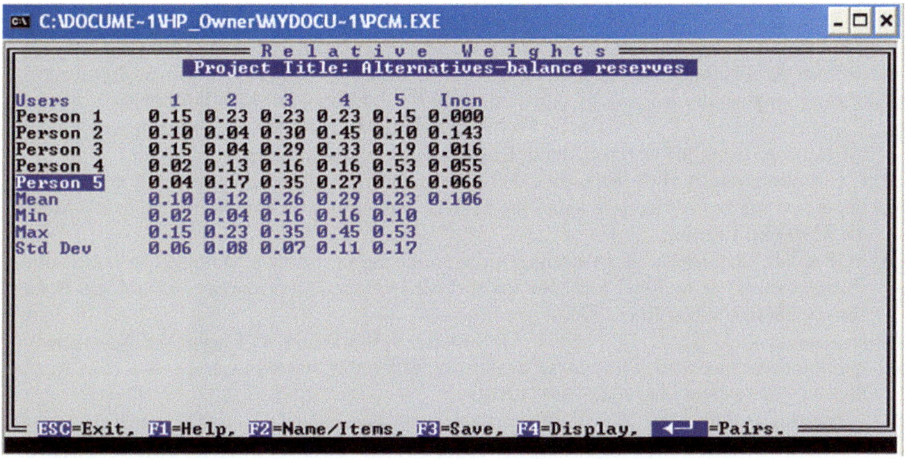

C:\DOCUME~1\HP_Owner\MYDOCU~1\PCM.EXE

Relative Weights

Project Title: Alternatives-balance reserves

Users	1	2	3	4	5	Incn
Person 1	0.15	0.23	0.23	0.23	0.15	0.000
Person 2	0.10	0.04	0.30	0.45	0.10	0.143
Person 3	0.15	0.04	0.29	0.33	0.19	0.016
Person 4	0.02	0.13	0.16	0.16	0.53	0.055
Person 5	0.04	0.17	0.35	0.27	0.16	0.066
Mean	0.10	0.12	0.26	0.29	0.23	0.106
Min	0.02	0.04	0.16	0.16	0.10	
Max	0.15	0.23	0.35	0.45	0.53	
Std Dev	0.06	0.08	0.07	0.11	0.17	

ESC=Exit, F1=Help, F2=Name/Items, F3=Save, F4=Display, ←┘=Pairs.

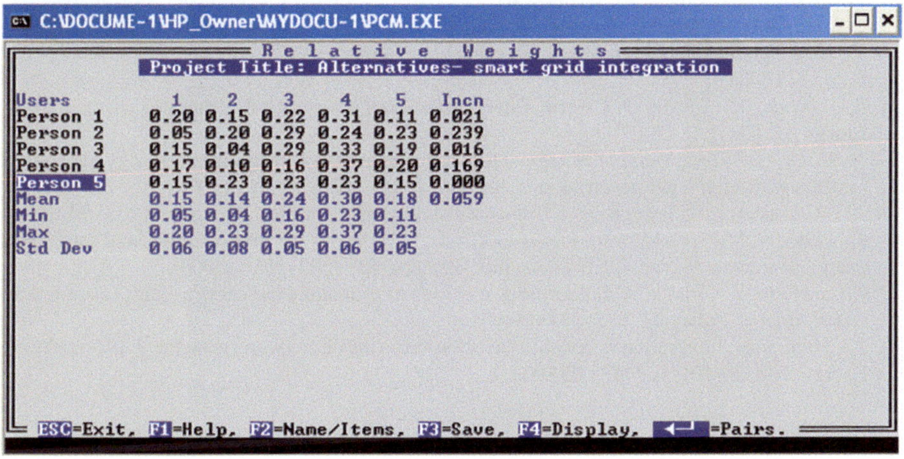

C:\DOCUME~1\HP_Owner\MYDOCU~1\PCM.EXE

Relative Weights

Project Title: Alternatives- smart grid integration

Users	1	2	3	4	5	Incn
Person 1	0.20	0.15	0.22	0.31	0.11	0.021
Person 2	0.05	0.20	0.29	0.24	0.23	0.239
Person 3	0.15	0.04	0.29	0.33	0.19	0.016
Person 4	0.17	0.10	0.16	0.37	0.20	0.169
Person 5	0.15	0.23	0.23	0.23	0.15	0.000
Mean	0.15	0.14	0.24	0.30	0.18	0.059
Min	0.05	0.04	0.16	0.23	0.11	
Max	0.20	0.23	0.29	0.37	0.23	
Std Dev	0.06	0.08	0.05	0.06	0.05	

ESC=Exit, F1=Help, F2=Name/Items, F3=Save, F4=Display, ←┘=Pairs.

References

1. J.B. Welch, A. Venkateswaran, The dual sustainability of wind energy. Renew. Sustain. Energy Rev. **13**, 1121–1126 (2009)
2. P. Gerdsri, D.F. Kocaoglu, HDM for developing national emerging technology strategy and policy supporting sustainable economy: a case study of nanotechnology for Thailand's agriculture, in *Portland International Conference on Management Engineering and Technology (PICMET 2008)* (2008), pp. 344–350
3. Oregon Department of Energy Oregon's Power Mix (2012), http://www.oregon.gov/ENERGY/Oregons_Electric_Power_Mix.shtml
4. S. Cantono, G. Silverberg, A percolation model of eco-innovation diffusion: the relationship between diffusion, learning economies and subsidies. Tech. Forecast. Soc. Chan. **76**, 487–496 (2009)

5. D. Diakoulaki, C. Zopounidis, G. Mavrotas, M. Doumpos, The use of a preference disaggregation method in energy analysis and policy making. Energy **24**, 157–166 (1999)
6. S. Roth, S. Hirschberg, C. Bauer, P. Burgherr, R. Dones, T. Heck, W. Schenler, Sustainability of electricity supply technology portfolio. Ann. Nucl. Energy **36**, 409–416 (2009)
7. I. Chatzimouratidis, P.A. Pilavachi, Multicriteria evaluation of power plants impact on the living standard using the analytic hierarchy process. Energy Policy **36**, 1074–1089 (2008)
8. I. Chatzimouratidis, P.A. Pilavachi, Sensitivity analysis of the evaluation of power plants impact on the living standard using the analytic hierarchy process. Energy Convers. Manag. **49**, 3599–3611 (2008)
9. M. Beccali, M. Cellura, M. Mistretta, Decision-making in energy planning. Application of the electre method at regional level for the diffusion of renewable energy technology. Renew. Energy **28**(10), 2063–2087 (2003)
10. N. Zografakis, E. Sifaki, M. Pagalou, G. Nikitaki, V. Psarakis, K.P. Tsagarakis, Assessment of public acceptance and willingness to pay for renewable energy sources in Crete. Renew. Sustain. Energy Rev. **14**, 1088–1095 (2010)
11. J. Zoellner, P. Schweizer-Ries, C. Wemheuer, Public acceptance of renewable energies: results from case studies in Germany. Energy Policy **36**(11), 4136–4141 (2008)
12. J.-J. Wang, Y.-Y. Jing, C.-F. Zhang, J.-H. Zhao, Review on multi-criteria decision analysis aid in sustainable energy decision-making. Renew. Sustain. Energy Rev. **13**, 2263–2278 (2009)
13. M. Jefferson, Sustainable energy development: performance and prospects. Renew. Energy **31**, 571–582 (2006)
14. R.U. Ayres, H. Turton, T. Casten, Energy efficiency, sustainability and economic growth. Energy **32**, 634–648 (2007)
15. M. Black, G. Strbac, Value of storage in providing balancing services for electricity generation systems with high wind penetration. J. Power Sources **162**, 949–953 (2006)
16. J.M. Carrasco, L.G. Franquelo, J.T. Bialasiewicz, E. Galvan, R.C.P. Guisado, M.A.M. Prats, J.I. Leon, N. Moreno-Alfonso, Power-electronic systems for the grid integration of renewable energy sources: a survey. IEEE Trans. Ind. Electron. **53**, 1002–1016 (2006)
17. F. Cavallaro, L. Ciraolo, A multicriteria approach to evaluate wind energy plants on an Italian island. Energy Policy **33**, 235–244 (2005)
18. T.J. Hammons, Integrating renewable energy sources into European grids. Int. J. Electr. Power Energy Syst. **30**(10), 462–475 (2008)

Chapter 4
Assessment of Wind Potential in Kalar Kahar Region by Comparing On-Site Data with NREL Wind Resource Map of Pakistan

Irfan Afzal Mirza, M. Shahid Khalil, Muhammad Amer, and Tugrul U. Daim

Abstract This chapter assesses the value of the wind resources in the Punjab area, Pakistan. Pakistan is an energy-deficient country having enormous potential of electricity generation from wind energy. Government of Pakistan (GoP) has taken firm measures to initiate wind power projects in Pakistan. A satellite mapping conducted by the United States Agency for International Development (USAID) in collaboration with the National Renewable Energy Laboratories (NREL), USA estimated a gross potential of 132 GW that can be exploited to generate wind energy all across the country. The most promising wind corridor of Gharo-Keti Bandar in southern coastal region offers a potential of 50 GW. The wind resource map developed by NREL also highlights wind potential in the upper Punjab and Balochistan. However, these regions are still not investigated and explored for wind potential due to unavailability of bankable wind data. Lack of credible wind resource data is considered the major reason towards underutilization of wind energy.

This chapter presents an investigation to validate the potential wind resources in the Punjab province. A wind strip is marked in the NREL map in the Punjab which shows good wind regime in the areas of the upper Punjab. To validate and verify the

I.A. Mirza
Renewable Resources (Pvt) Ltd., Karachi, Pakistan
e-mail: irfanmirza@renewableresources.com.pk

M.S. Khalil
University of Engineering and Technology, Taxila, Pakistan
e-mail: drmshahidk@gmail.com

M. Amer
Advanced Engineering Research Organization, Islamabad, Pakistan
e-mail: amer1992@gmail.com

T.U. Daim (✉)
Department of Engineering and Technology Management, Portland State University, 1900 SW 4th Avenue 50-02, Portland, OR 97201, USA
e-mail: tugrul@etm.pdx.edu

T.U. Daim et al. (eds.), *Policies and Programs for Sustainable Energy Innovations*, Innovation, Technology, and Knowledge Management,
DOI 10.1007/978-3-319-16033-7_4

NREL assessment, a suitable site in the Kalar Kahar region of the Punjab province was selected and an IEC (International Electrotechnical Commission) standard wind measuring mast was installed. Wind data was gathered for a period of 2 years and has been analyzed using Wind Atlas Analysis and Application Program (WAsP) software. This script presents the information about the site, mast installed, wind data received, and treatment of wind data as per international practices. The results of wind resource assessment were compared with the results shown in the NREL map and percentage difference is calculated. On the basis of comparison, conclusions and recommendations are derived for future deployment of wind energy projects in the Punjab region.

4.1 Energy in Pakistan

4.1.1 Overview of Pakistan

Pakistan emerged as a new country on 14 August 1947, after the division of former British India. It is located in South Asia, bordering the Arabian Sea on the south, India on the east, Iran and Afghanistan on the west, and China in the north. The country has coastline of approx. 1,100 km long with the Arabian Sea in the south. From Gwadar Bay in south-eastern corner, the country extends more than 1,800 km to the Khunjerab Pass on China's border. It is a densely populated country and covers 796,095 sq.km [1]. The total population is approximately 176 million [2]. The country has a literacy rate of 53 % [1].

The national economy is based on agriculture: wheat, cotton, rice, and sugar cane are the major crops. The country also has an expanding industry. Cotton, textiles, sugar, cement, and chemicals play important roles in the national economy. The textiles sector accounts for most of Pakistan's export earnings [3]. Over the 2004–2007 period, GDP growth in the 5–8 % range was spurred by gains in the industrial and service sectors, and between the 2001–2007 period, poverty levels decreased by 10 % as the government steadily increased the development budget [3]. However, the economic growth slowed down during the 2008–2009 period due to several reasons including the global financial crisis and severe electricity shortfalls [3].

Pakistan is divided into three major geographic areas: the northern highlands, the Indus River plain and the Balochistan Plateau. The northern highlands contain the Karakoram, Hindu Kush, and Pamir mountain ranges, which contain some of the world's highest peaks, including five out of the fourteen mountains with height above 18,000 m. The Balochistan Plateau lies in the west and the Thar Desert in the east. The 1,609 km long Indus River and its tributaries flow through the country from the Kashmir region to the Arabian Sea. There is an expanse of alluvial plains along it in Punjab and Sindh [4].

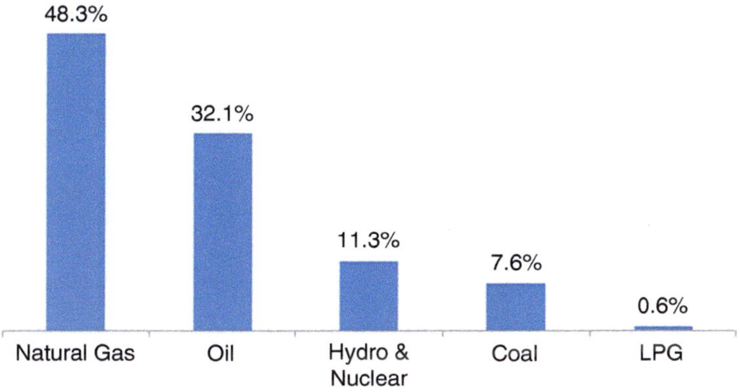

Fig. 4.1 Share of Pakistan's primary energy supplies by various sources during financial year 2008–2009 [6]

4.1.2 Overview of National Energy Sector

In this modern era, energy is a key element required for sustainable development and prosperity of a society. Pakistan is a developing country requiring sustainable sources of energy to foster a sustained economic growth and social development in the society [5]. The total primary energy supply (TPES) of Pakistan was 62.6 million tons of oil equivalent (MTOE) during the financial year 2008–2009 [6]. Energy resources like natural gas, oil, hydro and nuclear, coal, and liquefied petroleum gas (LPG) contribute to 48.3, 32.1, 11.3, 7.6, and 0.6 % of the primary energy supplies, respectively [6]. The share of the primary energy supplies by various sources is shown in Fig. 4.1.

Pakistan has a very limited fossil resource base, and the country's indigenous energy resources are insufficient to provide its economy with the necessary energy supplies [5]. Oil is the key resource of energy for electricity generation, and its import has put a heavy burden on the national economy [7, 8]. The national energy sector is heavily dependent on imported fossil-fuel. Presently, large hydropower dams are the only major renewable energy resource in the country for electricity generation. Usually, construction of large hydro dams results in a major relocation of people and changes in land use for the areas in which the dams are built. These projects have become controversial in Pakistan in recent years due to water shortages and significant impacts on the rivers, ecosystems, and surrounding communities. The large dams were developed in the 1970s, but the pace of new hydropower generation facilities has significantly slowed down over the last three decades due to the above-mentioned reasons [9]. The share of emerging renewable energy resources such as the use of wind energy for electricity generation is negligible in the country despite abundant renewable energy resources. It is shown in Fig. 4.2 that the share of electricity generated from wind energy is around 0.2 %, with the installed capacity of only 40 MW by the end of 2011 [10].

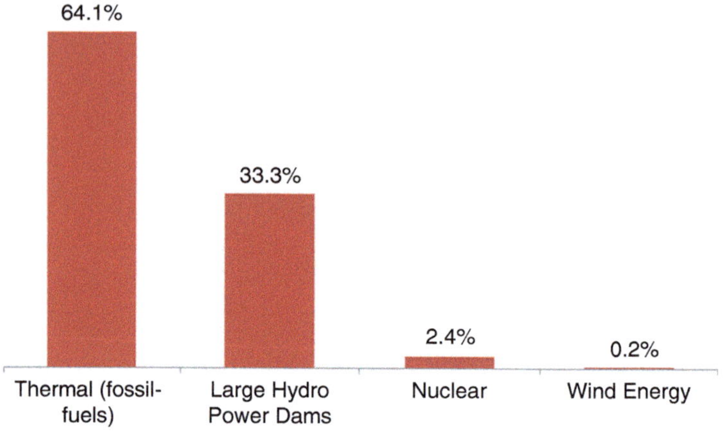

Fig. 4.2 Share of Pakistan's electricity from various sources [10]

The residential sector consumes almost 50 % of electricity produced in the country, followed by the industrial sector (26.7 %), agriculture sector (13.0 %), and the commercial sector (7.5 %) [8]. Given the current growth trends, it is expected that the demand of the domestic sector will further increase in the near future. The living standard is improving and there are more people living in large cities as compared to rural areas and rate of urbanization is 3.1 % [3]. There has been an increase in electricity consumers due to increase in the GDP, rapid urbanization, and extension of the national grid to include rural areas.

The power sector of Pakistan is dominated by two vertically integrated giants: Water and Power Development Authority (WAPDA) and Karachi Electric Supply Corporation (KESC). These two entities control the national electricity transmission and distribution network and generate almost 70 % of the country's power. Independent Power Producers (IPPs) produce 30 % of the country's electricity. In order to introduce a competitive environment and attract the private sector participation, in the year 2000, the Water and Power Development Authority (WAPDA) was restructured and unbundled into the following 12 separate units [5, 8]:

- Eight Distribution Companies (DISCOS)
- Three Thermal Generation Companies (GENCO)
- National Transmission and Dispatch Company (NTDC)

4.1.3 Energy Crisis in the Country

Pakistan is an energy-deficient country facing problems due to a shortage of energy, especially electricity. The country's electric sector is in a crisis because electricity demand continues to exceed supply, and it results in extended periods of blackouts (load shedding). It is widely recognized as a severe obstacle to growth and poverty reduction in the country [8]. The electricity deficit of the country was over 4,000 MW in the year 2008, and it is estimated to reach over 8,000 MW by the end 2011 [9]. Therefore, rotating blackouts throughout the country are also necessary to overcome this shortage. The load-shedding has caused significant damage to the national economy and the closure of industry; resulting in loss of production and jobs.

An increase in the electricity demand is directly linked to the growth of the country's economy. Research indicates that every one percent of GDP growth in Pakistan requires an increase in electricity supply of 1.25 %. Keeping in view the sustained growth in all sectors of the economy in the coming years, it is expected that the future demand for electricity will be more than 20,000 MW in the near future. Thus, an increase in electricity supply is required to sustain the economic growth [5]. This problem of electricity shortage will be further aggravated in the future because the national energy demand is also increasing at an average annual rate of 5.67 % [6].

The electricity crisis has forced the government to make decisions like early market shutdown, power cutoff to the industry, and two holidays per week for all businesses. These measures are negatively affecting all economic and business activities in the country. The extended periods of blackouts almost suspend the social life of people. There were also some riots over the power shortages in Pakistan. The per capita electricity consumption for Pakistan is 475 kWh, which is almost six times less than the average electricity consumption in the world [11]. Access to electricity is essential to provide modern health services, improve agricultural productivity, obtain the full benefits of improved educational systems, and build an economic base that can participate in today's globalized economy [5]. The shortage of electricity and frequent blackouts constrain economic development and disrupt health, education, and other services. Moreover, unreliable electricity service also undermines the cold chain vital to the distribution of medicines and perishable foods, and negatively affects public health.

Due to the shortage of electricity in Pakistan, the industrial sector has been badly affected and overall exports of the country have been reduced. Unreliable power supplies and frequent blackouts have encouraged the industries and businesses to install their own power supplies such as diesel generators. However, high operating costs of these generators raises the cost of local products and erodes their competitiveness within the region [5]. The shortage of conventional energy resources in Pakistan, when coupled with hiking energy prices worldwide, highlights a need to explore wind power in order to overcome the energy crisis in the country. Therefore, it is crucial for the country to formulate a diverse energy strategy and increase the share of sustainable energy resources by exploiting renewable energy technologies (RETs).

4.1.4 Importance of Wind Energy

RETs are the fastest growing energy resources in the world and various projections indicate that these resources will have a huge contribution in the future [12–14]. Pakistan mainly depends upon the conventional energy resources and there is not much effort for the exploitation of RETs for electricity generation. Due to over dependence on imported fossil-fuel, more than 60 % of the foreign exchange is spent for the import of energy [7]. Oil import is a significant burden on the national exchequer and foreign reserves.

The government is trying to increase the indigenous energy supplies and renewable energy sector has been identified as an important target area. Renewable resources have enormous potential and can meet many times the current national energy demand. These resources can enhance diversity in the national energy mix, secure long-term sustainable energy supplies, reduce atmospheric emissions, create new employment opportunities, and offer possibilities for growth of the domestic manufacturing industry [15].

Among all RETs, wind is the most mature, rapidly deployable, clean, and affordable energy resource. In the decade leading up to 2009, there has been an average annual growth rate of 30 % for the installed wind energy capacity in the world [16]. According to the World Wind Energy Association (WWEA), the global market for wind energy is gaining momentum and 40.5 GW of new wind capacity was installed in the year 2011 in more than 50 countries [17]. It indicates that wind energy is a rapidly growing, mature, and proven technology. Electricity is being generated from wind energy at a cost around 8 US cent/kWh in some Asian countries [18]. International Energy Agency (IEA) estimates that the investment cost of wind power is expected to further reduce in future as a result of technology development and economies of scale by 23 % for onshore and 38 % for offshore projects [16]. There is an increasing trend towards wind energy deployment [19–21]. Therefore, wind energy is a technically feasible alternative of renewable energy available for Pakistan at a competitive cost. The country's 1,100 km long coastline is ideal for the installation of wind farms to generate electric power. The growth of wind energy in the neighboring countries like China and India has been remarkable during the last decade. Pakistan is sharing the same coastal line of the Indian Ocean with India. Concerns about the security of energy supplies have led many countries in this region to diversify their energy mix through their indigenous wind resources.

Import of natural gas could be seen as a viable resource to overcome the depleting domestic reserves. However, the natural gas import has significant issues such as need for substantial capital investment in the infrastructure, security issues in the region, and physical terrain concerns. Moreover, it will further increase dependence on imported energy, and there is also price uncertainty over the future supply. Thus, wind energy has the potential of becoming a strong contributor in the national electricity mix.

A recent economic survey by the government indicates that more than 40,000 villages in the country do not have access to electricity [22]. Wind energy can be utilized to provide electricity to those villages. The deployment of wind energy projects can electrify these villages, improve living standards of the communities in those areas, and contribute towards the national economic growth [2].

Utilization of the indigenous wind energy resources can significantly help the country to overcome this severe energy crisis, improve living standards of the society, diversify the national energy mix, contribute to the national economic growth, improve rural economy, reduce the energy import bill, and ensure environment sustainability. The major hurdle in the development of wind power in the country is the unavailability of reliable and bankable wind data. Due to the current developments in the region of Gharo and Jhimpir, many project developers have installed their own wind masts to get bankable wind data. Unfortunately, wind power development is not planned for any other region of the country. Moreover, there is not any effort to install any wind measuring mast to assess the potential of other regions of the country. The energy crisis in the country demands the Government to accelerate its efforts towards development of indigenous and environment friendly power solutions to meet its future energy demands and to ensure sustainable environment. Therefore, it is critical to investigate and validate the potential wind resources in different regions of Pakistan in order to exploit the enormous potential of wind energy resources available in the country.

4.1.5 Wind Energy Resources in Pakistan

Wind power has been used from the ancient times for grinding grains, sailing ships, and pumping water for irrigation purposes. Wind power technology is the fastest growing renewable energy resource in the world [16]. The worldwide installed capacity of wind farms reached 254 GW by the end of June 2012 [17]. Several European countries are obtaining more than 10 % of electricity from wind power [11, 16]. It indicates that wind power technology is a rapidly growing, mature, and proven technology. The capacity and height of wind turbines have increased with time [16]. Generally, wind speed is higher and more stable at height. Increased height of the wind turbine allows increasing length of the turbine blades, so it captures more power due to larger area through which the turbine can extract energy (known as swept area of the rotor). Additionally, the rotor can be installed higher to take advantage of the higher wind speed [23].

Pakistan has a tropical desert climate with yearly precipitation of less than 250 mm. It is hot and dry in most of its areas, with a relative high average annual temperature. There are four distinct seasons: a cool, dry winter from December through February; a hot, dry spring from March through May; the summer rainy season, or southwest monsoon period, from June through September; and the retreating monsoon period of October and November. Rainfall varies greatly from year to year, and patterns of alternate flooding and drought are common [24]. The country is under a great influence of monsoon from the Indian Ocean, which brings both precious rain and abundance of wind energy resources. The thermal depression of South Asia and monsoon winds shape the country's southern coastal areas and northern mountain areas into a land rich in wind resources.

Various studies indicate that there is a good potential for generating electricity from wind energy along the coastline and many other regions of Pakistan [2, 7, 25–28]. Pakistan Metrological Department (PMD) has installed several wind data collection centers along the coastline and northern areas of the country. The wind data is obtained from 47 towers along the coastline. The collected wind data indicates that wind speeds from 5 to 7 m/s persist in the coastal regions and many valleys in the Northwest region of the country at a height of 50 m [29].

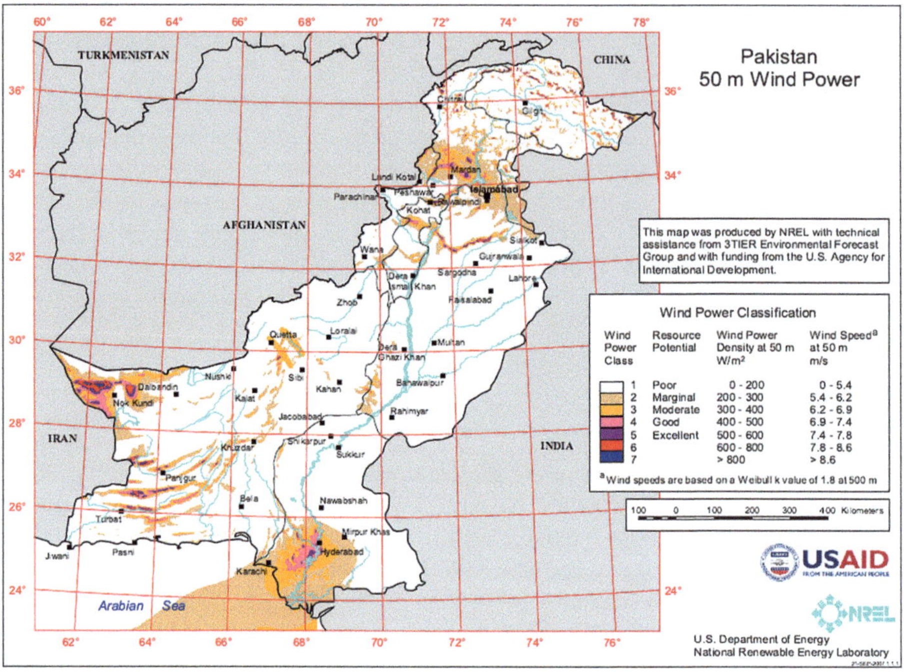

Fig. 4.3 Wind map of Pakistan [30]

In an effort to access the global potential of wind resources, the US National Renewable Energy Laboratory (NREL) and 3TIER Environmental Forecast Group developed 50 m wind map of Pakistan, shown in Fig. 4.3. This high-resolution wind map also indicates that many regions of country have good potential for generating wind energy [30]. Sheikh mentioned that many potential sites for wind energy generation in Pakistan have capacity factor of more than 25 %, which is internationally considered suitable for the installation of economically viable commercial wind farms [25]. Multiple studies indicate 50 GW of wind energy potential along the coastal areas of the country with an average wind speed of more than 7 m/s at 80 m height [26, 31]. Therefore, the coastal areas of Pakistan, especially the wind corridor around Gharo region is ideal for generating electricity from wind energy.

An in-depth analysis of the wind data is very useful to identify the best prospective areas and screen out less promising areas. The following areas in several regions of the country have good-to-excellent wind resources [32]:

- Southeastern Pakistan especially:
  - Hyderabad to Gharo region in southern Indus Valley
  - Coastal areas south of Karachi
  - Hills and ridges between Karachi and Hyderabad

Table 4.1 Good-to-excellent wind resource [32]

Wind resource	Wind class	Wind power (W/m^2)	Wind speed (m/s)	Land Area (Km^2)	Electricity potential (MW)
Good	4	400–500	6.9–7.4	18,106	90,530
Excellent	5	500–600	7.4–7.8	5,218	26,090
Excellent	6	600–800	7.8–8.6	2,495	12,480
Excellent	7	>800	>8.6	543	2,720
Total				26,362	131,800

- Northern Indus Valley especially:
  - Hills and ridges in northern Punjab
  - Ridges and wind corridors near Mardan and Islamabad
- Southwestern Pakistan especially:
  - Near Nokkundi and hills and ridges in the Chagai area
  - Makran area hills and ridges
- Elevated mountain summits and ridge crests, especially in northern Pakistan.

It is estimated that approximately 26,400 km^2 (3 % of Pakistan's total area) has class 4+ (good-to-excellent) wind resource for utility-scale applications. It has a potential of generating approximately 132,000 MW of electricity from wind (assuming 5 MW/km^2) [32]. Moreover, almost 9 % of the country's land area has a Class 3 or better wind resource. A summary of the wind resource at 50 m height along with potential of generating electricity from the available wind resources in the country are presented in Table 4.1.

Despite the availability of wind resources in abundance, there is not much progress made for the utilization these resources in the country. Presently, the installed capacity of wind power is only 40 MW in the country [10]. In the renewable energy policy announced by the government in the year 2006, surety has been given for the purchase of electricity generated by wind farms. Moreover, a unique concept of "wind risk" has also been incorporated to immune the investors and project developers from the risk of variability of wind resource (wind speed). This concept has been incorporated to overcome the fear related to the reliability and accuracy of the available wind data and insulate the investor from resource variability risk. This risk is absorbed by the power purchaser (government). The wind risk concept will ensure that the government will make monthly payments for the purchase of power in accordance with the benchmark wind speed tables [33]. The benchmark wind speed is determined for each project site on the basis of the independently monitored wind data. Subsequently, electricity generation levels corresponding to the benchmark wind speed are calculated. If less power is generated in a particular month due to wind speed lower than the benchmark wind speed, the government will make monthly payments to wind farms according to the benchmark wind speed data [34]. The principle behind the wind risk concept is to make the wind farm developers and investors immune to the wind speed variability factor, which is beyond their control.

However, project developer will be fully responsible for factors within their control such as availability of the wind farms [33]. The renewable energy policy also offers other benefits including some tax exemptions and waiver of import duties for the equipment required for renewable energy projects.

4.1.6 Need to Assess Wind Resources

Lack of credible wind resource data is considered the major reason towards underutilization of wind energy in Pakistan [35]. Due to limited availability of bankable wind data, it is important to collect wind data from the potential locations for installation of wind farms. Generally, wind data of remote locations suitable for installation of wind farms are not available. It is due to the reason that meteorological department only install weather stations at specific locations such as airports, ports, and areas with high density population and these locations are avoided during wind farms siting [36]. Literature highlights that for wind energy resource assessment applications, minimum of 1 year of wind measurements is required for resource assessment and to build wind climatology for a certain site [36–38]. A 1-year period of wind measurements can provide a reasonable indication of wind potential including percentage of uncertainty from 5 to 15 % [39]. In this chapter, data analysis for the period of 2 years is done using WAsP software. The results of the analysis were then compared with resource identified in NREL wind map of Pakistan.

4.2 Site Selection Approach

4.2.1 Site Selection Criteria

Selection of site for the installation of wind mast is a complex job. Various factors are involved that need to be addressed before finalizing the location, where the mast is to be installed. The likely future development of wind power projects in the area are also needed to be catered during installation of wind mast. Following are the critical parameter to install the mast on best suitable site [37, 40–48]:

- Land availability is the most critical factor. It should be ensured that the rights of the land selected for the site can be secured. The land owners, local authority, statutory stakeholders, and local communities must be consulted before the selection of site.
- Ecological assessment must be done to avoid disturbance to critical habitat in that region. Effects on any agricultural or recreational site in the selected region are also calculated. The land must be flat and easily accessible for logistics, construction, and O&M of the mast.
- Grid connection should be in close proximity to the installed mast likely to be cost-effective for the desired size of project development in the future.

- Prevailing wind speeds should be adequate. The basic meteorological data or wind maps available of the area can be used to assess the wind frequencies. The area with wind preferably in one direction should be chosen for installation of mast.
- Building permits need to be taken from the aviation authorities.
- Ground conditions must be checked and verified through soil investigation or geotechnical studies.

It is important to strictly follow the above-mentioned site selection criteria for the installation of the met mast for better results.

4.2.2 Location of the Mast

The site for the installation of met mast was selected by keeping in view the criteria mentioned above. The US National Renewable Energy Laboratories under the USAID assistance program has carried out the wind resource study of Pakistan and developed a mesoscale map showing the wind speed available at 50 m altitude [49]. This map was studies at the time of site selection which shows high wind power potential in Kalar Kahar region of the Punjab. The meteorological data of wind available for the Kalar Kahar region was also analyzed to verify the high wind speeds in the region. Based on the pre-analyses of NREL map and meteorological data, different locations of Kalar Kahar region was visited by authors. During visits, authors also collect the basic information from the local residents of the area regarding the wind conditions of the area. Land availability options were evaluated in different locations of Kalar Kahar. Land ownership issues were handled through consultation with statutory authorities and local communities to get the documented land rights to avoid any legal issues. Authors also got permission from Civil Aviation Authority (CAA) Pakistan for the installation of the mast. After several site visits, the location of the mast was finalized and land was obtained on lease.

The wind mast was installed in the Kalar Kahar region, located at 25 km south west of Chakwal near motorway M-2 in February 2010. The selected site for the installation of the mast is an elevated area consists of a plateau. There are small hills having approx. average height around 300–600 m. They attain the average height of 900 m with several peaks rising up to 1,200 m. Small hills of bare rock rise steeply above the surface. There are few large hills in the region. Surrounding areas are agricultural farmlands covered with seasonal crops, whereas the population is approximately 2 km away towards the north east of the site. This is the first wind mast installed in the Punjab province, where the anticipated potential of wind power generation is 5–10 GW. Satellite imaginary of the site is given in Fig. 4.4 followed by the images of mast surroundings in Fig. 4.5. The coordinates of the installed wind mast are 32.6579333 N, 72.66645E. During installation of wind mast, international industry standards (IEC 61400-12-1) were strictly followed to record reliable and bankable wind data.

Fig. 4.4 Site satellite imaginary of Kalar Kahar mast

Fig. 4.5 Surrounding areas of Kalar Kahar mast

4.3 Research Methodology

The objective of the research is to identify a windy area in the Punjab province, where future wind power projects can be installed. Detailed wind resource assessment is considered essential for estimation of the wind power potential and evaluation of the most promising sites for wind farm development [50, 51]. To achieve this objective, a wind mast was erected at the selected site according to the site selection criteria given in Sect. 2.1. The wind mast equipment was imported from NRG USA and all the equipment was installed as per industry standard IEC 61400-12-1. The complete wind measurement system comprises five anemometers for measuring wind speed, two wind vanes for wind direction measurement, one temperature, and one humidity sensor. The installation heights of these sensors are given in Table 4.2 in Sect. 3.1. Wind speed is the most important site indicator; therefore, anemometers were installed on three different heights to determine wind shear characteristics. Wind Atlas Analysis and Application Program (WAsP) was used which is one of the most widely used wind resource assessment model [37, 52–54]. Horizontal and vertical extrapolation of wind data can be performed with WAsP. It contains a complete set of models to calculate the effects on the wind of sheltering obstacles, surface roughness changes, and terrain height variations [50].

Once the system is fully commissioned, the data was retrieved through GSM network after every 10 min. Wind energy resource assessment applications require accurate wind data [36, 37, 55–59]. The collected data was retrieved and stored on the servers. Data is then validated manually and through wind data retrieval software to screen the data for any anomaly or error and the missing data is reported. The reason for erroneous data like faulty sensor, loose wire connection, broken wires, etc. were immediately detected and rectified to improve the reliability of the data. Data validation was done by verification of data records, time sequence, and range test, i.e., comparison of available data with upper and lower limit values. Literature recommends taking out the erroneous or missing measurements and refining the wind date [53]. Relational checks were made by verifying relationships between different physical parameters. It was a critical step because accuracy of the WAsP predictions depends on the accuracy of the measurement station [53].

Table 4.2 Mast information

Sensor/Serial number	Height (m)
Speed/121263	79
Speed/121274	79
Speed/120677	60
Direction	77
Direction	30
Temperature	5
Speed/120678	50
Speed/120679	30
Speed/120685	10

Table 4.3 Sensor information

S No	Sensor	Units	Height (m)	Possible records	Valid records	Recovery rate (%)
1	Direction 30 m	degree	30	73,350	70,920	96.69
2	Direction 77 m	degree	77	73,350	70,920	96.69
3	Speed 10 m	m/s	10	73,350	70,920	96.69
4	Speed 30 m	m/s	30	73,350	70,920	96.69
5	Speed 50 m	m/s	50	73,350	70,920	96.69
6	Speed 60 m	m/s	60	73,350	70,920	96.69
7	Speed 79 m A	m/s	79	73,350	70,920	96.69
8	Speed 79 m B	m/s	79	73,350	70,920	96.69

After the data validation, the data is processed to be used in wind data software. Validated 10 min data subsets were converted into an hourly average data base using WAsP software. Average wind speeds and annual wind speed values are then calculated from the software. Wind rose is developed for the wind direction data received from the site. Temperature data was also received and monthly averages were calculated. The complete data set of 2 years was analyzed to conclude the results of the research. It is according to the recommendations given in the literature, where emphasis is given to have minimum of 1 year of wind data for accurate resource assessment [36–38].

4.3.1 Mast Information

The equipment installed on the mast was supplied by NRG Systems. NRG has been into presence since 1982 and considered the most renowned make in the wind measuring equipment and can be found in more than 135 countries, serving electric utilities, wind farm developers, research institutes, government agencies, and universities [60]. NRG Systems manufactures complete wind measurement systems includes tilt-up towers, instruments, and sensors that measure and analyze wind speed, direction, and other environmental data important to sitting and operating wind energy projects. Details of the mast installed on site are given below:

Eight sensors are installed on the mast. Details of the sensors installed on the mast are given below in Table 4.3.

4.3.2 Data Recovery Rate and Faulty Data

Large-scale data sets are likely to have some errors. Data errors may be caused by a number of reasons including human mistakes and instrumental malfunctioning and other environmental factors. In order to make the given data reliable enough to be used for the wind resource assessment, it is important to process it to address its

Table 4.4 Data recovery

Lot No	Duration
01	23rd May 2010
02	1st January 2011 to 13 January 2011

Table 4.5 Missing data

Sensor	Correlation with	R^2
Speed 79 m A	Speed 79 m B	0.998
Speed 79 m A	Speed 60 m	0.990
Speed 60 m	Speed 50 m	0.996
Speed 50 m	Speed 30 m	0.980
Speed 30 m	Speed 10 m	0.947

errors and shortcomings. The available wind data sets have therefore been processed for quality control. Data received from the mast has recovery rate of 96.7 %. Calculated recovery rate of the data is given above in Table 4.4.

There were few anomalies in the received data. Fault temperature data observed in the data is ignored. There are two events when data is not received from all sensors. Details of these observations are given in Table 4.5. Data is investigated for any failure of battery, data logger failure, and data transmittance through GSM module. There are no clear observations to make reason of data not received from logger. As per mast maintenance team, the reason is due to insufficient storage memory.

4.4 Wind Resource Assessment

The following analyses are carried out to assess the wind resource at Kalar Kahar region in the Punjab province.

4.4.1 Shadow Analysis

Shadow count the effect of the mast installation on the measured results normally caused due to shelter of any anemometer from any direction. Data received from the mast is analyzed to investigate for any shadow. There is no shadow observed on the top anemometers. The shadow analysis is shown in Fig. 4.6.

4.4.2 Drift Analysis

Drift generally occurs due to anemometers cyclic bearing loading, contaminations and wear caused them to operate in off design fashion. Drift analysis is carried out on all speed sensors. It is observed that there is no drift on any speed sensor and all

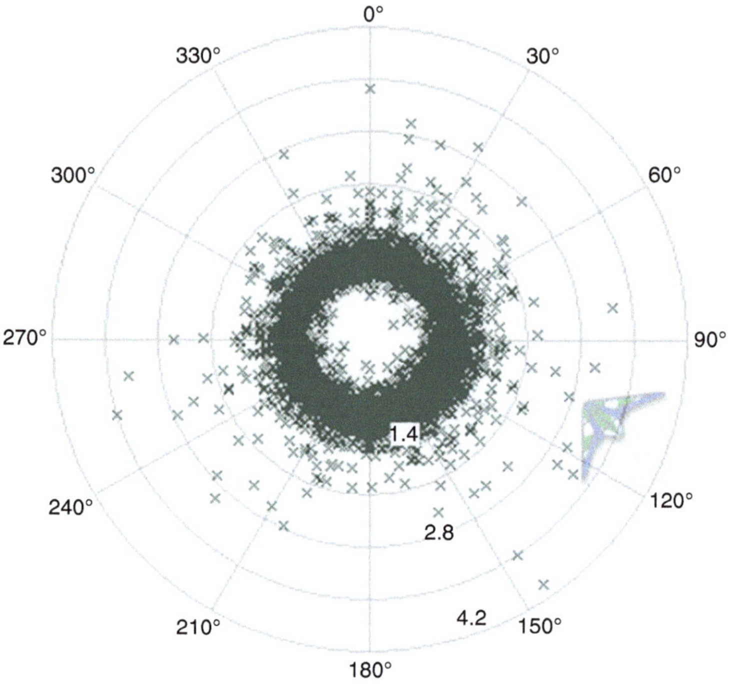

Fig. 4.6 Shadow analysis

speed sensors are working in good condition. It has been indicated in Fig. 4.7 that all sensors are performing in normal fashion.

4.4.3 Sensor Symmetrical Behavior

After having drift analysis, it was observed that all sensor are working fine with due time. For further investigations of the installed sensor, correlation of sensors at the consecutive heights is carried out. Purpose of this analysis was to find out any abnormality of the sensor functionality. A strong correlation between consecutive heights shows that all speeds are working in good conditions (Table 4.6).

4.4.4 Monthly Wind Speeds

Based on all above data checks, it can be concluded that at this stage the finalized data is in good quality and results can be extracted for any decision about site based on wind characteristics to install wind power project. Monthly mean of the site is

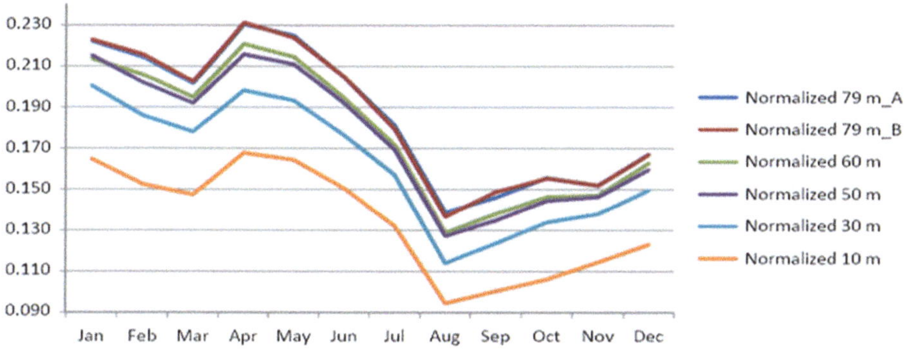

Fig. 4.7 Normalized speed data of the mast installed at Kalar Kahar

Table 4.6 Symmetric data trend check

Year	Month	Mean wind speed 79 m	Mean wind speed 79 m	Mean wind speed 60 m	Mean wind speed 50 m	Mean wind speed 30 m	Mean wind speed 10 m
2010	Feb	5.6639	5.7266	5.4334	5.2916	4.8477	4.0529
2010	Mar	5.2637	5.3072	5.0966	5.0157	4.6345	3.8177
2010	Apr	6.0812	6.1217	5.8354	5.7249	5.2699	4.5138
2010	May	6.144	6.1992	5.9001	5.7993	5.3528	4.5604
2010	Jun	5.4097	5.4646	5.178	5.1035	4.6942	4.0148
2010	Jul	4.7006	4.6501	4.4431	4.3984	4.0863	3.4285
2010	Aug	3.509	3.4575	3.2555	3.219	2.8823	2.3892
2010	Sep	3.6937	3.7653	3.4929	3.4123	3.1275	2.5389
2010	Oct	3.9402	3.9335	3.702	3.6553	3.3872	2.6849
2010	Nov	3.845	3.8447	3.7212	3.7021	3.4904	2.8952
2010	Dec	4.2316	4.2256	4.1206	4.0398	3.7834	3.1147
2011	Jan	5.6257	5.6441	5.4067	5.4496	5.0733	4.1691
2011	Feb	5.2883	5.303	5.0775	5.0067	4.6241	3.7438
2011	Mar	4.9378	4.942	4.7608	4.6973	4.3744	3.6397
2011	Apr	5.5803	5.5838	5.3453	5.1999	4.76	3.9748
2011	May	5.2636	5.1579	4.9849	4.9037	4.4529	3.7762
2011	Jun	4.9497	4.8966	4.6638	4.6161	4.2356	3.601
2011	Jul	3.9213	3.8857	3.751	3.6332	3.33	2.8382
All data		4.8929	4.8955	4.6755	4.6048	4.2462	3.5417
Mean of monthly means		4.7172	4.7192	4.5054	4.4453	4.1047	3.4078

Table 4.7 Monthly mean wind speed

Year	Month	Mean wind speed 79 m	Mean wind speed 79 m	Mean wind speed 60 m	Mean wind speed 50 m	Mean wind speed 30 m	Mean wind speed 10 m
2010	Feb	5.6639	5.7266	5.4334	5.2916	4.8477	4.0529
2010	Mar	5.2637	5.3072	5.0966	5.0157	4.6345	3.8177
2010	Apr	6.0812	6.1217	5.8354	5.7249	5.2699	4.5138
2010	May	6.144	6.1992	5.9001	5.7993	5.3528	4.5604
2010	Jun	5.4097	5.4646	5.178	5.1035	4.6942	4.0148
2010	Jul	4.7006	4.6501	4.4431	4.3984	4.0863	3.4285
2010	Aug	3.509	3.4575	3.2555	3.219	2.8823	2.3892
2010	Sep	3.6937	3.7653	3.4929	3.4123	3.1275	2.5389
2010	Oct	3.9402	3.9335	3.702	3.6553	3.3872	2.6849
2010	Nov	3.845	3.8447	3.7212	3.7021	3.4904	2.8952
2010	Dec	4.2316	4.2256	4.1206	4.0398	3.7834	3.1147
2011	Jan	5.6257	5.6441	5.4067	5.4496	5.0733	4.1691
2011	Feb	5.2883	5.303	5.0775	5.0067	4.6241	3.7438
2011	Mar	4.9378	4.942	4.7608	4.6973	4.3744	3.6397
2011	Apr	5.5803	5.5838	5.3453	5.1999	4.76	3.9748
2011	May	5.2636	5.1579	4.9849	4.9037	4.4529	3.7762
2011	Jun	4.9497	4.8966	4.6638	4.6161	4.2356	3.601
2011	Jul	3.9213	3.8857	3.751	3.6332	3.33	2.8382
All data		4.8929	4.8955	4.6755	4.6048	4.2462	3.5417
Mean of the monthly means		4.7172	4.7192	4.5054	4.4453	4.1047	3.4078

calculated on the basis data of 24 months; the maximum monthly mean is 4.72 m/s at a height of 79 m, whereas lowest monthly mean wind speed is 3.41 m/s at height of 10 m given in Table 4.7.

All monthly mean wind speeds are plotted in Fig. 4.8. Analysis of long-term wind data reveals interesting patterns. For example, it is observed that wind observed during the May 2010 is higher than observed during May 2011. Annual wind variability and cyclic patterns can be better observed based on long-term data.

4.4.5 Annual Wind Speeds

The annual wind speed data for the years 2010 and 2011 are shown in Table 4.8.

From Fig. 4.9, it can be understood that due to vertical mixing forced by solar irradiation, the wind speeds at the different heights tend to equal around

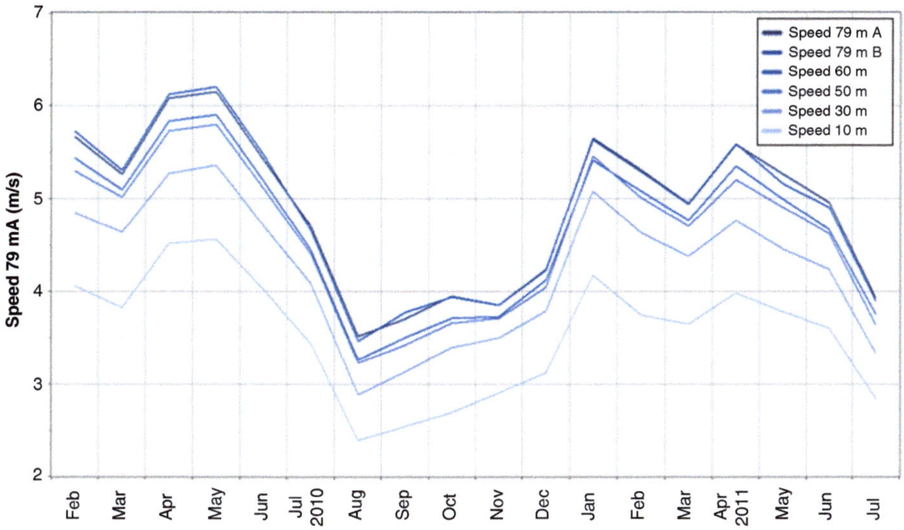

Fig. 4.8 Monthly mean wind speeds

Table 4.8 Annual wind speed

Year	Mean wind speed 79 m	Mean wind speed 79 m	Mean wind speed 60 m	Mean wind speed 50 m	Mean wind speed 30 m	Mean wind speed 10 m
2010	4.7246	4.7415	4.5161	4.4452	4.1039	3.4232
2011	5.2043	5.1806	4.9706	4.9002	4.5096	3.7608
Avg.	4.8929	4.8955	4.6755	4.6048	4.2462	3.5417

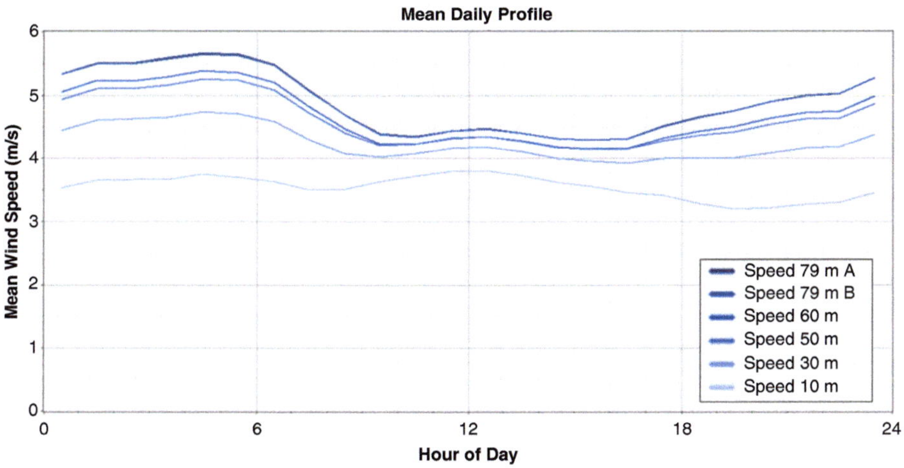

Fig. 4.9 Annual diurnal mean wind speed

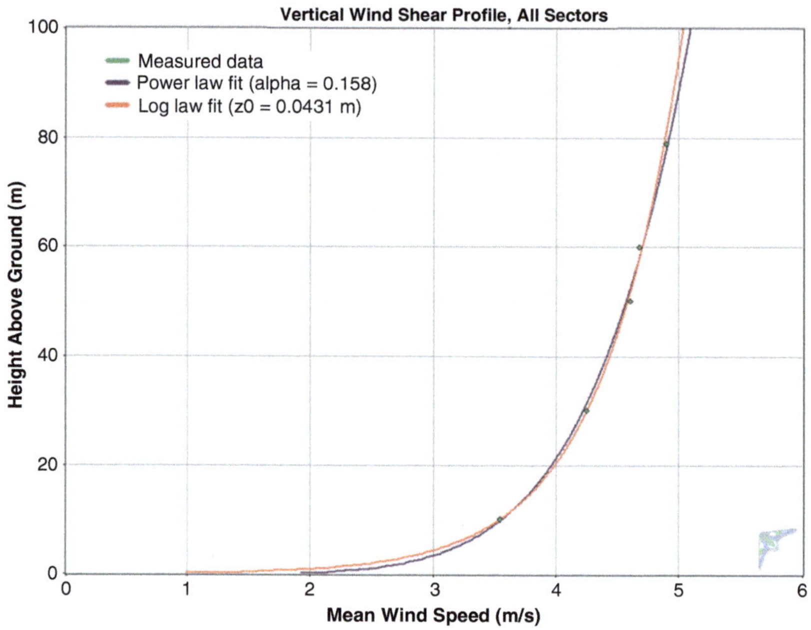

Fig. 4.10 Vertical shear profile

noon time, which provides the daily minimum wind speed for the top level 79 m around that time.

4.4.6 Wind Shear Profile

The vertical wind shear and monthly wind shear profiles for the wind data recorded during the period of February 2010 to July 2011 have been computed. The results derived from the recorded wind data are presented in Figs. 4.10, 4.11, and Table 4.9.

The lowest wind shears are observed during the months of October to January due well-mixed atmospheric boundary layer and less temperature gradient.

4.4.7 Wind Direction

Wind direction is a major factor for selecting land for a wind farm project. Wind rose is used to estimate the wind direction in reference to wind speed. Dominant wind is observed from North side. Considering Figs. 4.12 and 4.13 where it can

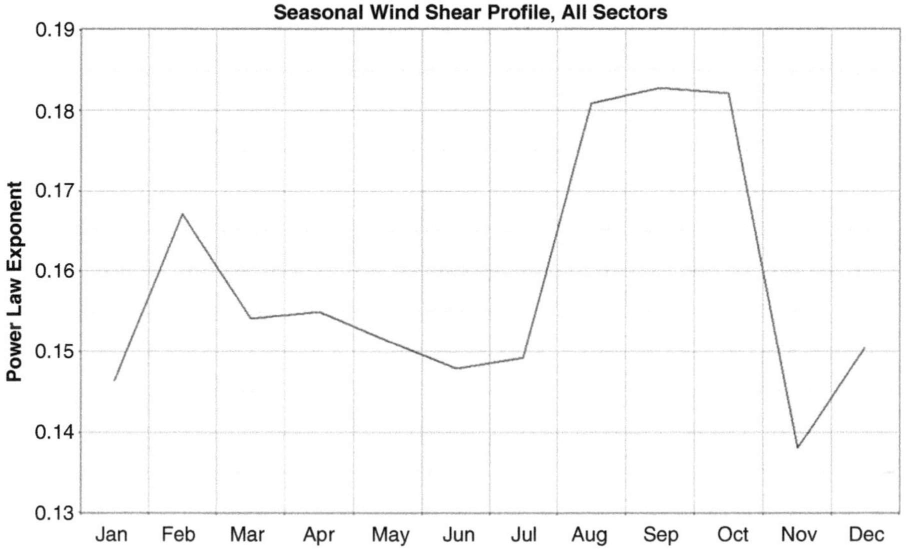

Fig. 4.11 Seasonal wind shear profile

Table 4.9 Seasonal wind shear profile

Months	Power law exponent
January	0.146
February	0.167
March	0.154
April	0.155
May	0.151
June	0.148
July	0.149
August	0.181
September	0.183
October	0.182
November	0.138
December	0.15
Monthly mean	0.159

be observed that it will be July when wind direction is from south east which is different from the dominant wind direction.

Finally, from Table 4.10, it can be seen that dominant wind is from the north direction.

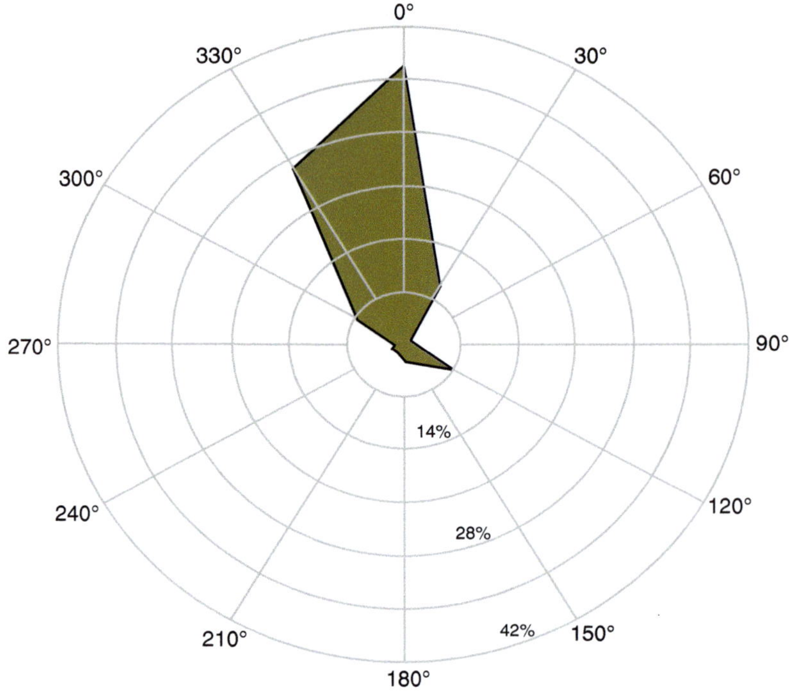

Fig. 4.12 Annual wind speed at a height of 70 m

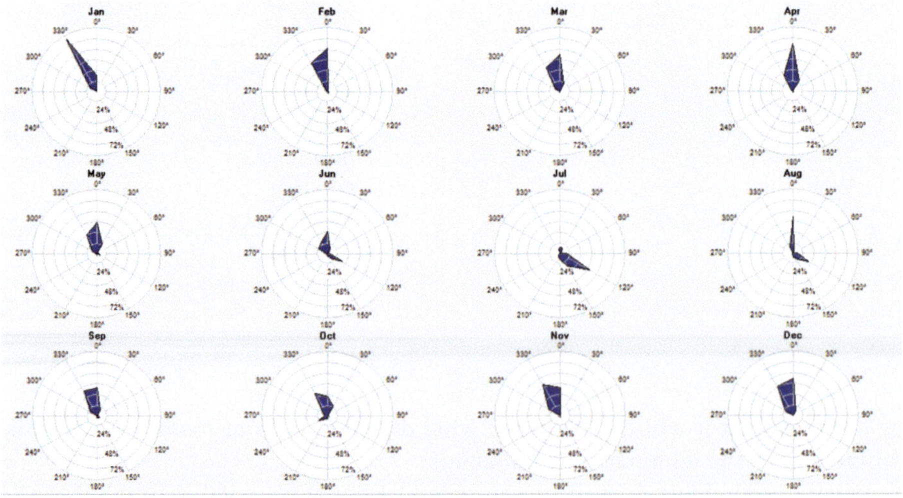

Fig. 4.13 Monthly wind rose

Table 4.10 Directional mean wind frequency

Sector No	Direction (77 m) Sector	Frequency (%)	Mean WS at 79 m Height (m/s)
1	345–15	18.66	6.5903
2	15–45	5.29	5.6589
3	45–75	1.92	3.6687
4	75–105	2.99	3.825
5	105–135	11.17	4.6379
6	135–165	9.46	3.7777
7	165–195	10.37	3.4254
8	195–225	7.85	3.3557
9	225–255	6.54	3.5828
10	255–285	3.03	3.6322
11	285–315	6.09	5.0149
12	315–345	16.63	6.2227

4.5 Comparison of Actual Site Data and NREL Mapping

Resource mapping of the Pakistan was done by the NREL funded by USAID in 2007. Mapping was done considering an elevation of 500 m and weibul k value of 1.8. Based on the values, the mountainous range of Chakwal is mapped as the region with the highest wind energy potential in the Punjab province. On the basis of wind mapping, wind mast was installed in the Chakwal region, where wind class 02 is identified. Power density identified for the same location is in the range of 200–300 W/m^2. Wind speed is monitored for more than 1 year. Based on the monitoring period of more than a year, the results showed deviation from the results with observed power density of 105 W/m^2 and wind class 01. Weibul k value observed is 1.804. Ground measurements are always more trusted as compared to remote sensing. However, it is required to put up more wind mast at new locations keeping considerations of the overall complex terrain of the region.

4.6 Conclusions

This chapter demonstrated to assess and promote the harnessing of wind power to generate electricity in Kalar Kahar region of the Punjab province. Recording of accurate time series wind data at different heights is one of the basic requirements. The analysis shows that the average wind speed of the area is around 4.8 m/s. Wind power class having a power density of 131 W/m^2 at height 79 m is although not recommended for MW class wind turbines, but it is ideal for small-scale applications. On the basis of the results, it is concluded that the observed wind data have relatively low wind potential as per international wind power classification.

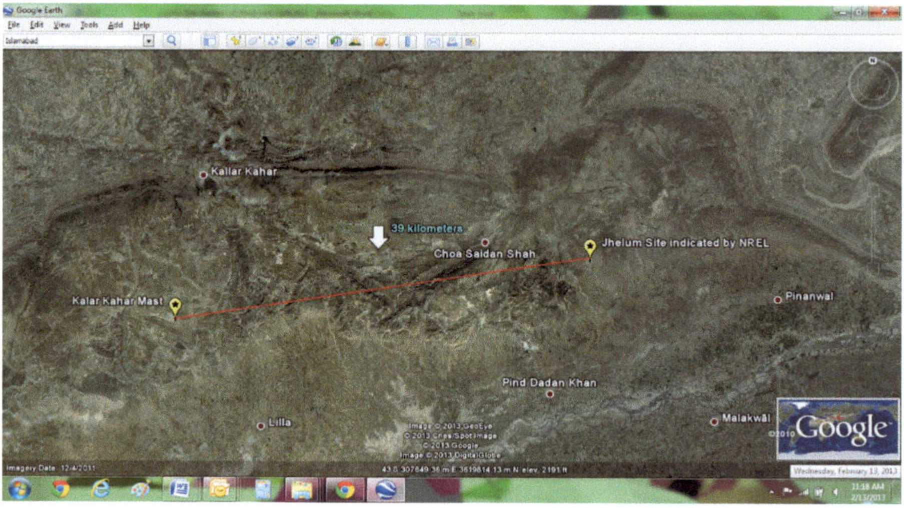

Fig. 4.14 Proposed locations for installation of mast as per map of NREL and USAID

However, through installation of suitable wind turbines, economical generation of electricity is possible for this remote location.

The results of the analysis were compared with the wind map of Pakistan developed by the US National Renewable Energy Laboratories (NREL). The NREL map shows high wind areas in the Punjab province. It was found that the installed mast is around 39 km away from the locations identified in the map of NREL. Based on the analysis, results and comparison with the NREL map, new locations of mast were identified, and it is recommended that the mast may be uninstalled from present site and reinstalled to any other potential site to set up a power project. The proposed new location of the wind mast is shown in Fig. 4.14. Moreover, on the basis of low potential but consistent winds in the area, where the wind mast is installed; it is proposed to use micro wind turbines for small applications like street lights and electrification of villages. It is also recommended that before planning any large-scale project in this area, micro-siting analysis must be completed for the economic viability.

References

1. GoP, Official website Government of Pakistan. www.pakistan.gov.pk. (2011, September 13)
2. U.K. Mirza, N. Ahmad, K. Harijan, T. Majeed, Identifying and addressing barriers to renewable energy development in Pakistan. Renew Sustain Energy Rev **13**, 927–931 (2009)
3. CIA, The World Factbook. Washington, DC: Central Intelligence Agency, Office of Public Affairs (2011)
4. To identify the wind potential areas. Pakistan Meteorological Department, Government of Pakistan. http://www.met.gov.pk/Projects/Wind_Project.html. (2012, December 24, 2012)

5. G. Weynand, Energy sector assessment for USAID/Pakistan. United States Agency for International Development http://www.usaid.gov/pk/downloads/eg/PEDP.pdf. (2007, December 14, 2011)
6. GoP. Pakistan Energy Year Book, Hydrocarbon Development Institute of Pakistan, Ministry of Petroleum and Natural Resources, Islamabad, Pakistan (2009)
7. M.A. Sheikh, Energy and renewable energy scenario of Pakistan. Renew Sustain Energy Rev **14**, 354–363 (2010)
8. C. Trimble, N. Yoshida, M. Saqib, Rethinking electricity tariffs and subsidies in Pakistan. The World Bank Report Number: 62971-PK. http://www-wds.worldbank.org/external/default/WDSContentServer/WDSP/IB/2011/08/31/000386194_20110831050158/Rendered/PDF/629710ESW0whit0020110Final00PUBLIC0.pdf. (2011, March 14, 2012)
9. M. Asif, Sustainable energy options for Pakistan. Renew Sustain Energy Rev **13**, 903–909 (2009)
10. AEDB, Alternate Energy Development Board, Government of Pakistan. http://www.aedb.org/. (March 18, 2010)
11. IEA, Key World Energy Statistics, International Energy Agency. http://www.iea.org/textbase/nppdf/free/2009/key_stats_2009.pdf. (2009, May 25, 2011)
12. EIA, International Energy Outlook 2011, Energy Information Administration, U.S. Department of Energy. www.eia.gov/ieo/pdf/0484(2011).pdf. (2011, March 30, 2011)
13. M. Jefferson, Sustainable energy development: performance and prospects. Renew Energy **31**, 571–582 (2006)
14. M. Amer, T.U. Daim, Application of technology roadmaps for renewable energy sector. Technol Forecast Soc Change **77**, 1355–1370 (2010)
15. T. Muneer, M. Asif, Prospects for secure and sustainable electricity supply for Pakistan. Renew Sustain Energy Rev **11**, 654–671 (2007)
16. H. Chandler, J. Bjerg, T. Kerr, U. Remme, B. Beck, C. Philibert, T. Rinke, Wind Energy Technology Roadmap, International Energy Agency (IEA). www.iea.org/G8/docs/Roadmaps_g8july09.pdf. (2009, March 23, 2011)
17. WWEA, The World Wind Energy Association: Half-year Report 2012. http://www.wwindea.org/webimages/Half-year_report_2012.pdf. (2012, Nov 17, 2012)
18. K.Q. Nguyen, Wind energy in Vietnam: resource assessment, development status and future implications. Energy Policy **35**, 1405–1413 (2007)
19. T.U. Daim, I. Iskin, X. Li, C. Zielsdorff, A.E. Bayraktaroglu, T. Dereli, A. Durmusoglu, Patent analysis of wind energy technology using the patent alert system. World Patent Inform **34**, 37–47 (2011)
20. I. Iskin, X. Li, C. Zielsdorff, A.E. Bayraktaroglu, Patent alert system for wind turbine technology, in *Portland International Center for Management of Engineering and Technology (PICMET)*, Portland, Oregon, USA, 2011
21. T.U. Daim, D. Lim, F.A. Gomez, J. Schwarz, S. Jovanovic, Storage technologies for wind power in the Columbia River Gorge. Int J Sustainable Energy **33**, 1–15 (2014)
22. GoP, Pakistan Economic Survey 2006–2007, Government of Pakistan, (2007)
23. T.U. Daim, M. Amer, R. Brenden, Technology road mapping for wind energy: case of the Pacific Northwest. J Clean Prod **20**, 27–37 (2012)
24. H. Aksoy, Z.F. Toprak, A. Aylek, N.E. Anal, Stochastic generation of fairly mean wind speed data. Renew Energy **29**, 2111–2131 (2004)
25. M.A. Sheikh, Renewable energy resource potential in Pakistan. Renew Sustain Energy Rev **13**, 2696–2702 (2009)
26. I.A. Mirza, N.A. Khan, N. Memon, Development of benchmark wind speed for Gharo and Jhimpir, Pakistan. Renew Energy **35**, 576–582 (2010)
27. U.K. Mirza, N. Ahmad, T. Majeed, K. Harijan, Wind energy development in Pakistan. Renew Sustain Energy Rev **11**, 2179–2190 (2007)
28. A.M. Chaudhry, R. Raza, S.A. Hayat, Renewable energy technologies in Pakistan: prospects and challenges. Renew Sustain Energy Rev **13**, 1657–1662 (2009)

29. PMD, Wind Power Potential of Sindh, Pakistan Meteorogical Department, (2007)
30. NREL, Wind Map of Pakistan. National Renewable Energy Laboratory. http://www.nrel.gov/wind/pdfs/final1-1-1pwr50pk.pdf. (2010, Nov 12, 2010)
31. A. Ghayur, Role of satellites for renewable energy generation technologies in urban regional and urban settings, in *International Conference on Advances in Space Technologies (ICAST)*, Islamabad, Pakistan, 2006, pp. 157–161
32. D. Elliott, Wind resource assessment and mapping for Afghanistan and Pakistan, SARI-Energy, National Renewable Energy Laboratory. http://www.nrel.gov/international/pdfs/afg_pak_wind_june07.pdf. (2007, August 14, 2011)
33. GoP, Policy for Development of Renewable Energy for Power Generation, Government of Pakistan. http://www.aedb.org/. (2006)
34. M. Amer, T.U. Daim, Selection of renewable energy technologies for a developing country: a case study of Pakistan. Energy Sustain Dev **15**, 420–435 (2011)
35. N. Keyes, Mapping the renewable energy revolution. Renewable Energy World. http://www.renewableenergyworld.com/rea/news/article/2013/06/mapping-the-renewable-energy-revolution?cmpid=SolarNL-Saturday-June22-2013. (2013, June 20, 2013)
36. S. Al-Yahyai, Y. Charabi, A. Gastli, Review of the use of Numerical Weather Prediction (NWP) models for wind energy assessment. Renew Sustain Energy Rev **14**, 3192–3198 (2010)
37. S. Ganesan, S. Ahmed, Assessment of wind energy potential using topographical and meteorological data of a site in Central India (Bhopal). Int J Sustainable Energy **27**, 131–142 (2008)
38. AWS Scientific Inc. Wind resource assessment handbook; fundamentals for conducting a successful monitoring program. New York: AWS Scientific Inc.; 1997, TAT-5-15283-01
39. S. Frandsen, C.J. Christensen, Accuracy of estimation of energy production from wind power plants. Wind Eng **16**, 257–671 (1992)
40. R. van Haaren, V. Fthenakis, GIS-based wind farm site selection using spatial multi-criteria analysis (SMCA): evaluating the case for New York State. Renew Sustain Energy Rev **15**, 3332–3340 (2011)
41. G. Joselin Herbert, S. Iniyan, E. Sreevalsan, S. Rajapandian, A review of wind energy technologies. Renew Sustain Energy Rev **11**, 1117–1145 (2007)
42. A.L. Drewitt, R.H. Langston, Assessing the impacts of wind farms on birds. Ibis **148**, 29–42 (2006)
43. J.F. Manwell, J. MacLeod, S. Wright, L. DiTullio, J.G. McGowan, Hull Wind II: a case study of the development of a second large wind turbine installation in the Town of Hull, MA, in American Wind Energy Association. Windpower 2006 Conference, (2006)
44. B. Ozerdem, H. Turkeli, Wind energy potential estimation and micrositting on Izmir Institute of Technology Campus, Turkey. Renew Energy **30**, 1623–1633 (2005)
45. L.M. Druyan, Review article wind climate studies for wecs siting. J Climatology **5**, 95–105 (1985)
46. K. Altaii, R. Farrugia, Wind characteristics on the Caribbean island of Puerto Rico. Renew Energy **28**, 1701–1710 (2003)
47. B. Ozerdem, S. Ozer, M. Tosun, Feasibility study of wind farms: A case study for Izmir, Turkey. J Wind Eng Ind Aerodyn **94**, 725–743 (2006)
48. A. Kusiak, Z. Song, Design of wind farm layout for maximum wind energy capture. Renew Energy **35**, 685–694 (2010)
49. D. Renné, Wind and solar maps for Pakistan and Afghanistan, in Renewable Energy Symposium by NREL, Islamabad, Pakistan, (2007)
50. N.G. Mortensen, L. Landberg, I. Troen, E. Lundtang Petersen, *Wind atlas analysis and application program (WAsP)* (Risoe National Lab, Roskilde, Denmark, 1993)
51. T.L. Acker, S.K. Williams, E.P. Duque, G. Brummels, J. Buechler, Wind resource assessment in the state of Arizona: inventory, capacity factor, and cost. Renew Energy **32**, 1453–1466 (2007)
52. B. Jimenez, F. Durante, B. Lange, T. Kreutzer, J. Tambke, Offshore wind resource assessment with WAsP and MM5: comparative study for the German Bight. Wind Energy **10**, 121–134 (2007)

53. C. Quine, Estimation of mean wind climate and probability of strong winds for wind risk assessment. Forestry **73**, 247–258 (2000)
54. E. Migoya, A. Crespo, Á. Jiménez, J. García, F. Manuel, Wind energy resource assessment in Madrid region. Renew Energy **32**, 1467–1483 (2007)
55. D.M. Hardy, J.J. Walton, Wind energy assessment, in Alternative Energy Sources, Volume 4, (1978, pp. 1835–1863)
56. A. Dorvlo, D. Ampratwum, Wind energy potential for Oman. Renew Energy **26**, 333–338 (2002)
57. A. Ucar, F. Balo, Evaluation of wind energy potential and electricity generation at six locations in Turkey. Appl Energ **86**, 1864–1872 (2009)
58. K. Radics, J. Bartholy, Estimating and modelling the wind resource of Hungary. Renew Sustain Energy Rev **12**, 874–882 (2008)
59. P. Martin, Some wind characteristics and applications of wind power on the east coast of the Arabian Peninsula. Solar Energy **34**, 9–11 (1985)
60. NRG, NRG Systems. http://www.nrgsystems.com/allproducts.aspx. (November, 12, 2012)

Chapter 5
Technological Assessment of Emerging Technologies in Conversion of Municipal Solid Waste to Energy

Tugrul U. Daim, Lokesh Honnappa, Mamatha Murthy, Claudiu Rusnac, and Chakaphan Pornsatit

Abstract The use of biomass to generate heat, energy, and petroleum substitutes such as bio-oil or bio-crude has showed much promise as a tool for reducing our reliance on imported oil and reducing the world's total carbon output through carbon recycling. This chapter provides a case of technology assessment of biomass conversion technologies assuming a hypothetical organization, Green Tech. The chapter outlines the steps on how Green Tech went from defining a problem to performing a gap analysis, defining requirements, identifying selection criteria, and finally performing a cost–benefit analysis on three biomass conversion technologies.

> We can make Oregon the national leader in renewable energy and renewable product manufacturing… Development of renewable energy will lessen our reliance on fossil fuels, protect Oregon's clean air and create jobs.—Governor Kulongoski, 2003 [1].

Nothing is more associated with Oregon than its natural resources and how we protect, enjoy, and utilize these resources is inextricably connected to the way we generate and supply energy to Oregonians and our economy. Recognizing this, Governor Kulongoski in 2003 [1] promoted diversity of renewable energy resources. Because some renewable energy fuels like bio-fuels are freely accessible and are environment friendly, they help stabilize electric rates and reduce our dependence on petroleum/natural gas. Further, investments create jobs, displacing the use of fossil fuel generation and avoiding numerous pollutants and global warming gases.

T.U. Daim (✉) • M. Murthy • C. Pornsatit
Portland State University, Portland, OR, USA
e-mail: tugrul@etm.pdx.edu

L. Honnappa
Lam Research, San Jose, CA, USA

C. Rusnac
Xerox Corp, Rochester, NY, USA
e-mail: crusnac@gmail.com

T.U. Daim et al. (eds.), *Policies and Programs for Sustainable Energy Innovations*, Innovation, Technology, and Knowledge Management,
DOI 10.1007/978-3-319-16033-7_5

The use of biomass to generate heat, energy, and petroleum substitutes such as bio-oil or bio-crude has showed much promise as a tool for reducing our reliance on imported oil and reducing the world's total carbon output through carbon recycling. Various combustion and gasification processes have been proven to be effective methods for converting biomass into useful chemical and oil derivatives as well as other carbon products.

For biomass fueled power plants, reliance on forest and agricultural waste means that a continuous supply of fuel may be uncertain. Generation of energy requires large quantities of biomass. It becomes highly important to look for alternate biomass resources and develop a facility to handle multiple biomass fuel types. Municipal solid waste (MSW) is one such product used as a feedstock, which is available abundantly and is cost competitive with natural gas power generation.

Green Tech, Inc., a renewable energy company, is in the process of expanding its energy generation by including MSW as the feedstock in addition to the homogenous feedstock. Homogenous feedstock comprises of arboricultural activities, yard waste, and wood waste. The current technology—conventional fluid bed reactor design, however, is not designed to accept MSW as the feedstock.

This chapter outlines the process that Green Tech used to assess technologies that could be a suitable replacement for its existing process that includes MSW as a heterogeneous feedstock.

5.1 Introduction

5.1.1 Problem Definition

World oil demand is growing substantially faster than production (refinery) capacity. Currently, the USA consumes 19.15 million barrels per day of oil, which is more than 25 % of the world's total [2]. As a result, it shows that the USA produces one-fourth of the world's carbon emission, which may be a contributing factor to climate change.

Also, USA has spent more than $250 billion annually to import oil [3]. In 2010, it is also estimated that the USA imported 10.27 million barrels per day of oil while it produced around 9.69 million barrels of oil per day [4]. As a result, more than 50 % of the US oil consumption is imported. By expanding the existing portfolio of energy, the USA could be more self-sufficient and provide energy locally.

Burning of fossil fuel has had significant environmental, political, and economical consequences. From environmental aspect, burning fossil fuel generates greenhouse gases (GHG), which consist of carbon dioxide, nitrous oxide, etc. Even though GHGs are beneficial in terms of maintaining earth's temperature, they could be harmful if produced too much. Also governments have to regulate and find the way to control these gases to ensure safety of the people. Economically, people are concerned about fossil fuels. For example, rising of the oil prices in many countries is an obvious reason why people are searching for other alternatives.

Furthermore, fossil fuel prices are highly volatile. With the conversion from MSW to energy, the energy manufacturers do not have to worry about fluctuating price of fossil fuel since MSW is easily accessible everywhere.

The USA spend significant amount of resources disposing of MSW. In 2010, the USA spent more than $40 billion dollars to dispose its annual production of 250 million tons of garbage [5]. Also, according to our research, Oregon in particular produced about 2.4 million tons in 2008 and disposed 50 % of its garbage to landfill [6]. However, landfill space in the USA is depleting rapidly. The number of operating landfills in the USA has declined over the last two decades, falling from 7,924 in 1988 to 1,754 in 2007 [7].

Finally exploring new reserves of fossil fuels has become more risky and costly thus, MSW-to-energy conversion technology could be a solution.

5.1.2 Solutions

One solution to this problem is conversion of biomass into energy products using various processes such as chemical, bio-chemical, and thermochemical conversion process. In this paper, we focus on MSW as a feedstock. The main advantage of biomass conversion is that the conversion process has zero net CO_2 emission (Fig. 5.1) [8].

5.1.3 Why Biomass?

Biomass is abundant in nature and is freely accessible and can be easily converted into usable energy products in all forms that people need. Thus, it helps to reduce our dependency on fossil fuel and natural gas. Figure 5.1 shows end products that could be produced by using biomass as a fuel. According to a report by Columbia University [10] it shows that almost 90 % of total energy consumption comes from fossil fuel and natural gas. Thus, biomass is an excellent alternative for energy production.

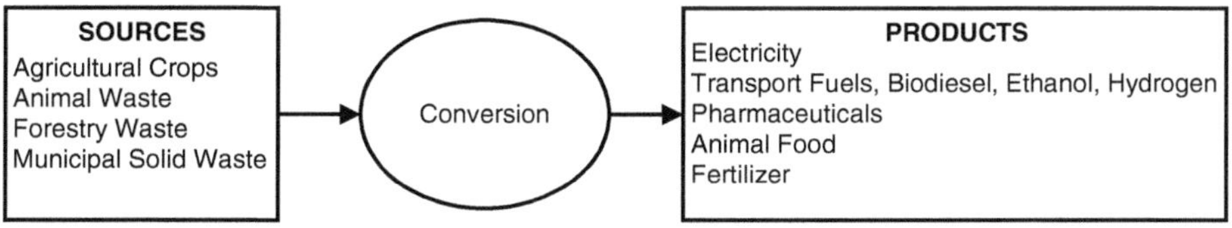

Fig. 5.1 Biomass conversion process [9]

Biomass, being a renewable energy source, reduces the costs of operation and maintenance which are major costs of the project other than capital cost. Moreover, renewable energy produces little or no waste products, so it causes little impact on the environment. Finally, renewable energy projects can be economically useful to many areas since it increases the use of local services [11].

By 2008, the USA was in second place of the highest level carbon emission countries, which represents 18.11 % of the total [12]. Most scientists believe that wide ranges of biomass resources are beneficial due to its carbon neutral nature.

MSW, a biomass energy source, is easily accessible everywhere and can be directly combusted into energy with minimal processing. The technology presents the opportunity for both electricity production and an alternative to landfilling. Further MSW facilities are paid by the fuel suppliers to take the fuel (known as a "tipping fee").

5.1.4 Current State and Overview of a Hypothetical Company

Green Tech Inc. is a renewable energy company generating electricity using homogenous biomass in the Portland metro area. Homogenous feedstock comprises of arboricultural activities, yard waste, processed wood, and wood from forest. The plant with a current capacity of 100 t/day converts the homogenous biomass using conventional fluid bed reactor technology into energy.

Conventional fluid bed reactor is a thermochemical conversion process, which provides a medium in which rapid heating of the biomass particle takes place [14]. Adding an inert medium to the bed, the technology provides a controlled fluidized environment and uniform temperature over a wide range of biomass feed rates [14]. However, Green Tech Inc. is facing certain challenges to use the same technology to process MSW.

- Burning of waste with partial supply of oxygen produces synthetic gas (also called as Syngas), which is mainly composed of carbon monoxide and hydrogen. The gas needs to be cleaned or purified before further processing.
- The process requires upfront processing of feedstock.
- The output is sensitive to input meaning the conventional process relies on high calorific value material.
- The current process is not optimized and designed to handle a heterogeneous feedstock like MSW.
- The conventional conversion facility is not self-sustaining. Green Tech cannot use the gases emitted during the process to heat the reactor.
- The current process is very inefficient in terms of conversion of energy from waste to electricity.

Thus, it is very important for Green Tech to look for a technology that satisfies company's goals and strategy.

5.1.5 *Future State*

Due to a number of problems that the company is currently facing, it needs effective and affordable solution to solve and improve the present operating system. To achieve an optimal MSW-to-energy system, there are a couple of requirements that the company is targeting.

1. The new technology should be easily integrated into the existing system.
2. The adopted technology should be able to process MSW effectively so as to generate better profits and have additional revenue stream in terms of tipping fees.
3. The selected technology should reduce or eliminate the process of feedstock preparation. By eliminating feedstock preparation process, it saves the company time and cost of operation.
4. The integrated technology should be efficient in terms of enhancing output products. In other words, it should be able to produce several by-products that are marketable such as bio-oil, bio-char, and other chemicals.

The next section illustrates the possible by-products from the considered technologies.

5.1.5.1 Bio-oil

Bio-oil is a complex oxygenated compound comprised of water, water-soluble compounds such as acids, esters, and water-insoluble compounds. It is a dark brownish viscous liquid resembling fossil crude oil [15].

Bio-oil can be used as a substitute for fossil fuels to generate heat, power, and chemicals. Boilers and furnaces can be fueled by bio-oil in the short term whereas turbines and diesel engines can be fueled by bio-oil in longer term. Plus, transportation fuels like methanol liquid can be drawn from bio-oil by using the bio-oil as a feedstock instead of the biomass. Furthermore, there is a wide range of chemicals that can be extracted or derived from the bio-oil such as resins, acetic acid, sugars, feedstock chemical industry, etc. [16].

Bio-oil has potential uses as a fuel for production of heat and electricity so it should be marketed to energy industries. It may also have additional higher value as a feedstock for green chemical industries.

5.1.5.2 Bio-char

Bio-char is a solid material, which is a by-product of pyrolysis. It is rich in carbon and can endure in soil for thousands of years.

It can be utilized in two main applications. First, it can store unwanted CO_2 generated by combustion and decomposition of woody biomass and agricultural residue in the soil. As a result, it reduces GHG emissions, which are the cause of climate change. Also, it enhances soil fertility by providing sufficient nutrients for plant growth and water retention. Therefore, bio-char offers promise for its climate benefits and soil productivity [17].

Based on its capability to absorb CO_2 and improve soil fertility, the potential market could be from agricultural industries to energy industries.

5.1.5.3 Slag

Slag, a by-product of the gasification process, occurs in several forms depending on its cooling process. Air-cooled slag which is a black glassy rock can be processed into bricks, synthetic gravel or asphalt, and other materials. On the other hand, slag becomes rock wool if compressed air is blown through a stream of molten slag. Rock wool looks similar to gray cotton candy and is light. It is an efficient insulation material, twice as effective as fiberglass. Since it is lighter than water and very absorbent, it could effectively be used to help contain and clean oil spills in the ocean. Cleanup crews could spread rock wool over and around an oil spill. The rock wool would float on the water while soaking up the oil [18].

Based on what slag is capable of, it can potentially be marketed to construction and water treatment organizations.

5.2 Methodology

The research in this chapter was done in three phases to answer the question "What technology is most efficient for processing MSW to energy?" (1) Intensive literature review to identify all technologies that can be used to process MSW to energy; (2) establish evaluation criteria at higher level and filtering technologies based on these criteria; (3) apply cost–benefit analysis method to determine the most suited and capable technology for providing a successful, long-term project at Green Tech Inc. Fig. 5.2 shows the process followed during our research.

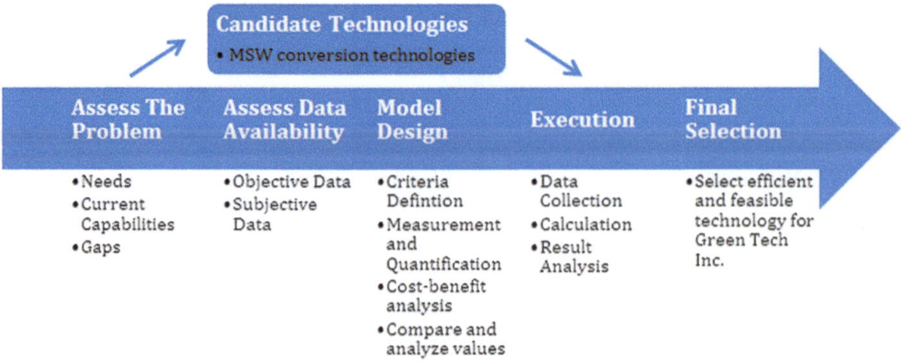

Fig. 5.2 Research process

5.3 Gap Analysis

5.3.1 Technical

With the increase in governmental regulations over the years, it has never been so important to minimize greenhouse gas emissions from industries. A new bill however has relaxed these emission control regulations until 2014, after which emission values are expected to be either par or less than the current limit. Similar to the gas emissions is the ash formation, which should be less than 3 %. Ash is a nonmarketable by-product and disposing it is a problem. Thus, it is in the best interest for Green Tech Inc. to incorporate a technology that either produces the same or less than the current emission capability.

MSW landfills are the largest source of human-related methane in the USA and it accounts for 34 % of these emissions. Thus, the plant should be capable of diverting more waste from landfill and produce energy at minimal cost.

Further, the company currently produces about 685 kWh/t of electricity; however, rest of the by-products produced are negligible and is not marketable. Thus, the new technology should have a potential for additional revenue stream in terms of tipping fee and more marketable by-products.

Finally, the technology selection should be such that the conversion process accepts the waste with minimal or no preparation and manage to process irrespective of feedstock's moisture content. Currently, feedstock preparation involves drying and grinding biomass into smaller particle size. Also it is essential to utilize heterogeneous feedstock such as medical waste, hazardous/toxic materials and still produce multiple useful products. Table 5.1 shows all technical gaps.

5.3.2 Organizational

There are several requirements that the organization is pursuing. First, the organization wants to add additional revenue streams. It is very significant from every organization's point of view as they need to survive in today's competitive environment. So they try to find the ways to generate profits as much as they could.

There are several ways to do so. The utilization of MSW as a feedstock provides additional revenue stream in the form of tipping fees. Also, Green Tech currently doesn't have any business relationship established with waste management companies. Thus, it is crucial to develop good relationship with waste management companies and secure access to waste so as to compete with other competitors.

The new technology which will be adapted to utilize MSW should be easily integrated into existing infrastructure and process. The requirement to select a technology should be easily retrofittable with minimal equipment change or process change. Moreover, the organization wants the minimal investment and technologies which has proven emission level by EPA regulation. It is obvious for every company

Table 5.1 Technical gaps

Requirements	Capabilities	Gaps
Minimal carbon emission	With in the EPA mandated emission limit CO—0.299 lb/t CO_2—1,970 lb/t [19]	Technology not proven with the MSW to maintain emission rate with—in current limit
Minimal waste going to landfill	Can convert waste into energy (electricity). In 2008, about 50 % total waste was land filled [16]	
Increase production efficiency and product mix	Currently generating 685 kWh/t [20] of biomass electricity and by-products are negligible	Technology needs to be proven for commercialization which can increase the electricity generation by at least 15 % or improves/introduces new marketable by-product
Reduce (less than 3 %) or eliminate ash formation	Current ash content 5–20 %	No proven technology that reduces ash content to less than 3 %, no proven technology which does not utilize preprocessing
Should be insensitive to input waste	Current system is capable of processing feedstock irrespective of the type of biomass but yield varies with calorific value of the input waste	No proven technology that can utilize hazardous/toxic materials, medical wastes, asbestos, tires, etc.—with closed loop system
No feedstock preparation	Feedstock drying and grinding in to smaller particle size	No proven reactor design that can take feedstock without any preparation
The process should take heterogeneous feedstock	Can process homogeneous feedstock	

that paying less is better, but they have to ensure that whatever technology they want to adopt is approved by involved regulator.

Finally, the organization wants an effective storage and transportation of by-products. This is required because the current plant does not have an effective storage yet for by-products that will come out from the process. Plus, by-products such as bio-oil, bio-char, and syngas require effective transportation (Table 5.2).

5.3.3 Personal

As part of the plant retrofit project to incorporate MSW, retraining the existing and new employees on the processes and procedures becomes necessary. Existing employees have knowledge of the current processes/procedures that can be tailored to incorporate the updated system or procedures. The scope of the training overhaul will be taken into consideration depending on the technology that is implemented.

Table 5.2 Organizational gaps

Requirements	Capabilities	Gaps
Additional revenue stream	Access to waste in Portland Metro Area	Increasing competition—need to develop relationship with waste management companies
New technology should be easily integrated into existing infrastructure and process	The existing systems can be easily upgraded to increase efficiency	Uncertainty in governmental regulations
Minimal investment and technologies which has proven emission level by EPA regulation	Need to prove candidate technologies that do not violate EPA standards	
Need an effective storage and transportation of by-products	Currently can handle gas output effectively	No existing storage facility and transportation infrastructure

Table 5.3 Personal gaps

Requirements	Capabilities	Gaps
Training existing employees on new process and technology	Employees have knowledge of existing processes	Train employees on new technology
Health and safety of employees in conversion process	Know safety policy/ process for existing technology	Bio-oil has known carcinogens—update policy/ process for handling

Further, personnel safety needs to be evaluated; training employees on how to properly handle bio-oil becomes necessary, as bio-oil has known carcinogens. Employees need to have a good understanding of the toxicity levels and appropriate controls that need to be in place to protect plant personnel [21]. Employee training needs to incorporate updated safety information regarding handling of all MSW process outputs (bio-oil, char, and slag) (Table 5.3).

5.4 Technology Evaluation and Selection

5.4.1 Technology Requirements

Using the current biomass conversion at Green Tech as a baseline and the output from the gap analysis, technology requirements were developed to identify a future state. The technology requirements were used as primary screening parameter to identify potential candidate technologies. In Table 5.4 is a list of technology criteria that was derived from the gap analysis.

Table 5.4 Technology requirements

Requirement	Description
Should process heterogeneous feedstock	The current biomass process only supports wood waste as a feedstock. Since the evaluation is identifying MSW as a feedstock, the new reactor design should be able to support heterogeneous feedstock
Should be easily integrated into existing system/process	In order to minimize major capital investment, retrofitting the plant design is necessary. The future state should be able to leverage existing processes and some of the existing equipment
No or minimum feedstock preparation	The current process requires extensive feedstock preparation, which includes feedstock grinding and feedstock drying to eliminate excessive moisture. Future reactor design should either eliminate or minimize feedstock preparation
Should generate 15 % more electricity than current output	If future reactor design can generate electricity, the output should have a net increase
100 % carbon conversion process	The future reactor design should recycle excessive waste and or production to act as a fuel source
Reduce (less than 3 %) or eliminate ash formation	Current state produces ASH as a by-product that is not marketable. Future reactor designs should either eliminate or reuse by-products
Output should be insensitive to input	The current process is sensitive to input. Future reactor designs should be able to accept heterogeneous feedstock and produce multiple outputs
Increase production efficiency and product mix	Expanding the product mix will act as additional revenue sources for Green Tech

5.4.2 Technology Selected for Evaluation

Sixteen biomass technologies that are proven to process MSW were identified that met some or all of the technical requirements and were used as a base for further consideration [22–28]. Table 5.5 lists all selected technologies that were considered for evaluation.

In order to complete the evaluation, we adopted nine (Appendix 1) evaluation criteria. The criteria were established as minimum screening parameters, with the objective that each technology would be required to meet most or all of the criteria in order to be further considered for future procurement. The criteria were structured to assess the feasibility and viability of a MSW conversion plant that meets all the established requirements defined in the gap analysis.

5.4.2.1 First Level Evaluation Matrix

See Table 5.6.

Table 5.5 Biomass technologies that were selected for evaluation

C#	Technology	C#	Technology
1	Updraft gasification	9	Circulating fluid bed reactor
2	Downdraft gasification	10	Biomass catalytic cracking
3	Circulating fluid bed reactor	11	Aerobic digestion
4	Plasma arc gasification	12	Anaerobic digestion
5	Vacuum pyrolysis	13	Fermentation
6	Ablative fast pyrolysis	14	Hydrolysis
7	Rotating cone pyrolysis	15	Micro turbine technologies
8	Bubbling fluidized bed reactor	16	Esterification

Table 5.6 Biomass technologies that were evaluated by the criteria

Candidate technologies	C1	C2	C3	C4	C5	C6	C7	C8	C9	Total
Updraft gasification	1	0	0	1	1	0	1	1	1	6
Downdraft gasification	1	1	1	0	1	0	0	1	1	6
Circulating fluid bed reactor	1	1	1	0	1	0	0	1	1	6
Plasma arc gasification	1	1	1	1	1	1	1	1	0	8
Vacuum pyrolysis	1	0	1	1	0	1	1	1	0	6
Ablative fast pyrolysis	1	0	1	0	1	0	0	0	1	4
Rotating cone pyrolysis	1	0	1	0	1	0	0	0	1	4
Bubbling fluidized bed reactor	1	0	1	1	1	0	1	1	1	7
Circulating fluid bed reactor	1	0	1	1	1	0	1	1	1	7
Biomass catalytic cracking	1	0	1	0	1	0	0	1	0	4
Aerobic digestion	1	0	0	0	0	0	0	0	1	2
Anaerobic digestion	1	1	0	1	0	0	0	1	0	4
Fermentation	1	0	0	0	0	0	0	0	1	2
Hydrolysis	1	0	0	0	0	0	0	1	0	2
Micro turbine technologies	1	1	0	1	0	1	0	0	1	5
Esterification	1	0	1	0	0	0	0	1	1	4

5.4.3 First Level Criteria Technology Selection

After evaluating 16 biomass technologies through the criteria process, 3 technologies were selected for further analysis. Below is a description of each technology that will be further evaluated using cost–benefit analysis.

5.4.3.1 Plasma Arc Gasification

Plasma arc gasification is a waste disposal technology that turns garbage into usable by-products without burning it by using electrical energy and the high temperatures created by an electrical arc gasifier. Temperatures as high as 7,200–12,600 °F are reached in the arc column. At this range of temperature, most types of waste are broken into basic elemental components in a gaseous form. The organics of waste solids (carbon-based materials) are converted to a synthesis gas (syngas) whereas inorganic materials and minerals produce a rock-like glassy by-product (slag) [29].

There are three main by-products of plasma arc gasification:

1. Syngas: a mixture of hydrogen and carbon monoxide. Most of the produced syngas could generate the electricity that powers the plant. The remaining could be sold to utility companies [30].
2. Slag: a solid residue resembling obsidian. Once molten slag is cleaned of contaminants, it can be funneled into brick or paving stone molds and then air-cooled into ready-to-use construction material [30].
3. Heat: Heat from the molten slag helps to maintain the temperature within the furnace. Some of the heat from gases can be used to convert water into steam, which in turn can turn steam turbines to generate electricity [30].

The plasma arc gasification has been considered as an effective waste-to-energy technology because of several facts. First, it is capable of breaking down all kinds of MSW due to its high-temperature operation. Second, it requires minimal or no feedstock preparation. Moreover, it produces useful by-products that could be applicable. Finally, it is an environmentally friendly waste-to-energy technology that produces less greenhouse gas than other thermal conversion technologies since there is no burning process occurred [31]. The technology is in early stage of development and we notice that there are only two manufacturers in the USA—Westinghouse [32] and Geoplasma [33].

5.4.3.2 Fluidized Bed Reactor Pyrolysis

Fast pyrolysis is a process similar to circulating bed reactor (CFB). Here small particles of biomass waste (less than 2–3 mm) are rapidly heated to high temperatures (500–550 °C) in the absence of oxygen, vaporized, and then condensed into liquid fuel. Products of the process are typically 70 wt% of liquid bio-oil, 15 % solid char, and remaining noncondensable gases (NCG). Most importantly, the process has no waste since both pyrolysis oil and char have significant commercial application and value, while noncondensable gases are recycled and produce approximately 75 % of the energy required for the pyrolysis process. Because of their long history of service and inherently simple operating design, this type of reactor is considered to be very reliable and virtually trouble free as a system capable of conducting fast pyrolysis of biomass [21] (Fig. 5.3).

5.4.3.3 Circulating Fluidizing Bed Reactor Pyrolysis

Fast pyrolysis is a process by which small particles of biomass waste (less than 2–3 mm) are rapidly heated to high temperatures (500–550 °C) in the absence of oxygen, vaporized, and then condensed into liquid fuel.

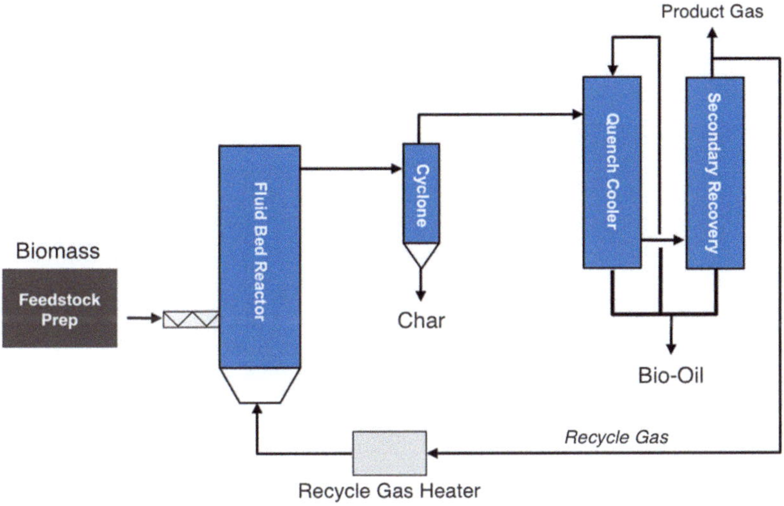

Fig. 5.3 Fluidized bed reactor pyrolysis [21]

CFB operates similarly to traditional fast-pyrolysis reactor designs in converting MSW to bio-oil and other minor by-products. The design is slightly more complicated as the process involves moving large quantities of sand into the reactor. Sand flow rate is also 10–20 times greater than the biomass feed rate significantly increasing the energy cost to operate CFB reactor designs.

Feed size needs to be taken into consideration with the CFB system designs. Particles only reside in the high heat transfer pyrolysis zone for only 0.5–1 s before it is entrained over to the char combustion section. For relatively large particles, this would not be enough time to transport heat to the interior of the particle. Consequently, if larger feed particles are used, the oil yield will be reduced due to combustion of incompletely pyrolyzed particles [21] (Fig. 5.4).

5.4.4 *Second Level Evaluation: Economic Analysis*

The second level evaluation was to do cost–benefit analysis on all three candidate technologies selected in the previous section; this involved calculating NPV, IRR, and payback period and selecting the most efficient one. In order to perform the financial analysis, we made a couple of assumptions after which initial investment and operating cost were calculated for each technology. Appendices 2, 3, 4 lists all general, production, and financial assumptions.

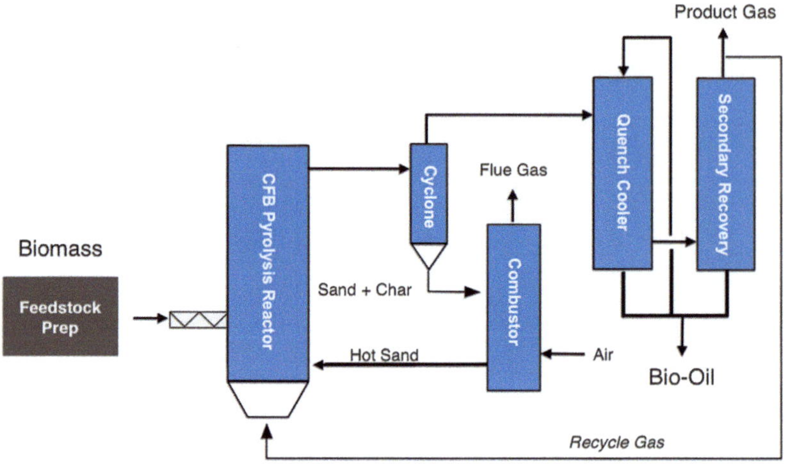

Fig. 5.4 Circulating fluidized bed reactor pyrolysis [21]

5.4.4.1 General Assumptions

Plant and financial assumptions are listed in Appendix 2. The plant is assumed to operate during 317 scheduled operating days with an 85 % utilization rate. The utilization rate is a crucial factor as it determines how much feedstock is needed and the quantity of outputs that can be produced and sold. The operating time parameters include annual days of downtime, annual operating hours, onstream percentage, and downtime costs. Further, the plant fuel will be derived from the outputs of each technology. The maintenance costs for the plant are assumed to be 2 % of the total equipment cost.

5.4.4.2 Capital Costs

Using similar papers from NREL [27, 28, 34–36] and companies that are using the above-mentioned technologies, each major piece of equipment was sized and estimated [37, 38]. For nonstandard equipment, other methods such as comparison with other similar equipment were employed. Installation charges were applied based on the economic analysis performed by the companies [37, 38] trying to use these technologies and most recent NREL papers [27, 28, 34–36]. Appendix 3 provides breakup for equipment required to retrofit each technology.

A contingency factor of 20 % was applied to project the total equipment costs. This factor was designed to account for the uncertainty in the analysis and if any miscellaneous equipment left out of the analysis. Thus, the total equipment cost summed all equipment cost, its installation charges, and the contingency factor.

Using this total equipment cost, the total project investment (TPI) was derived using the bottom-up costing approach. Table 5.7 outlines all the costs used.

Table 5.7 Capital costs

	Cost		
Description	Plasma arc gasification	Fluidized bed reactor	Circulating fluid bed reactor
Total equipment cost	$11,766,000.00	$2,704,000.00	$3,054,000.00
Site development cost/ warehouse	$0.00	$270,400.00	$305,400.00
Total installed cost (TIC)	$11,766,000.00	$2,974,400.00	$3,359,400.00
Installation cost (20 % of equipment cost)	$2,353,200.00	$594,880.00	$671,880.00
MISC start up cost (5 % of total installed cost)	$588,300.00	$148,720.00	$167,970.00
Project contingency (20 % of TIC)	$2,353,200.00	$594,880.00	$671,880.00
Total project investment (TPI)	$14,707,500.00	$4,312,880.00	$4,871,130.00

5.4.4.3 Revenue

Appendices 4, 5, and 6 show the revenue generated for first year. For simplicity purpose we assumed that the feedstock processed for 5 years will be constant @ 30,600 t/year.

Plasma Arc: Electricity selling price was derived based on the current selling price by BPA [39] in Oregon which is about $ 47.00 per MW. The slag price was assumed to be $1.21 per kilogram based on analysis done by city of Marion, Iowa [40]. The total revenue generated by Plasma Arc for 1 year is 6.8 million; this is based on the net electricity output of 865 kWh [33].

The selling price of bio-oil was assumed to be $50.50/barrel and $21/kg for bio-char [37]. Fluidized bed reactor with yield of 70 % bio-oil [35] and 15 % [35] bio-char is estimated to generate the total revenue of $8.9 million for each year. Circulating fluidized bed reactor with the yield of 75 % bio-oil [35] and 10 % bio-char [35] is estimated to produce total revenue of 9.3 million. Appendix 4, 5, and 6 list all important assumption made for revenue calculation.

5.4.4.4 Operating Costs

- Labor Costs: The wages were decided according to the personnel. A payroll burden of 33 % was included for calculations. It was assumed to require 16 personnel for plasma arc gasification process, 17 for fluid bed reactor process, and 18 for circulating fluid bed reactor process during all scheduled operating hours.
- Maintenance Costs: The maintenance costs for the plant are assumed to be 2 % of the total equipment cost.

- Insurance Costs: The insurance costs for the plant is assumed to be 11 % of total initial investment. Property insurance premiums were based on 0.4 % of asset value for buildings and 0.7 % of the building contents value.
- Taxes and Depreciation: For the biomass plant, taxes paid on net cash flow (minus depreciation) were incorporated to determine NPV, IRR, and payback period. The model developed for this project assumed that federal taxes are paid according to Internal Revenue Service (IRS) Form.

Double declining depreciation of capital over the useful operating life of 15 years is assumed for the purpose of reporting taxable income to the IRS. No special financing and grant programs or accelerated depreciation are used in the analysis, nor are production credits or employment credits.

To obtain the final total expenses, we sum all of the operating costs represented above with the depreciation and loan interest for each technology. As a result, plasma arc gasification by calculation has the most expensive operating costs ($307,735), which are approximately twice as much as the other two technologies ($139,564 and $152,947).

5.4.4.5 Financial Analysis Results

The economic viability of all three technologies was evaluated based on cost–benefit analysis comparing Net Present Value (NPV), Payback Period, and IRR for all three technologies. Appendices 4, 5, and 6 show the financial analysis for all three technologies. A life of 15 years was considered for all three technologies.

Retrofitting plasma arc gasification technology at Green Tech Inc. shows total revenue of $6.8 million with a net profit of $360K. However, the technology shows a longer payback period than 5 years with a negative 17 % internal rate of return. The NPV for the project is negative $10 million.

However going with the fluidized bed reactor, pyrolysis technology involves a capital cost of $4.3 million producing a net profit of $2.4 million including the revenue due to tipping fee. The time recovery period for the investment would be 2.37 years with a 31 % rate of return, a value greater than hurdle rate of 7 %. The NPV is positive with a value of $2.5 million. The circulating fluid bed reactor technology, on the other hand, shows similar results with a closer profit of $2.7 million compared to the fluidized bed reactor with a payback period of 2.34 years (Table 5.8).

Due to the closeness in values between fluidized bed reactor and circulating fluid bed reactor technology, incremental cost analysis method was used to select the most suitable and feasible one. According to its results, circulating fluid bed reactor was found to be a good option for Green Tech Inc. as the extra amount invested earns a return that exceeds the IRR (Table 5.9).

Table 5.8 Financial summary

Candidate technologies		NPV	Payback period	IRR
A	Plasma arc gasification	($10,252,291.82)	>5 years	−17.11 %
B	Fluidized bed reactor pyrolysis	$2,542,582.64	2.37 years	31.44 %
C	Circulating fluidized bed reactor pyrolysis	$2,974,404.44	2.34 years	32.13 %

Table 5.9 Incremental cost analysis

	Alt. B	Alt. C	Alt. C−Alt. B
Initial cost	−$4,312,800	−$4,871,130	−$558,250
Net annual income	$2,364,167	$2,772,185	$408,018
IRR on total cash flow	31.44 %	32.13 %	67.54 %

5.5 Conclusion

The research process in this chapter sufficiently helped Green Tech to establish a technology assessment method to select a technology that can be procured for implementation and provide a sound profit for the organization. This paper outlines the steps in how Green Tech went from defining a problem to performing a gap analysis, defining requirements, identifying selection criteria, and finally performing a cost–benefit analysis on three technologies that met all or majority of the defined criteria.

It is important to note that in order to make a conclusion about what technology that was suitable, a hypothetical organization was needed to define the research boundaries. Since boundaries were set, majority of the decision-making process focused on identifying a technology that was cost effective and met the technical requirements that were defined early on. Structuring research in this method had a positive output that allowed Green Tech to make a technology decision, but was limited to assessing technology from other perspectives. Using environmental, organizational, and other perspectives might have yielded different results.

After reviewing 16 different waste conversion technologies that process MSW, we were able to come up with 3 different technologies that are in the best interest of Green Tech. Further based on cost–benefit analysis, circulating fluid bed pyrolysis technology was proposed which is environmental friendly with a net zero CO_2 emission and has no ash as the by-product. Also, with minimal investment, the technology can be easily retrofitted to existed process.

Future research on this topic could be expanded to include the following:

- Adding additional criteria for more granular definition of the technology.
- Use alternative models such as analytic hierarchy process (AHP) and compare the results. This would validate other important factor to selecting technology

and not only selecting a technology from a financial perspective. Other perspectives such as environment and political impacts can be taken into consideration.
- Perform a 15-year cost analysis rather than the current 5-year cost model. Expanding the cost analysis to 15 years might yield different results in the final technology selection.

Identifying a suitable alternative to the current biomass technology at Green Tech is the first step. In order to implement the technology selected through the technology assessment process, several additional steps need to be performed and need further evaluation:

1. Determine license fees for technology. Pyrolysis technology is a patented process owed by Dynamotive [40] and needs to be evaluated and considered whether additional license fees need to be taken into consideration as an additional procurement or operational cost.
2. Identify vendors who manufacture the reactors. Research of all vendors who manufacture circulating fluid bed reactors. Other components that were identified for replacement need vendors identified.
3. Get exact quote for equipment. The current cost estimates defined in Section 5.4.4 are based upon published papers and technology assessments done in various locations throughout the US Establishing relationships with vendors and getting exact quotes for the equipment is necessary to get an exact cost projection. The cost model needs to be adjusted accordingly.
4. Identify strategic partners. It is necessary to develop the appropriate relationship for both feedstock inputs (MSW) and biomass process outputs. These partnerships are critical to Green Tech as there is a need for constant supply of MSW and the sale of output as bio oil has a limited shelf life.

Appendix 1: Baseline Evaluation Criteria

Plant and Financial Assumptions

C#	Criteria	Description
C1	Process MSW (heterogeneous feedstock)?	The current process currently only supports homogenous feedstock comprised of arboricultural activities, yard waste, processed wood, and wood from forest products. The need is to expand and retrofit the plant to expand by processing MSW as MSW will be a source of revenue
C2	Generate electricity? If yes, >685 kWh/t	The current process supports the generation of electricity and is a current source of revenue for Green Tech. The new MSW technology should support additional electricity generation
C3	Produce bio-oil/ bio-char?	Current process output currently is only electricity. By diversifying output to bio-oil, and bio-char, Green Tech. will have additional revenue sources

C#	Criteria	Description
C4	Technology proven for MSW?	Reactor designs should be able to support MSW. This criterion allows Green Tech to take advantage of a new revenue source (MSW) and diversify its feedstock
C5	Easily integrated to existing system	To minimize initial capital investment, procuring a system and that can be retrofitted into the existing system is critical. Leveraging existing systems, like chillers, quencher, and other plant components will minimize overspending
C6	Need feedstock preparation?	Minimal or no feedstock preparation is critical to minimizing the investment in additional processing equipment and the space needed to process the feedstock. Since MSW can vary in size, this is critical factor to processing various feedstock sizes
C7	Can process medical waste, hazardous waste, etc.	This is not a critical must-have criterion, but might prove to be beneficial as this may be a business differentiator and set Green Tech apart from competitors
C8	Carbon emission level less than or equal to current level?	Existing regulation does not specify a maximum carbon emission level, but rather than wait until regulation is imposed, Green Tech should select a technology that has minimal carbon emissions
C9	Minimum investment < $4 million	Green Tech. is a small corporation with limited funds. Retrofitting the existing plant will cut down on initial capital investment. Doing a cost–benefit analysis will identify the most suitable technology for procurement and implementation

Appendix 2: General Assumptions

Plant and Financial Assumptions

Plant assumptions	Plasma arc gasification	Fluidized bed reactor	Circulating fluid bed reactor
Plant capacity (tpd)		100	
Plant availability (85 %)		317 days/year	
Dry feedstock consumed (t/year)		30,600	
Yield (%electricity/%slag)		84/16	
Plant operating fuel	Electricity	Pyrolysis oil	
Plant life (years)		15	
Study period (years)		5	
Maintenance cost (% equipment cost)		2 %	
Financial assumptions			
Hurdle rate		7 %	
Tax rate		30 %	
Tipping fees ($/t)		$58	

Appendix 3: Equipment Cost

Equipment	Qty	Unit cost	Cost
Plasma arc gasification			
Plasma ARC	1	$9,000,000.00	$9,000,000.00
Heat exchanger	1	$2,266,000.00	$2,266,000.00
Utility interconnect	1	$500,000.00	$500,000.00
Total equipment cost			$11,766,000.00
Fluidized bed reactor pyrolysis			
Feed stock handling and drying	1	$1,000,000.00	$1,000,000.00
Fluidized bed—pyrolysis system	1	$750,000.00	$750,000.00
Quench cooler	2	$352,000.00	$704,000.00
Heat recycle to gas heater	1	$250,000.00	$250,000.00
Total			$2,704,000.00
Circulating fluidized bed reactor pyrolysis			
Feed stock handling and drying	1	$1,100,000.00	$1,100,000.00
Fluidized bed—pyrolysis system	1	$1,000,000.00	$1,000,000.00
Quench cooler	2	$352,000.00	$704,000.00
Heat recycle to gas heater	1	$250,000.00	$250,000.00
Total equipment cost			$3,054,000.00

Appendix 4: Plasma Arc Gasification—Financial Summary

PLASMA ARC GASIFICATION

Plant Assumptions		Financial Summary (Year 1)			
Plant Capacity (tpd)	100	**Revenues**	**FY2012**	**Cash Flow**	
Plant availability (85%)	317 days/year	Electricity	$1,243,620.00	Cash from Sales	$6,797,520.00
Dry feedstock consumed (ton/year)	30,600	Slag	$5,553,900.00	Non operating other income	$1,774,800.00
Yield (%Electricity/%Slag)	84/16	Total revenue	$6,797,520.00	Sub total cash received	$8,572,320.00
Plant operating fuel	Electricity				
Plant life (years)	15	Direct cost of sale	$5,066,280.00	Cash Spent on Operations	$6,156,985.00
Study period (years)	5			Net Cash Flow	$2,415,335.00
Maintenance cost (% Equipment cost)	2%	Gross margin	$1,731,240.00		
		Other Income (tipping Fees)	$1,774,800.00		
Investment		Total Gross profit	$3,506,040.00	**Cost-Benefit Analysis**	
Capital cost ($)	$17,060,700.00	Gross margin%	25.47%	Rate of Return (IRR)	-17.11%
Production				Payback Period	>5 years
Net Electricity Production (KWHR/ton)	865	**Expenses**		NPV(RR=7%)	-$10,252,291.82
Slag production (kgs/ton)	150	Operating cost/expense	$2,991,429.00		
		Depreciation expense	$1,568,796.00	**Financial Assumption**	
Net Electricity Production (MW/year)	26460			Hurdle Rate	7%
Slag Production (Kgs/year)	4,590,000	PBITDA	$514,611.00	Tax rate	30%
		EBITDA	$2,083,407.00	Tipping Fees ($/ton)	$58
Price Assumption:		Taxes incurred	$154,833.00		
Electricity Selling price ($/MW)	47	**Net profit**	$359,778.00		
Slag selling price($/Kg)	1.21	Net Profit/Sales	5.30%		

Appendix 5: Fluidized Bed Reactor—Financial Summary

FLUDIZED BED REACTOR PYROLYSIS					
Plant Assumptions		**Financial Summary (Year 1)**			
Plant Capacity (tpd)	100	**Revenues**	**FY2012**	**Cash Flow**	
Plant availability (85%)	317 days/year	Bio-Oil	$8,205,846.00	Cash from Sales	$8,892,510.00
Dry feedstock consumed (ton/year)	30,600	Bio-Char	$686,664.00	Non operating other income	$1,774,800.00
Yield (%Bio-Oil/%Bio-Char)	70/15	Total revenue	$8,892,510.00	Sub total cash received	$10,667,310.00
Plant operating fuel	Pyrolysis Oil				
Plant life (years)	15	Direct cost of sale	$5,224,118.00	Cash Spent on Operations	$7,320,548.00
Study period (years)	5			Net Cash Flow	$3,346,762.00
Maintenance cost (% Equipment cost)	2%	Gross margin	$3,668,392.00		
		Other Income (tipping Fees)	$1,774,800.00		
Investment		Total Gross profit	$5,443,192.00	**Cost-Benefit Analysis**	
Capital cost ($)	$4,312,880.00	Gross margin%	41.25%	Rate of Return (IRR)	31.44%
Production				Payback Period	2.37 Years
Bio-Oil (Barrels/ton)	5.31	**Expenses**		NPV(RR=7%)	$2,542,582.64
Bio-Char(Kgs/ton)	136	Operating cost/expense	$2,065,810.00		
		Depreciation expense	$396,576.00	**Financial Assumption**	
Bio-Oil (Barrels/year)	162,486			Hurdle Rate	7%
Bio-Char(Kgs/year)	4,161,600	PBITDA	$3,377,382.00	Tax rate	30%
		EBITDA	$3,773,958.00	Tipping Fees ($/ton)	$58
Price Assumption:		Taxes incurred	$1,013,215.00		
Bio-Oil selling price($/Barrel)	50.5	**Net profit**	$2,364,167.00		
Bio-Char selling price($/Kg)	0.17	Net Profit/Sales	26.59%		

Appendix 6: Circulating Fluidized Bed Reactor—Financial Summary

CIRCULATING FLUDIZED BED REACTOR PYROLYSIS					
Plant Assumptions		**Financial Summary (Year 1)**			
Plant Capacity (tpd)	100	**Revenues**	**FY2012**	**Cash Flow**	
Plant availability (85%)	317 days/year	Bio-Oil	$8,777,304.00	Cash from Sales	$9,234,743.00
Dry feedstock consumed (ton/year)	30,600	Bio-Char	$457,439.00	Non operating other income	$1,774,800.00
Yield (%Bio-Oil/%Bio-Char)	75/10	Total revenue	$9,234,743.00	Sub total cash received	$11,009,543.00
Plant operating fuel	Pyrolysis Oil				
Plant life (years)	15	Direct cost of sale	$4,983,467.00	Cash Spent on Operations	$7,260,062.00
Study period (years)	5			Net Cash Flow	$3,749,481.00
Maintenance cost (% Equipment cost)	2%	Gross margin	$4,251,276.00		
		Other Income (tipping Fees)	$1,774,800.00		
Investment		Total Gross profit	$6,026,076.00	**Cost-Benefit Analysis**	
Capital cost ($)	$4,871,130.00	Gross margin%	46.04%	Rate of Return (IRR)	32.13%
Production				Payback Period	2.34 Years
Bio-Oil (Barrels/ton)	5.68	**Expenses**		NPV(RR=7%)	$2,974,404.44
Bio-Char(Kgs/ton)	90.6	Operating cost/expense	$2,065,810.00		
		Depreciation expense	$396,576.00	**Financial Assumption**	
Bio-Oil (Barrels/year)	173,808			Hurdle Rate	7%
Bio-Char(Kgs/year)	2,772,360	PBITDA	$3,960,266.00	Tax rate	30%
		EBITDA	$4,356,842.00	Tipping Fees ($/ton)	$58
Price Assumption:		Taxes incurred	$1,188,080.00		
Bio-Oil selling price($/Barrel)	50.5	**Net profit**	$2,772,186.00		
Bio-Char selling price($/Kg)	0.17	Net Profit/Sales	30.02%		

References

1. Oregon Department of Energy, *Oregon's renewable energy action plan.* Action Plan, Apr 2005
2. The World Factbook, Central Intelligence Agency
3. U.S. net imports, U.S. Energy Information Administration, 28 July 2011
4. Waste—non-hazardous waste—municipal solid waste, EPA, 17 Nov 2011
5. Metro: how much are we recycling? Metro|Making a great place [Online], http://www.oregonmetro.gov/index.cfm/go/by.web/id=24920. Accessed 10 Dec 2011
6. D. Thornley, Myth5: U.S. forests and landfill space are shrinking, Center for Energy Policy and the Environment, 2009
7. L.-S. Fan, ARPA-E's 37 projects selected from funding opportunity announcement #1, Ohio State University
8. Renewable Energy/Biomass Power, CREM Ltd [Online], http://www.crem-ltd.com/renewables_biomass.html. Accessed 15 Nov 2011
9. Sylvia Prieler and Guenther Fischer, *Agricultural By-Products Associated with Biofuel Production Chains, Elobio*, Energy Research Center of the Netherlands (ECN), December 2009. [Online]. Available: http://ec.europa.eu/energy/intelligent/projects/sites/iee-projects/files/projects/documents/elobio_impact_of_biofuels_on_food_and_feed_markets.pdf Accessed: 12-Nov-2011
10. The advantages and disadvantages of renewable energy, School Energy Monitoring
11. List of countries by carbon dioxide emissions—Wikipedia, the free encyclo-pedia, Wikipedia [Online], http://en.wikipedia.org/wiki/List_of_countries_by_carbon_dioxide_emissions. Accessed 12 Nov 2011
12. Woody biomass properties, Oregon State University, 12 Mar 2010
13. M. Abdollahi, C. Guy, J. Chaouki, Biomass gasification in rotating fluidized bed
14. S Sadaka, A.A. Boateng, Pyrolysis and bio-oil, University of Arkansas
15. Pyrolysis oil applications, Biomass Technology Group, 2011
16. Biomass pyrolysis, Altenergymag, Feb 2009
17. J. Strickland, Plasma converter byproducts
18. Refuse combustion, EPA
19. L.J. Circeo, Plasma arc gasification of municipal solid waste
20. M. Ringer, J. Scahill, V. Putsche, Large-scale pyrolysis oil production: a technology assessment and economic analysis, National Renewable Energy Laboratory, Nov 2006
21. J. Ciferno, J. Marano, Benchmarking biomass gasification technologies for fuels, chemicals and hydrogen production, US Department of Energy, June 2004
22. A. Klein, N. Themelis, Energy recovery from municipal solid wastes by gasification, in *Presented at the North American Waste to Energy Conference (NAWTEC 11) 11 Proceedings, ASME International*, Tampa, FL, April 2003, pp. 241–252
23. Pyrolysis, EMRC [Online], http://www.emrc.org.au/pyrolysis.html. Accessed 12 Dec 2011
24. Waste conversion technologies: emergence of a new option or the same old story?
25. A. Rajvanshi, Biomass gasification
26. R. Swanson, J. Satrio, R. Brown, Techno-economic analysis of biofuels production based on gasification, National Renewable energy Laboratory, Technical NREL/TP-6A20-46587, Jan 2010
27. M. Wright, J. Satrio, R. Brown, D. Daugaard, D. Hsu, Techno-economic analysis of biomass fast pyrolysis to transportation fuels, National Renew-able energy Laboratory, Technical NREL/TP-6A20-46586, Jan 2010
28. G.C. Young, Plasma arc the leading light? Waste Manage World
29. Plasma arc gasification: turning garbage into gas, Earthnet: renewable energy sustainable future, 28 Jan 2008
30. Plasma arc benefits, C.A.I. Technologies LLC, 2007

31. What is plasma gasification?|Westinghouse plasma, Westinghouse Plasma Corporation—a division of Alter NRG Corp [Online], http://www.westinghouse-plasma.com/technology/what-is-plasma-gasification. Accessed 22 Nov 2011
32. Aitkin County Plasma Gasification Study: does it make sense to move forward with a full feasibility analysis? Aitkin County Economic Development & Forestry
33. M. Ringer, V. Putsche, J. Scahill, Large-scale pyrolysis oil production: a technology assessment and economic analysis, National Renewable Energy Laboratory, Technical NREL/TP-510-37779, Jan 2006
34. D. Peterson, S. Haase, Market assessment of biomass gasification and combustion technology for small and medium scale applications, National Renewable Energy Laboratory, Technical NREL/TP-7A2-46190, Jan 2009
35. K. Kazi, J. Fortman, R. Anex, Techno-economic analysis of biochemical scenarios for production of cellulosic ethanol, National Renewable energy Laboratory, Technical NREL/TP-6A2-46588, Jan 2010
36. R. Waterloo, B.A. Kingston, A preliminary look at the economics of a new biomass conversion process by dynamotive, 27 Mar 2009
37. C. Sorenson, A comparative financial analysis of fast pyrolysis plants in southwest Oregon
38. Current BPA Power Rates (pbl/rates), Bonneville Power Administration [Online], http://www.bpa.gov/Power/PSP/rates/current.shtml. Accessed 7 Dec 2011
39. B. Clark, Economic feasibility of a plasma arc gasification plant, City of Marion, Iowa, in *Presented at the 18th Annual North American Waste-to-Energy Conference, Orlando, FL*
40. Dynamotive Energy Systems» Fast Pyrolysis [Online], http://www.dynamotive.com/technology/fast-pyrolysis/. Accessed 8 Dec 2011

Chapter 6
Bio-fuel Adoption: Can Best Practices from Brazil Be Applied in the United States?

Abrahim Abdulsater, Asem Alluhibi, Bahare Saatchi, and Judith Estep

Abstract Bio-fuel energy can be defined as an alternative source of energy due to being sustainable in both producing energy at a lower price and avoiding exceeding carbon dioxide emission into the atmosphere. The most common worldwide types of bio-fuel are bio-diesel and ethanol. There are many methods to understand how and why technologies are accepted in a country. This report reviews bio-fuel adoption relative to political, cultural, technical, environmental, and economic perspectives for the two largest bio-fuel producers, Brazil and the United States. The research approach is to review the successful bio-fuel adoption in Brazil and ultimately to understand if any of these practices can be applied to the United States. The lessons learned from Brazil could be used by the United States to promote more widespread use of bio-fuel, with the long-term objective of being an "oil-independent" nation in the future.

The need to switch energy strategies from oil to a less risky alternative is the result of a rise in fossil fuel cost, incremental increase of global energy demand, expanding instability of regional policies that supply oil and the augmentation of carbon emission threat. Looking for such alternatives, scholars consider bio-fuel options as a promising substitution for oil [1].

> We can get fuel from fruit, from the shrub by the roadside, or from apples, weeds, sawdust, almost anything! There is enough alcohol in one year's yield of an acre of potatoes to cultivate that field for a hundred years. And it remains for someone to find how this fuel can be produced commercially—better fuel at a better price than we now know. (Henry Ford 1925)

A. Abdulsater
Intel Corp, Hillsboro, Oregon, USA

A. Alluhibi
Portland State University, Portland, OR, USA

B. Saatchi
Nike Corp, Beaverton, Oregon, USA

J. Estep (✉)
Bonneville Power Administration, Portland, OR, USA
e-mail: jaestep@bpa.gov

T.U. Daim et al. (eds.), *Policies and Programs for Sustainable Energy Innovations*, Innovation, Technology, and Knowledge Management,
DOI 10.1007/978-3-319-16033-7_6

Regarding these alternatives, many developed countries consider expansion of bio-fuel production as a strategy not only to reduce the fossil fuel demand [2] but also as a significant beneficial choice to decision makers for several reasons which include:

1. It can be refilled. Unlike oil that depends on exhaustible resources, bio-fuel is a reproducible agricultural resource.
2. Decreases the carbon emission so it is a promising solution to air pollution.
3. Increases farm income because farmers have an opportunity to produce different types of crops and consequently they may need fewer subsidies.
4. Secures less reliance on importing sources for energy.
5. Requires more labor in comparison with other technology so it is an opportunity for entrepreneurship.
6. Its characteristics are quite similar to gasoline or diesel so it requires less adjustment on engines.
7. It is simple to use and produce for the customer and producers, respectively [1].

From this list it can be seen that there are many aspects which would affect adopting alternative fuels. These include the political, cultural, technological, environmental, and economic (PCTEE) consequences [3–6]. This research reviews how bio-fuel has (or has not) been adopted in Brazil and the United States, against this criteria.

According to the history of bio-fuel adoption, Brazil is one of the first countries that adopted renewable energy and could successfully employ bio-fuel energy throughout the entire nation, such that it is known as a leader of alternative energy production [4]. Although the United States has adopted an aggressive strategy to enhance the use of ethanol as a bio-fuel, the nation still debates the efficacy of this fuel alternative among the politicians, the economists, and the environmentalists. The other debates are coming from the various interest groups and their lobbyists [5]. This research is about what best practices could be used, based on the successful adoption in Brazil.

6.1 Bio-fuel Adoption in Brazil

> Brazil will be the largest supplier of renewable energy in the twenty-first century. We will no longer talk about prospecting petroleum; we will talk about planting petroleum. [3]

In 2006, Brazil ranked 35th among the world's most competitive nations. Relatively speaking, this mediocre rating is not representative of how successful Brazil has been in the development and adoption of alternative fuels, specifically ethanol. Relative to the United States, Asian countries, and the European nations, who have debated the benefits and adoption methods of alternative fuels, Brazil has been very effective in developing and integrating this fuel source into their culture. They have been so successful that in 2007, over 83 % of the vehicles sold in Brazil were

Flexible Fuel Vehicles (FFVs) [3]. Brazil's success serves as a global model for the production, distribution, and use of ethanol fuel [4].

Brazil's alternative fuel development began in 1930 and incrementally increased to the level it is today. There were many PCTEE reasons that can be attributed to its success. In general, like most other agricultural nations, Brazil prescribed to the "energy farming" method as a way to improve the quality of life and the economy in a sustainable way [6]. Specifically, the objectives for their alternative fuel program included:

- Energy diversification and security
- Mitigation of air pollution
- Minimize greenhouse effect
- Expansion of agribusiness
- Opportunities for rural workers

Provided are how the PCTEE elements influenced their ability to meet these objectives and achieve the illustrious title of an oil-independent country.

6.1.1 Political Impact

The government relied heavily on incentives and subsidies to affect the development and adoption of alternative fuel, with the first policy established in 1931. This policy required the addition of ethanol to imported gasoline in an effort to reduce oil consumption. However, the government's effort and influence intensified in the 1970s, with escalating oil prices the need to achieve oil independence was a top priority. The government established Proalcool (Brazilian National Alcohol Program) with the objective of supporting ethanol production. Notably, Proalcool provided the equivalent of US$11 billion in federal incentives. These incentives were directed at research institutions, private firms, and the general public in an effort to increase ethanol efficiency, production, and adoption. In general, the basis of the incentives was to (1) create and maintain a market for ethanol, (2) increase production, and (3) foster technological development in the alternative fuel sector. These translated into cheaper credit for private firms, protection against ethanol imports, and tax breaks for the consumers.

For the consumer, the incentives were very attractive. Specifically, the price of ethanol was 59 % of gasoline; the government regulated gasoline prices so the ability to steeply cut the cost of ethanol in an effort to influence a consumer's purchase was feasible. Also, the taxes on ethanol fueled cars were significantly lower than gasoline powered (vehicles). In fact, the incentives were so attractive that by 1984, 96 % of the cars sold in Brazil were fueled by ethanol [4]. Such a large fleet of vehicles needed an infrastructure to support it—imagine having only a limited number of fueling stations and the impact this would have on planning a trip. Brazil addressed consumer's fueling concerns by ensuring that ethanol was available at every Petrobas station. Note: 64 % of Petrobas shares are owned by the Brazilian government [7].

However, beginning around 1985 it was becoming more costly to maintain the level of subsidies and incentives, relative to the cost of oil; oil prices fell to US$12–20 per barrel. Fewer incentives, lower oil prices, and a subsequent reduction in ethanol production, but still a high demand for the product due the large number of ethanol cars that flooded the market, set up the perfect storm for an ethanol supply crisis. In 1988, the Proalcool program lost credibility and was terminated. A new fuel was developed MEG—(methanol, ethanol, and gasoline) to handle the existing fleet of passenger vehicles. Finally, in 1999 all ethanol government regulations ended, except for their ability to mandate gasohol blending rates [4]. The government also slowed research and development efforts which logically had an impact on realizing efficiencies with the production of ethanol or related technological achievements (e.g., manufacturing of agricultural assets).

Then at the start of the twenty-first century oil prices began to steadily increase, once again calling attention to the risks and issues of being dependent on foreign oil imports. The Brazilian government's response was similar to that in the 1930—regulating the ratio of ethanol to gasoline mix.

The influence of the government, associated policies, and specifically the influence of former President Luis Ignacio Lula da Silva cannot be underestimated. During his presidency (2003–2010), Brazil witnessed significant ethanol production and adoption. "…the biodiesel program demonstrates the Brazilian government's capacity to coordinate policies that redirect resources and transform production and consumption across both the private and public sectors and within a broad set of government ministries and agencies" [8].

(Note: There have been significant changes which could redefine the future of bio-fuel in Brazil. The influences are mentioned but, because they are so new, both short- and long-term impacts have yet to be seen. Speculation on future impacts doesn't necessarily support a lessons learned research approach. Therefore, the recent changes in Brazil may be outside the scope of this report. However, the authors chose to include this information to provide a more comprehensive picture of the bio-fuel landscape. Inferences are drawn on their potential to change the use of bio-fuel. Future research is suggested to understand how they have impacted bio-fuel in Brazil, as they inevitably will).

Fast forward into the future and consider if the newly elected (January 2011) President, Dilma Rousseff will maintain similar policies? If not, how could the differences impact the continued use of bio-fuel in Brazil? Some articles suggest that since she was influential for bio-diesel adoption during President Lula's term as the Minister of Mines and Energy, she will continue her pro bio-fuel policies during her term [8]. Only time will tell. Another significant influence on the future of bio-fuel is a proven pre-salt hydrocarbon reserve off the coast of Brazil, discovered in 2007. As reported by the Brazilian National Petroleum Agency, the range of available oil reserves could contain between 50 and 80 billion barrels of crude equivalent [9]. Proven reserves have been reported by Brazil as 14 billion barrels of crude. To put this in perspective, this amount is equivalent to 3 years worth of global oil consumption [9]. The impact on the future of bio-fuel could be significant. Similar to when oil prices became competitive with bio-fuel in the 1980s, will Brazil revert to using oil as its primary fuel source? Future research is suggested to see the impact of both President Rousseff's bio-fuel policies and the reaction to the off shore reserve.

6.1.2 Cultural Impact

Brazil is not considered one of the most technologically advanced nations. However, the Brazilians have been extremely successful at integrating alternative fuel technology to the extent that they could be considered "oil independent." Certainly, the political influence and other factors contributed to this success. However, when similar policies are implemented elsewhere, the adoption is not as pronounced. In fact, the gap is so big, it is estimated that it would take decades for the United States to achieve the level of adoption seen in Brazil [3]. Why? As suggested by the article, "Beyond a better mousetrap: A cultural analysis of the adoption of ethanol in Brazil," Nardon, L; Aten, K. *Journal of World Business*, vol 43 (2008) the answer lies in the Brazilian culture itself. More specifically, such a successful adoption can be attributed to the term jeitinho, or logics of action of flexible adaptation used to deal with various problems. A brief historical perspective will help to clarify the term.

Brazil was original inhabited by the Portuguese, and they imported a large number of slaves to work in the sugarcane plantations. The combination of backgrounds developed into a unique cultural foundation. Historically, the population is described as "seeking for his soul in the dialectic profusion of his physical and spiritual components, who has to develop a flexible, labile, plastic personality in order to survive, live, and build a country" or a flexibility of body and spirit, allowing deviation from obstacles [3]. As result of the fundamental need to adapt and be flexible, the concept of jeitinho evolved.

Flexibility permeates every aspect of Brazilian lifestyle—from disputes on the soccer field to providing an effective response to fuel scarcity [3]. Considering this perspective, it is not surprising that the FFV or ethanol in general was such a success—it fits perfectly within the fundamental beliefs of their culture. Brazil's response to fuel scarcity follows a logic principle of flexible adaptation. The FFV is the ultimate legitimization of the logic of changing fuels to adapt to external circumstances. The power of deciding fuel mix has moved away from the government and into the final consumer's hands [3].

6.1.3 Technology Impact

In addition to technology impacts on agriculture and harvesting equipment as well as contributing to the efficiencies in sugarcane processing, by and large, one of the most significant contributions to the adoption of alternative fuel was the development of FFVs—vehicles that can run on ethanol, gasoline, or any combination of the two. This flexibility allows the consumer to adapt to changing markets/prices.

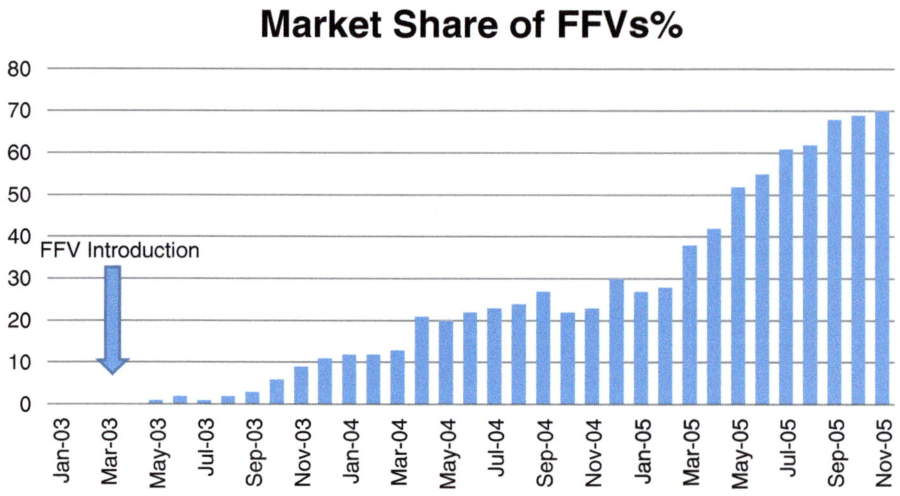

Fig. 6.1 Market share of FFVs in Brazil January 2003–October 2005 [6]

In addition, the vehicle's performance was comparable to a gas-only vehicle and similarly priced. It was unexpected how quickly this technology was accepted. Initial forecasts suggested that by 2006, the adoption rate would be ~60 %. However, Fig. 6.1 shows the actual market penetration in 2005 Q3 exceeded 60 % [3] and by 2007, 83 % of the cars sold in Brazil used this technology!

The FFV was the result of a partnership between Germany-based Bosch and an Italian firm Magneti Marelli. From a technology management perspective, the FFV could not have come at a better time. The development of the vehicle was able to apply lessons learned from previous efforts to integrate alternative fuels and recognized the benefits of providing a car that could run, efficiently, on any fuel. Their solution empowered consumers and provided a long-term solution to the ebbs and flows of the oil industry.

6.1.4 Environmental Impact

Sugarcane, for the production of ethanol, can be harvested either manually or mechanically; almost 80 % of the sugarcane harvested is done manually. In order to increase the yield by up to 30 % and lower transportation costs, the sugarcane is typically burnt prior to harvesting. The effects of the burn could negate the positive environmental impacts of using an alternative fuel—burning results in an increase in atmospheric pollution, accelerated soil degradation, pollution of the aquatic system, and loss of biodiversity. Also, the impact to the most vulnerable citizens should be considered. Depending on the time of year, a significant amount of smoke and suspended particles can remain in the air, posing a serious health threat to the elderly and children. As an example, in municipalities with >50 % of the land dedicated to sugarcane production, there is a 15 % increase in elderly respiratory illness and a 12 % increase in children. The respiratory illness is directly attributable to the suspended particles resulting from a pre-burn [2].

The article suggests a more comprehensive review of policy, rather than just the economic value of alternative fuels, and urges the government to restrict burning times or mandate complete mechanization of sugarcane harvesting. In response to public pressure, there have been some attempts to mitigate the impact of smoke and suspended air particles. Brazilian law 11,241 calls for a gradual increase of mechanization; however, the law has not been rigorously enforced and pushback is common. The law also states that manual harvesting in areas that can be potentially mechanized has to be converted by 2020, and phasing out manual harvesting completely, by 2030. Some legislation is written to protect the aquatic areas and soil by restricting harvesting in areas close to rivers/streams, but if enforced, would eliminate between 4 and 28 % of cultivation [2]. It remains to be seen if the legislation will have a positive impact on burning. As of 2006, based on the number of fires, expansion of manual cultivation is proceeding faster than mechanization [2]. As well, in Sao Paulo the government has forbidden burning from July until the middle of October, as well as prohibiting burning if the relative humidity falls below 20 %, environmental conditions that are conducive to respiratory illness [3].

Similar to adopting an ethanol policy, the government is implementing policy to minimize the negative effects on the environmental and health of its citizens.

6.1.5 Economic Impact

In 2004, Brazil was responsible for producing more than one third of the total ethanol produced in the world and was one of the main exporters [6]. In addition to providing a profitable exportable commodity, the impact of the ethanol program resulted in an increase of higher quality jobs for rural populations and successful technology transfers which stimulate wealth among the private sector.

6.1.5.1 Sugarcane as an Export

Sugarcane has two primary outputs—sugar production (as a food source) and ethanol. There is a significant external market for raw sugar and a modest one for ethanol exports [4]. Specific to ethanol, Brazil can have a fluctuating market. Figure 6.2 shows for the period 1990–1997, the increase in imports can be explained by remnants of the pure ethanol vehicles—as mentioned before during this period it was not cost-effective to produce ethanol due to competitive oil prices, yet there remained a supply due to the strong government ethanol policies and the public's acceptance of ethanol-only vehicles. Conversely, there appears to be a consistent export potential—in most years Brazil provided some export of ethanol, except for those years where fuel had to be imported to address the ethanol supply crisis in the early to mid 1990s [6].

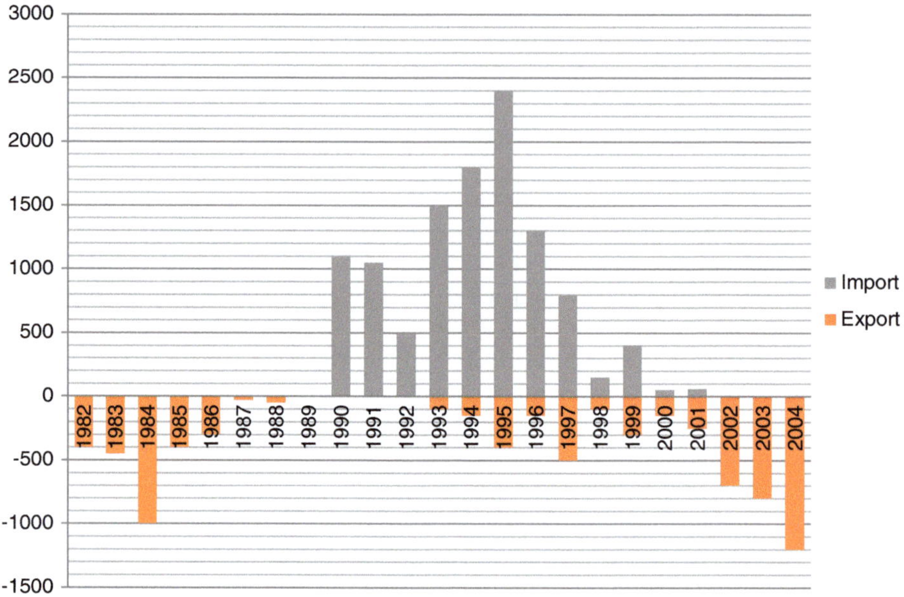

Fig. 6.2 Ethanol trade in Brazil 1982–2004 [6]

In addition to the government policies which affect export and import trends, the climate also has an impact. Brazil is subject to the affects of El Nino and La Nina effects. These correspond to lower sugarcane yields and hence higher import trends.

6.1.5.2 Increase in High-Quality Jobs for Rural Farmers

Referring to Dave Williams discussion, "Impact Your Community with your Technology," if we define community as the rural farm workers, then the development and adoption of ethanol had a significant impact on the creation of sustainable communities. Some references cite in excess of 800,000 rural jobs have been created as a result of ethanol adoption. In addition, there were 250,000 indirect jobs created [6].

Once again we see the influence of government legislation. A portion of the net sugarcane and ethanol prices are reserved for the assistance in improving services for rural farm workers (e.g., improvements in medical, dental, pharmaceutical, sanitary conditions).

6.1.5.3 Stimulating Wealth Through Successful Technology Transfers

Proalcool addressed the adoption of ethanol in two phases. The first phase focused on adjusting ratio of alcohol to gasoline mix in an effort to curtail oil imports. Phase two introduced the use of ethanol fueled cars. These cars were made by Brazilian manufacturers who purchased the technology from public research centers;

investing in research was a primary objective of Proalcool. The technology for the ethanol-based vehicles was developed in the 1970s and sold to the automakers who then continued to develop the technology [6].

6.2 Bio-fuel Adoption in the United States

> ...this is our generation's Sputnik moment. Two years ago, I said that we needed to reach a level of research and development we haven't seen since the height of the Space Race.... we'll invest in...clean energy technology—an investment that will strengthen our security, protect our planet, and create countless jobs for our people. (President Obama, State of the Union Address, January 25, 2011)

The United States has been paying the price of ongoing industrialization and the incremental demand for energy sources. Increased energy demand can also be seen as a result of the growing and aging population. In addition, the general public and government agencies awareness and concerns about the climate changes and their associated impact to the environment and population have played a big role in the search for alternative energy sources, mainly bio-fuel products.

Ethanol, as shown in Table 6.1, is not a totally new concept to the United States, over the past decades the nation has spent a great deal of effort in the area of bio-fuel research. The goal was and still is to limit its dependency on foreign oil, as well as control, and hopefully reverse the course of accelerated emissions of greenhouse gas (GHG). By-products of a national renewable energy sources would include increased employment as well as a lower level of GHG emissions, which both directly and indirectly impact the US population.

In the United States, ethanol and biodiesel have been the main forms of bio-fuel focused on by researchers; in fact, the US ethanol production in 2010 accounted for 57.5 % of the total global production Table 6.2 [10]. The sources vary from corn and other feedstock to algae, a promising source, which will not only eliminate the food vs. fuel competition but also minimize the need for arable land. Other product and by-products include biobutanol (with similar properties as gasoline), and alternative protein sources for aquaculture and livestock feed [11].

Table 6.1 A summary of the major US ethanol timeline events [31, 32]

1826	Development of first engine that uses ethanol and turpentine
1862	Taxation of ethanol by the Union Congress to cover the Civil War expenses
1896	First automobile for pure ethanol by Ford
1908	The Ford's Model T as the world first Flex-Fuel-Vehicle (FFV)
1920s	Gasoline became the major motor fuel with ethanol as a booster
1940s	First US Fuel ethanol plant was built by the US army
1940s–late 1970s	Low prices of gasoline fuel forced ethanol out of the market
1989–2000	Multiple regulations passed to control the US Motor gasoline
Late 1990s–present	US FFVs that can perform on a blend of up to E85 entered the market
2005–present	More regulations, subsidies, and research funding

Table 6.2 2010 World fuel ethanol production [33]

Continent	North and Central America	South America	Europe	Asia	Australia	Oceania	Africa
Millions of gallons	13,720.99	7,121.76	1,208.58	785.91	66.04	66.04	43.59
Nation	Brazil	European union		China		Canada	
Millions of gallons	6,921.54	1,176.88		541.55		356.63	

6.2.1 Political Impact

As the leader of the modernized world, the United States has an obligation to be among the first nations (if not the first) to think about alternative fuel. Indeed, during the last decades of the twentieth century, both federal and states legislative bodies have translated the worries about the nation's energy security and climate global changes into policies. The increasing energy demand, unpredictable crude oil prices, instability of oil-source regions, and the alarming rate of global warming have been among the main reasons why the United States is pursuing an aggressive investment in bio-fuel research and production.

According to a 2010 congressional research service, 22 programs and provisions have been established between the years of 1975 and the year of 2009. As of November of 2011, seven programs have expired; 10 will reach their end of life between December of 2012 and 2019, while the remaining five are not tied to an expiration date. Billions of dollars have been allocated for various programs supported by different governmental departments or agencies such as the Environmental Protection Agency EPA, the US Department of Agriculture DOA, the Department of Energy DOE, the Internal Revenue Service IRS, and the Customs and Borders Protection. To list few examples, we mention the 1975 Manufacturing Incentive for FFVs to stimulate the production/sales of FFVs, the 1980 Import Duty for Fuel Ethanol that imposed a tax on imported ethanol to promote and encourage national production. The 2005 Energy Policy Act, and the 2008 Food, Conservation and Energy Act are considered to be the major ones; therefore, we will briefly detail them [12].

The 2005 Energy Policy Act was administered by the EPA and has multiple sections; The §1501 (P.L. 109-58) Renewable Fuel Standard (RFS) which was revisited by the 2007 Energy Independence and Security Act set up the rules for gasoline and bio-fuel products blending; starting with 4.0 billion gallons in 2006 and reaching 36 billion gallons in 2022 with 21 billion gallons from a non-corn source [13]. Another section, Title XVII, or what is known as the DOE Guarantee Loan, was initiated to fund energy-related projects such as bio-fuel researches. The §942 (P.L. 109-58) Cellulosic Ethanol Reserve Auction allocated a total of $1 billion to the DOE spending in support of cellulosic bio-fuel production. The regulations were finalized in October 2009. Other sections were crafted too under the 2005 Energy Policy Act.

Advanced bio-fuels were also supported by The 2008 Food, Conservation and Energy Act, better known as the 2008 Farm Bill, under the §15,321 (P.L. 110-246) Credit for Production of Cellulosic Bio-fuel. This IRS program allows the cellulosic bio-fuel producers to claim a maximum tax credit of $1.01 per gallon. As generous as it sounds, this regulation was tied to others that would lead to the reduction of the final allowable claimed credit.

There are a wide range of beneficiaries which include farmers, rural small business, and bio-fuel producers. A complete list of the 22 programs can be found in the September 15, 2010 Congressional Research Service "Bio-fuels Incentives: A Summary of Federal Programs" by Brent D. Yacobucci [12].

The second form of governmental influence on bio-fuel adoption came at the state level. Multiple states legislators were attracted to the concept of achieving a national self-sustained energy status and used their influence within their own states. Since these policymakers are more familiar with the landscape of their state, the overall interaction of the community and their surroundings, and their various resources as well as their stakeholders, they could be more innovative and creative than the federal government to establish effective policies. The initiatives were classified and categorized under various policy labels which included the development of sustainable feedstock guidelines, establishing minimum RFS, researching and developing locally appropriate feed stocks, and conversion technologies. Specifically, a few examples are noted [14]:

1. The Pennsylvania's Alternative Fuels Incentive Grant assisted the funding of more than 50 projects valued at $17.8 billion since the start of 2006 and invested a yearly $5.3 million till 2011 in support of locally produced bio-fuel.
2. In the "2007 California State Alternative Fuels Plan," the Air Resources Board and Energy Commission realized the large unused biomass resource in the form of agricultural waste, forestry, and urban waste streams. California decided to benefit from them in the energy production arena in order to reach a Low-Carbon Fuel Standard.
3. In 2007, the "North Carolina's Strategic Plan for Bio-fuels Leadership" was published. The plan was drafted to reduce the annual fuel demand by 10 % and to replace it with local bio-fuels product by the year 2017.
4. In 2008, the Maryland's Agricultural Water Quality Cost-Share program was designed to pay landowners up to $85 per acre to plant their land for winter to minimize the soil erosion and nutrient runoff. The program aimed at two targets; increasing the harvest of feedstock and the carbon segregated in the soil.

A more comprehensive list of the state influenced legislatures can be found in the February 16, 2010 "Developing an Advanced Bio-fuel Industry: State Policy Options for Lean and Uncertain Times" published by the Environmental and Energy Study Institute [14].

The bio-fuel adoption path was never obstacle free, related to government policies. Even though the giant oil companies, such as ExxonMobil and Chevron, invested in some future bio-fuel technologies, their lobbyists' actions suggested otherwise. In reality, they mostly opposed or delayed passing of any mandates for alternative fuels. Their thinking was that it would detract from short-term gains and control of the oil industry [15].

6.2.2 Cultural Impact

The need for the creation of employment opportunities has perhaps the biggest impact on the population. Bio-fuels use has had a great impact on the US population. Specifically, the transfer of land from local owners to investors for large-scale industrial plantations has created income for rural farmers. Also, others have migrated from their local establishments to seek employment in these new plantations and associated agribusinesses [16]. These new job opportunities also affect the surrounding by creating ancillary services (medical, schools, and other social services). As a result, employees at these farms have an opportunity at a more prosperous livelihood.

Conversely, bio-fuel production may contribute to increasing the cost of food. The expected increase in corn prices is of particular concern, since more than 70 % of the corn grown in the United States is used to feed livestock [17]. Diverting corn into bio-fuel production may, as a consequence, increase the price of meat and grain products [17]. The increased use of bio-fuel energy carries many associated impacts, both positive and negative, related to social impacts. These effects need to be critically assessed to bring a more thorough understanding and formulate, the best suited strategies promoting positive development, creating jobs, while limiting the negative impacts such as an increase of food prices.

6.2.3 Technology Impact

The bio-fuel history goes back to the 1880s (refer to Table 6.1) when the first diesel cars were designed to run on peanut oil. Henry Ford produced bio-fuel cars as early as the 1908 T model. Hemp and peanut oil were the main resources of the bio-fuel sold by Standard oil, and it did account for 25 % of the total fuel sold. Unfortunately, the biodiesel industry collapsed in the 1930s under the massive and aggressive campaign of the petroleum industry. It took the industry, and the world, some 40 years and a couple of oil crises to realize that nonrenewable energy sources would at one point be depleted, not to mention the inability to control the foreign oil sources. Recently, due to public awareness and demand for more environmentally friendly products, the auto industry is beginning to support the market with bio-fuel, flex-fuel, and hybrid cars. In fact, all the US passenger cars sold since 2000 are regulated

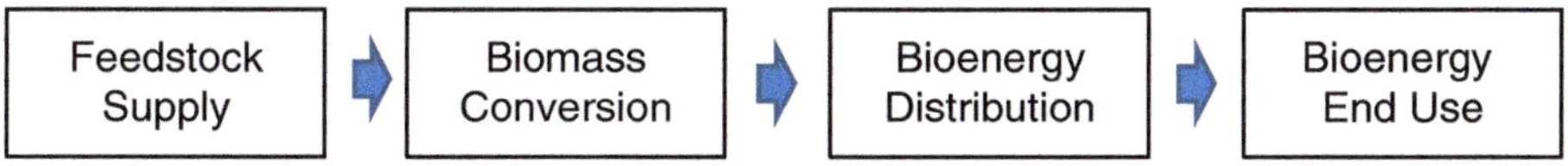

Fig. 6.3 Biomass-to-bioenergy supply chain [20]

and designed to run on a mix of gas and bio-fuel [18]. The bio-fuel technology was kick-started in the 1970s to the present day with US Energy Secretary Chu's latest position on bio-fuel research:

> ...Using consolidated bioprocessing, a research team led by James Liao of the University of California at Los Angeles for the first time produced isobutanol directly from cellulose. The team's work, published online in Applied and Environmental Microbiology, represents across-the-board savings in processing costs and time, plus isobutanol is a higher grade of alcohol than ethanol... [19]

As of today, multiple processes to manufacture ethanol exist or are in their final stages of development. A large number of private and governmental R&D laboratories are working to decrease manufacturing costs and increase efficiency. A typical cycle is represented by Fig. 6.3 [20].

The supply ranges from peanut oil, hemp, corn, panicum (switch grass) and all kind of plants and plant-derived materials. The processing and conversion could be biochemical, thermochemical (which may complement each other), or by gasification [21].

Another technological front, which is moving at a rapid rate, is biomass algae. This technology is significant because it relies on non-food biomass. Using this technology would address the depletion of food sources (corn) for ethanol production. Via a continuous harvest process, carbon dioxide (CO_2), the major GHG, emitted by power plants, oil and gas refineries, and cement factories provide high growth rates for the marine microalgae [22]. This bio-fuel production procedure, from photosynthetic microbes, started in the early 1980s and more than US$25 million was invested on various programs by the US National Renewable Energy Laboratory (NREL) and was terminated in the late 1990s by judging the program non-economical, non-feasible. But Cellana scientists, supported by the founders, managed to attract investments for another US$20 million to refund the research of the Aquatic Species Program (ASP). In a 4-year period (1998–2001), the results were shocking and the conclusion that was based on a large-scale pilot operation proved the NREL wrong. In a very detailed article, the coauthors postulate a target of fossil-free independence by 2020 [23]. This accomplishment was finally recognized by the current administration on May 5, 2009. President Obama and Secretary of Energy announced a US$800 million investment in new research on bio-fuel, including the algae as an alternative and renewable source of biomass feedstock. A National Algal Bio-fuels Technology Roadmap Workshop was convened by the department of energy (DOE) in December 2008 [24].

Two more technologies are worth mentioning for the very positive and promising results they display: The Hydrothermal Liquefaction (a thermochemical technique conducted by Professor Yuanhui Zhang at the University of Illinois) that successfully produced bio-crude oil from waste material mimicking the nature's process beneath the earth's crust. Professor Zhang is very optimistic about his process especially if algae are used as the feedstock [25].

The second is the cellulosic technology [26]—this technology is in its final stages of development. This technique, like the other advanced bio-fuel techniques, will not rely on food material as feedstock instead it will be using abundant agricultural wastes and any other wood-like material to generate about 60 billion gallons/year an approximate 30 % of 2030 gasoline consumption once commercialized [26].

In parallel with the achievements of the advanced bio-fuels, the current technologies for the commercial corn ethanol production are improving on all fronts from the corn field to the refinery [27].

6.2.4 Environmental Impact

The NREL was the first body to venture into bio-fuel research, concentrating primarily on cellulose ethanol. Environmental impacts associated with bio-fuels can be associated to the nature of its production. In recent years, there has been a lot of emphasis on the technology and challenges of bio-fuel systems and their relative efficiencies with respect to energy and carbon dioxide emissions. This research has paid less attention to the impact on the environmental issues associated with the development of large-scale bio-fuel production [28]. The researchers suggested that the average future GHG emission from corn ethanol and gasoline or diesel fuel could be similar, however, there is uncertainty associated in these estimation [28]. In general, ethanol doesn't provide any quality advantage in local air over gasoline [28].

According to Kojima and Johnson 2005, the liquid bio-fuel production was a little over 1 % of the global renewable energy and a little shy of 1 % of the global crude oil. This statement suggests that nations started considering issuing policies that will encourage the involvement in alternative fuel research and the impact may be felt within a decade [1]. These policies fueled the concerns of environmentalists which feared a shift toward more corn-oriented farming to take advantage of these policies and realize more economic gain. To achieve a better crop, farmers may tend to increase the use of fertilizers and pesticides. The increased use of these chemicals may cause additional damage to the environment. Also, there is a fear that additional farmland will be needed. At what cost? Cellulosic ethanol requires wood-like biomass which would impact forestation and there are already debates about how to use corn crops—for food or ethanol production. The potential of expanding farming/forestation to manufacture ethanol directly affects the amount of GHG presence in the environment [28].

6.2.5 Economic Impact

The US government invested a significant amount of money in domestic bio-fuels production hoping that they will one day be the primary source of energy. This way, a nonrenewable fuels independency will be possible—the diversification of the energy market would be of a positive benefit to the producers and consumers. This investment is a direct response to forecasted shortages and increased prices of fossil fuels. The Agro-biotechnology journal estimates that by 2015 the US corn market is expected to support 15 billion gallons of ethanol which will only be enough for less than a quarter of the overall US population fuel demand. Also, 12.3 billion bushels of corn would be available for the food industry and export market [29]. However, the economic potential associated with bio-fuel is dependent on energy prices and policies regulating the researches and production of renewable energy. Day-to-day fluctuations of oil prices make it hard to predict ethanol's economic potential in the United States [1].

6.3 Discussion

Although Brazil is not as developed a nation as the United States, it still could successfully adopt alternative fuel technology to the extent that it is known as an "oil-independent" nation. In comparison, the United States hopes to be fossil fuel independent by 2020, yet it is one of the prosperous countries in world. The research attempted to understand this paradox by examining the political, cultural, technical, economic, and environmental impacts surrounding bio-fuel adoption. In general, studying the history of alternative fuels in Brazil and the United States shows that the countries employed completely different policies at the same period of time. In the 1930s, Brazil established the first policy of alternative fuel while the US consumption of biodiesel collapsed due to a total absence of any marketing campaigns. Instead US investors pursued the petroleum industry.

Regarding the source of bio-fuel, each country uses a different source. Brazilian feedstock relies primarily on sugarcane while the United States cultivates corn to produce ethanol. By comparing sugarcane and corn characteristics (Table 6.3), it is understood how sugarcane characteristics help Brazil to produce ethanol through easier processes and increased yields vs. corn, which is used by the United States [30].

However, in 2007, the United States surpassed Brazil in ethanol production. This can be the result of enforcing some environmental restrictions in Brazil—the government has forbidden burning from July until the middle of October, as well as prohibiting burning if the relative humidity falls below 20 %. These restrictions focus on environmental conditions that are conducive to respiratory illness [3, 27].

One reason that Brazil could adopt alternative fuel relatively easy is Brazil's historic ability to be flexible. This cultural basis is known as Jeitinho. Brazil has a cultural reference of adaptability and willingness to change vs. the United States,

Table 6.3 Sugarcane and corn characteristics comparison [30]

Brazil-sugarcane	The United States-corn
The sugar (source) in sugarcane can be converted directly into ethanol	The starch in corn is first converted into sugar. Then the sugar is converted into ethanol
Sugarcane is planted every 6 years using cuttings	Corn is planted every year using seeds
Sugarcane provides five cutting over 6 years and then is replanted	Corn is harvested once each year
Sugarcane yields about 35 ton per acre (entire plant) per harvest acre	Corn yields about 8.4 ton per acre (entire plant) per harvested acre
Sugarcane yields about 4.2 ton of sources per acre (10–15 % of sugarcane yield)	Corn yields 4.2 ton of corn graph per acre (150 bushels) or 2.4 ton of starch
An acre of sugarcane produces about 560 gallons of ethanol (35 ton yield)	An acre of corn produces about 420 gallons of ethanol (150 bushel yield)
Sugarcane feedstock is cheaper to grow than corn per gallon of ethanol	Corn feedstock is more expensive to grow than sugarcane per gallon of ethanol
Sugar-ethanol can be produced cheaper than corn-ethanol	Corn-ethanol is more expensive to produce than sugarcane-ethanol
The by-product of ethanol production is bagasse	The by-product of ethanol production is distillers grains with soluble that is used as livestock feed
The energy source for ethanol production is bagasse	The energy source for ethanol production is natural gas coal, coal, and diesel
About nine million acres are used for ethanol production	About 180 million acres are used for ethanol production
Brazil has great potential for expanding sugarcane acreage without limiting the acreage of other crops	US expansion of corn acreage will come at the expense of reduced soybean and other crop acres
No subsidies for ethanol	Subsidy reduction from $0.51 per gallon to $0.45
No import tariff on ethanol	A $0.54 per gallon import tariff

where people seem reluctant and resistant to changes. Related to the cultural adaptability, this may not be an area where the United States can apply the lessons learned from Brazil.

Based on their previous bio-fuel experience, beginning in the 1930s, Brazil could largely develop the required technological infrastructures to use ethanol throughout the nation. Brazilian car manufacturers produced FFVs that can run on gasoline, ethanol, or any combination of the two. This flexibility allows the consumer to adapt to changing markets/prices. Also in Brazil all gas stations are well equipped for storing and distributing both gasoline and ethanol. In comparison, the United States does not use FFV in the same manner as in Brazil; US FFVs only run on E10, a mix of ethanol and gasoline, nor does the United States have an infrastructure to support wide-scale production or distribution. It can be inferred that the inaccessibility of bio-fuel is causing less acceptance in the United States (as compared to Brazil).

Both Brazil and the United States have concerns about the environmental effects of bio-fuel, but in this respect both countries have taken different measures to deal with those concerns. In Brazil, as mentioned earlier, the government has legitimated some regulations for producing bio-fuel, while the United States is expanding research on algae as an alternative source for producing bio-fuel; the process of manufacturing alternative fuel from this source is less harmful to the environment.

In Brazil bio-fuel has proved to be economically beneficial, both as an export and national fuel source. The United States is struggling with a wide-scale adoption so economic impacts are less certain. One economic impact that is readily seen in the United States is on the price of corn—it's a balance between using corn as a food source and using it to manufacture bio-fuel.

6.4 Conclusions

This research focused on bio-fuel adoption for the two largest ethanol producers—Brazil and the United States. Both countries were well studied regarding political, cultural, technological, environmental, and economic issues. A summary of their response for each perspective is summarized in Table 6.4. Each country had/has a strong political influence which encouraged research in alternative fuels. Similarly, each had supportive policies and regulations such as assigning tax credit and increasing investment.

In general, this research identified two areas where the United States might better understand the success in Brazil and determine how these technologies could be applied. Prior to the manufacturing process there does not appear to be significant opportunities to utilize best practices. It can be inferred that the technologies used to harvest and manufacture ethanol from sugarcane are too disparate from those required to manufacture ethanol from corn—it would be like comparing apples to oranges. Also, the cultural foundation of jeitinho cannot be useful to the US adoption. However, post production there appear to be opportunities to better understand the success in Brazil and apply to the United States.

Two areas where the United States can learn from Brazil involve technological and environmental (infrastructure) perspectives. It would appear that having a fully FFV, that is one that can operate on any ratio of ethanol to gasoline, was critical in the widespread adoption in Brazil. This technology accommodates adjustments to fuel blends as a result of fluctuating prices or facilitates a more gradual transition from gasoline to ethanol. More specifically identifying the obstacles to manufacturing and implementing FFVs in the United States is suggested for future research. Also, the United States does not have a sufficient infrastructure to support the delivery and consumption of bio-fuel. Parallel to other bio-fuel research, the United States should consider the infrastructure success in Brazil and determine necessary steps to have one (infrastructure) in place to meet the objectives previously stated: to be oil independent by 2020.

Table 6.4 Brazil and US comparison regarding bio-fuel approach

<table>
<tr><th>Aspects</th><th>Brazil</th><th>The USA</th></tr>
<tr><td rowspan="10">Political</td><td>Alternative fuel first policy-1931</td><td>Tax credit 1.01$/gallon</td></tr>
<tr><td>Proalcool to support ethanol production</td><td>Adoption at the state level</td></tr>
<tr><td>Investing on developing ethanol (efficiency, production, and adoption)</td><td>Biodiesel collapsed due to new petroleum industry-1930s</td></tr>
<tr><td>Cheaper credit for private firm</td><td>Support bio-fuel market by car manufacturers-1970s</td></tr>
<tr><td>Ethanol accessibilities gas station</td><td>800 million investment</td></tr>
<tr><td>Tax breaks for customer</td><td rowspan="5">Oil-independent-2020</td></tr>
<tr><td>Switch over to oil–oil price reduction in 1985</td></tr>
<tr><td>Switch back to bio-fuel-twenty-first century</td></tr>
<tr><td>Strategy for Alternative fuel vehicle production (96 %)</td></tr>
<tr><td>Oil-independent-now</td></tr>
<tr><td rowspan="3">Cultural</td><td rowspan="3">Jeitinho-historical flexibility in adopting change</td><td>Transferring lands from farmers to investors</td></tr>
<tr><td>Work for investors</td></tr>
<tr><td>Change rural life into more civilized life</td></tr>
<tr><td rowspan="4">Technological</td><td>Flexible Fuel Vehicle (FFV)- 100 % of either fuel</td><td>Use of peanut oil in diesel car-1880</td></tr>
<tr><td>Previous experience of alternative fuel</td><td>FFV-combination of gas and bio-fuel</td></tr>
<tr><td rowspan="2">Sugarcane supply [30]</td><td>Corn supply [30]</td></tr>
<tr><td>Algae for bio-fuel production</td></tr>
<tr><td rowspan="4">Environmental</td><td>Environmental damage due to burning sugarcane</td><td>Environmental damage due to excess of farming consequences like:</td></tr>
<tr><td>Health threat</td><td>Soil fertilizing</td></tr>
<tr><td rowspan="2">Prohibiting burning actions</td><td>Pesticides</td></tr>
<tr><td>Vegetable cleaning and burning</td></tr>
<tr><td rowspan="5">Economical</td><td>2004, 33 % ethanol production in the world</td><td>2010, 57 % ethanol production in the world</td></tr>
<tr><td>Profitable export</td><td>Increase food price</td></tr>
<tr><td>Two productions: sugar and ethanol</td><td rowspan="3">Increase hunger</td></tr>
<tr><td>High-quality jobs</td></tr>
<tr><td>Technology transfer—wealth among private sector</td></tr>
</table>

6.5 Future Research

Any one of the perspectives considered in this report could be developed further, to more specifically understand the issues. Often one perspective had an effect on another (e.g., mandating 100 % mechanized harvesting in Brazil would decrease the number of jobs for rural workers). Therefore, general future research could include a discussion on the interrelationships between the perspectives, as they pertain to bio-fuel adoption.

The report identified potential areas for additional research, which if further understood, may facilitate a more widespread adoption in the United States.

- FFV: Vehicles that can operate on 100 % ethanol, 100 % gasoline, or any combination of the two. This technology allows Brazil to respond to fluctuating pricing policies. Analogous to this type of technology, the United States is researching the use of smart meters to be used in energy conservation. If the population is ready for flexibility with their energy use, then perhaps they might be amenable to flexibility with their vehicles?
- Assuming the United States will meet the 2020 target for oil independence, what alternative fuel source will be used? There are already significant debates about corn and cellulosic ethanol. However, the report identified algae as a promising substitute. Future research is recommended to more fully understand the potential of this fuel source. Could this be a golden opportunity to solve the bio-fuel answers for the United States?

Finally, while not directly related to applying lessons learned but discussed in this report, it will be very interesting to watch the impact of a new President and offshore oil reserves on the future of bio-fuel in Brazil.

References

1. D. Rajagopal, D. Zilberman, Review of environmental, economic and policy aspects of biofuels, Energy and Resources Group, University of California Berkeley, 2007
2. M. Uriarte et al., Expansion of sugarcane production in Sao Paulo, Brazil: implications for fire occurrence and respiratory health. Agric. Ecosyst. Environ. **132**, 48–56 (2009)
3. L. Nardon, K. Aten, Beyond a better mousetrap: a cultural analysis of the adoption of ethanol in Brazil. J. World Bus. **43**, 261–273 (2008)
4. L.L. Nass, P.A.A. Pereira, D. Ellis, Biofuels in Brazil: an overview. Crop. Sci. Soc. Am. **47**, 2228–2237 (2007)
5. P. Alvarez et al., Fundamentals of a Sustainable U.S. Biofuel Policy, James A. Baker III Institute for Public Policy of Rice University, 2010
6. A. Papageorgiou, Ethanol in Brazil, PREMIA, Greece, Report, 2005
7. Wikipedia Contributors, Plagiarism, Wikipedia, The Free Encyclopedia (2011), http://en.wikipedia.org/wiki/Petrobras. Accessed 15 Nov 2011
8. M.S. Longevin, The Brazilian Biodiesel Program [Online] (2010), http://www.ensec.org/index.php?option=com_content&view=article&id=273:brazilian-biodie

9. Respol, Brazil Special [Online] (2000–2011), http://www.repsol.com/es_en/corporacion/prensa/publicaciones/especial-brasil/
10. Wikipedia Contributors, Plagiarism, Wikipedia, The Free Encyclopedia (2011), http://en.wikipedia.org/wiki/Ethanol_fuel. Accessed 7 Nov 2011
11. Cellena, Biofuels [Online] (2011), http://cellana.com/products-overview/
12. B.D. Yacobucci, Biofuels Incentives: A Summary of Federal Program [Online] (2010), http://www.stoel.com/files/R40110.pdf
13. C. Berg's, World Ethanol 2008: Ethanol in 2008/09—Light at the End of the Tunnel? in *F.O. Licht 4 th Annual Conf. Sugar and Ethanol Brazil, Germany*, 2008
14. EESI, Developing an Advanced Biofuels Industry [Online] (2010), http://files.eesi.org/state_biofuel_021610.pdf
15. Y. Alizadeh, Giant oil companies and dynamic capabilities approach: how much "peak oil" and "global warming" make them dynamic? Portland International Conference on Management Engineering and Technology (PICMET 2011) (2011), pp. 1–10
16. D. Arvizu, Bio-fuels Research and Development to Reduce Reliance on Imported Petroleum, National Renewable Energy Laboratory, USA, 2007
17. M.A. Altieri, E. Bravo, The ecological and social tragedy of crop-based biofuel production in the Americas [Online] (2007), http://www.foodfirst.org/node/1662
18. Mean Green, Learn About the History of Biofuel Cars [Online] (2011), http://www.meangreenbiofuels.com/learn-about-the-history-of-biofuel-cars/
19. U.S. DOE, Energy Department Announces New Advance in Biofuel Technology [Online] (2011), http://apps1.eere.energy.gov/news/progress_alerts.cfm/pa_id=497
20. U.S. DOE, Biomass Feedstocks [Online] (2011), http://www1.eere.energy.gov/biomass/biomass_feedstocks.html
21. U.S. DOE, Processing and Conversion [Online] (2010), http://www1.eere.energy.gov/biomass/processing_conversion.html
22. E. Miller, NELHA's Cellana at Forefront of Emerging Biofuel Industry [Online] (2011), http://www.westhawaiitoday.com/sections/news/local-news/nelhas-cellana-forefront
23. M.E.Huntleyand,D.G.Realge,CO_2MitigationandRenewableOilfromPhotosyntheticmicrobes: a New Appraisal [Online] (2009), http://www.scribd.com/doc/11570295/Co2
24. J. Ferrell, V.S. Reed, National Algal Biofuels Technology Roadmap, Technical Report, U.S. DOE, College Park, MD, 2010
25. B. Sims, Waste Not [Online] (2011), http://biorefiningmagazine.com/articles/5840/waste-not
26. NREL, Cellulosic Ethanol [Online] (2007), http://www.nrel.gov/biomass/pdfs/40742.pdf
27. RFA, From Farm to Biorefinery: Ethanol Production Efficiency Improves [Online] (2011), http://www.ethanolrfa.org/news/entry/from-farm-to-biorefinery-ethanol-production-efficiency-improves/
28. K.E. Phipps et al., The Opportunities and Risk of Large-Scale Production of Bio-fuels [Online] (2007), http://woods.stanford.edu/docs/biofuels/BiofuelsCommentary.pdf
29. M. Schlincher, Biofuels in the US: Today and in the Future, Agro-biotechnology Management & Economic, Manhattan, USA, 2007
30. D. Hofstrand, Brazil Ethanol Industry [Online] (2008), http://www.agmrc.org/renewable_energy/ethanol/brazils_ethanol_industry.cfm
31. Fuel-Testers, Ethanol Fuel History [Online] (2006–2009), http://www.fuel-testers.com/ethanol_fuel_history.html
32. B. Cole, A Timeline of the Ethanol Industry [Online] (2006), http://minnesota.publicradio.org/projects/2006/09/energyproject/timeline.shtml
33. RFA, World Fuel Ethanol Production [Online] (2005–2011), http://ethanolrfa.org/pages/World-Fuel-Ethanol-Production

Chapter 7
Assessing Alternatives for District Heating

Baraa Abudawod, Rupha Natarajan, Namitha Shetty, and Tugrul U. Daim

Abstract In order to accommodate the growing demand for hot water and the possibility of using an alternative district heating source, Portland State University (PSU) was trying to identify and evaluate future fuel sources for its campus through an objective process. This paper is focused toward developing an evaluation model to identify the most feasible fuel option for PSU's district heating purposes. The study evaluates three fuel alternatives using the Hierarchical Decision Model (HDM) together with the Technology Valuation (TV) Model. The three fuels evaluated are natural gas, marine diesel oil, and pyrolysis oil. It is determined from the model using expert responses that natural gas is the preferred alternative. The highest weighting for the criteria was associated with cost while the lowest weighting was associated with environment.

This chapter demonstrates an assessment approach of fuel alternatives of commercial heating system. HDM and TV Model are used to evaluate three fuels for heating system of PSU's campus. The campus consists of approximately 60 buildings on 50 acres of land. The main heating system that PSU currently relies on consists of two heating plants with seven natural gas fired boilers. In addition, a 2.5 MW diesel fired turbine was installed in the university's newest building in 2006. The campus also has seven small natural gas fired boilers that serve individual buildings in the area for PSU residents. On average, PSU's heating system is required 8 months of the year for approximately 14 h a day, 6 days a week. The evaluation model utilizes different factors and expert's subject judgments.

B. Abudawod
GlaxoSmithKline, Jeddah, Saudi Arabia

R. Natarajan • T.U. Daim (✉)
Portland State University, Portland, OR, USA
e-mail: tugrul@etm.pdx.edu

N. Shetty
TransUnion, Chicago, IL, USA

T.U. Daim et al. (eds.), *Policies and Programs for Sustainable Energy Innovations*, Innovation, Technology, and Knowledge Management,
DOI 10.1007/978-3-319-16033-7_7

7.1 Assessment Tools

7.1.1 Hierarchical Decision Model

The Hierarchical Decision Model (HDM) is a quantitative model that is used to evaluate different alternatives in order to make decisions. It is a four-level hierarchical model; those four levels are the objective level, criteria level, subcriteria level, and technology alternatives level [1]. The criteria are broken into a set of subcriteria/factors and each of these subcriteria/factors contributes as a part of each criterion's weight. Pairwise comparisons are completed for the criteria and subcriteria levels in order to find the contribution of the each criterion and subcriterion/factor to the overall objective. Then, the weights for the criteria and subcriteria are used to find the contribution of each technology alternative to the overall objective.

The HDM model is a robust model that facilitates the decision-making process. Calculating the weights for the criteria and subcriteria level makes it an easier process to make the right decision. One disadvantage of this model is that it mostly follows a quantitative analysis to evaluate the technology alternative. However, some criteria or subcriteria are more qualitative, hence it would be difficult to apply that into a quantitative-based model.

7.1.2 Technology Valuation Model

According to Gerdsri [2], one of the investors of the model, "The TV model enables the evaluation of emerging technologies though semi-absolute values instead of the relative values." The model is a product of three matrices: the criteria priorities, the relative importance of technological factors on each criterion, and the semiabsolute value of emerging technologies on each technological factor.

The TV value or the Technology Value is calculated using the equation below. In order to find the relative priority values of the criteria and the factors (subcriteria), pairwise comparison input sheets were provided to experts both in the facilities department for PSU and other industry experts identified by us as a team.

$$\mathrm{TV}_n = \sum_{k=1}^{k} \sum_{j_k}^{j_k} w_k \cdot f_{j_k,k} \cdot V\left(t_{n,j_k,k}\right)$$

where:

TV_n—technology value of technology, (n) determined according to a company's objective
w_k—relative priority of criterion (k) with respect of the company objective
$f_{j_k,k}$—relative importance of factor (j_k) with respect of criterion (k)

$t_{n,j_k,k}$—performance and physical characteristics of technology (n) along with factor (jk) for criterion (k)

$V\left(t_{n,j_k,k}\right)$—desirability value of the performance and physical characteristics of technology (n) along factor (jk) for criterion (k)

7.2 Methodology

In this chapter, the Hierarchical Decision Making Model is used to evaluate the three candidate fuels for Portland State University's (PSU's) district heating system (natural gas, marine diesel oil (MDO), and pyrolysis oil) in combination with the TV model (see Fig. 7.1). Therefore, the model quantifies the contribution of each fuel option to the overall objective, which is to find the best fuel option for PSU.

The first level in this model contains five criteria: cost, environment, availability, safety, and sustainability. For the criteria level, employees from the facilities and planning department at PSU were contacted in order to do the pairwise comparison for this level. Since the project was focused on PSU, employees from PSU had to be contacted in order to weigh the importance of each criterion for PSU. Three employees at PSU did pairwise comparisons in order to calculate the weights of the five criteria.

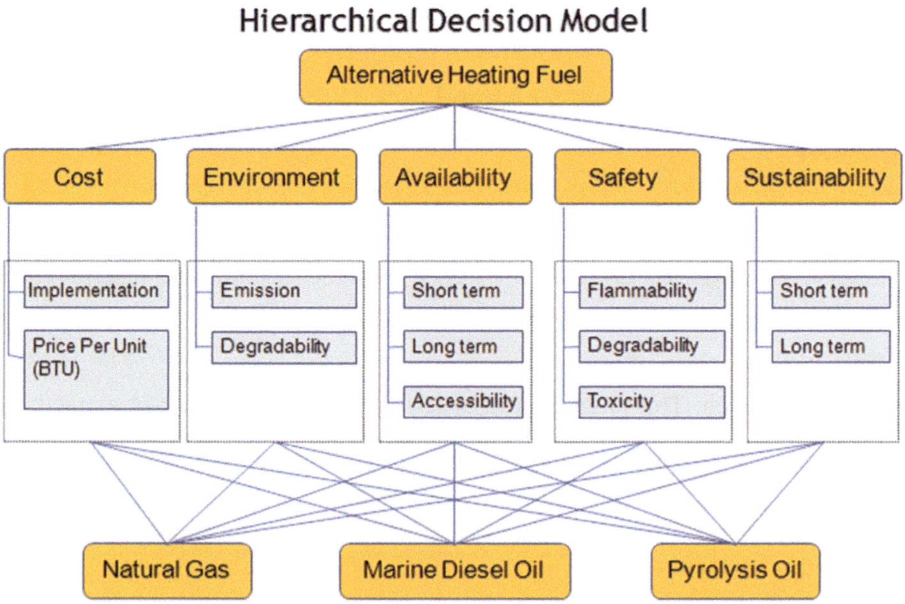

Fig. 7.1 HDM model

Consequently, the weights for the subcriteria level were calculated by doing pairwise comparison for the subcriteria/factors. The subcriteria level required more understanding and knowledge in the topic. Therefore, four industry experts with technical backgrounds in the oil and gas industry were contacted in order to do the pairwise comparisons for the subcriteria/factors.

After figuring out the weights for the criteria and subcriteria levels using the PCM software, it was possible to find the contribution of each fuel alternative to the overall objective of this project using the TV equation.

In this chapter, we try to evaluate between alternative fuel sources for the PSU campus heating system. We used an HDM in order to structure the decision into an objective, criteria, subcriteria, and alternatives. The following methodology was therefore followed:

1. Establish a HDM with criteria, subcriteria, and alternatives for the selection of the preferred fuel source (objective)
2. Create a survey with pairwise comparisons and distribute to experts in this field
3. Use the Pairwise Comparison (PCM) software to determine the weights for criteria and subcriteria from the expert responses
4. Evaluate the weights of the criteria and factors to determine their importance. The criteria and subcriteria with the highest weights are seen as the most important to the expert
5. Calculate the technology value for each alternative and determine the best alternative
6. Discuss the preferred alternative that should be pursued by PSU

7.3 Hierarchical Decision Model

7.3.1 Criteria and Factors

The criteria chosen for the HDM model were cost, environment, availability, safety, and sustainability. The description of each of these criteria and subcriteria is described below.

Cost

- Implementation: This is the cost of acquiring the fuel consisting of extraction, storage, and transpiration cost including processing, pipe construction, storage tank, import expenses, boiler adjustment, etc.
- Price per unit (BTU): This is the cost of delivery for the final consumer for usage as a fuel source. This might include some portions of the transportation cost embedded in them.

Environment

- Emissions: This includes the various chemical components that the fuel contributes to the atmosphere on combustion such as COx, NOx, SOx, particulates, and mercury including greenhouse gases if any. This also evaluates the impact the fuel has on acid rain generation.
- Degradability: This evaluates the long-term negative impact on the environment due to spills into the soil and water. This specifically evaluates the oil spill that has devastating impact on the environment and the amount of time the fuel takes to naturally degrade.

Availability

- Short term: This refers to the capacity of local distributors to meet the needs of PSU's boiler heating systems including local and external fuel resources. This includes the ability of the fuel option to support boiler heating for the next couple of years.
- Long term: This refers to the total volume of resources, including the estimated duration of time that these resources will be available in the long run. Special consideration should be given to fuel resources that can be defined as having a limited known quantity, i.e., fossil fuels.
- Accessibility: This refers to the ability and ease with which the distributor can meet the demands for PSU. Considers the rapidity and convenience of fuel delivery, as well as local storage versus distribution and feed systems. This capability to generate the fuel locally should be given higher preference.

Safety

- Flammability: This refers to the overall mixture of oxygen to material that can be ignited. A larger low to high concentration area should be considered more flammable. Special consideration should be given to mixtures less likely to combust at standard PSU temperatures currently available.
- Degradability: Most materials biodegrade over time. This is a measurement of the amount of time that given substance will biodegrade within a standard period. Fuel substances with high degradability should be given higher preference.
- Toxicity: Fuels may be more or less likely to burn the skin or cause irritation of eyes when comes in direct contact. This considers toxicity in terms of a possible fuel leak into the environment, or fuel-handler exposure both in the short and the long term.

Sustainability

- Short term: This refers the degree to which system operation may deplete the fuel resources in the short term. In the case of boiler fuels, fossil fuels cannot be recreated in a timely fashion, and therefore would rank low.
- Long term: This refers the degree to which system operation may deplete the fuel resources in the long term. Boiler fuels with higher sustainability would be biomass-generated fuels such as pyrolysis oil generated from organic and renewable feedstock.

7.3.2 Available Alternatives

The three alternatives chosen in the study was natural gas, MDO, and pyrolysis oil. Each of these alternatives is described below:

7.3.2.1 Natural Gas

Natural gas is a combustible mixture of hydrocarbon gases. While natural gas is formed primarily of methane, it can also include ethane, propane, butane, and pentane. It is colorless, shapeless, and odorless in its pure form. It is a known fact that natural gas is a fossil fuel and it does bring along its own advantages and disadvantages that are generic to any fossil fuel. This gas is colorless and odorless in its pure form but the final product delivered to the customers is highly processed to act as a usable fuel [3].

Analysts have projected that there is abundant availability of natural gas in North America. There is abundant availability for natural gas both in the short term as well as the long-term horizon [3]. Appendix 1 further projects the status of natural gas usage and availability for the State of Oregon [4].

The cost of natural gas is close to $12.5 per 1,000 cubic feet (including implementation and delivery costs), which is one of the lowest among its competitor fuel options. This has enabled natural gas to be a highly economic fuel option [5]. The prominent contributor to the cost of natural gas is the storage and transportation cost. The reasons being that; once the natural gas is extracted it has to be stored in large tanks underground before being processed for final use. The transportation of gas is done though pipelines laid underground which significantly increases the initial implementation cost while the delivery cost remains low as they are easy to transport and relatively safer as compared to other fuel options. Having that said, the US natural gas pipeline is highly integrated transmission and distribution grid that can transport natural gas to nearly any location including the Western Region. According to the US energy administration, ten interstate and nine intrastate natural gas pipeline companies provide transportation services to and within the Western Region. The largest capacity natural gas pipeline within the region is the El Paso Natural Gas Company system, which has the capability to transport up to 6.2 billion cubic feet (Bcf) per day [6].

Natural gas is lighter than air and its impact on the environment is minimalistic as it directly evaporates into the air without leaving any kind of residue behind. This makes natural gas 100 % degradable [7]. Natural gas is also nontoxic which makes it the safest fossil fuel when used by common people at home [7]. According to one of the recent research article published in 2012, the major commercial uses of natural gas include space heating, water heating, and cooling. According to the study, 36 % of the usage of natural gas is toward space heating, 20 % is toward lighting, and 8 % is used toward water heating. Natural gas is also used for infrared (IR) heating units, which are highly innovative and economic method of utilizing natural gas [8].

7.3.2.2 Marine Diesel Oil

Marine gas oil and MDO are manufactured when kerosene, light, and heavy gas oil fractions are added to distilled crude oil [9]. MDO is generally used as fuel in marine diesel engines, furnaces, boilers, etc. [10].

Below are the physical and chemical properties of marine diesel fuel [9]:

- Marine diesel is liquid in nature with clear, straw colored appearance with petroleum odor.
- It has sulfur content between 1 and 2 % along with 0.3 % water and ash content of 0.01 % m/m maximum.
- It has relative density of 81–92 at 150 °C and kinematic viscosity at 40 °C is 11 cst (centistokes).
- It has high thermal stability with high resistance to oxidation.

The energy generation of MDO in BTU/gal is up to maximum of 133,781 to minimum of 132,311 [11]. The current price of MDO in Houston is $865/Metric Ton (06/08/2012) [12]. One of advantages of using MDO is it has lower maintenance cost when compared to other fuels [10, 13].

MDO when burned emit large amount of toxic gases like carbon dioxide, nitrogen dioxide, sulfur dioxide, and harmful gases, which pollutes the atmosphere thus causing global warming and climate change. Appendix 5 shows the emission from MDO as compared to natural gas.

- SOx, COx—Sulfur dioxide and carbon dioxide are formed due to unburned parts of sulfur and carbon in combustion of MDO. Sulfur dioxide when combined with ozone layer forms smog and when combined with water vapor forms acid rain.
- CFC's—Chlorofluorocarbons is a poisonous air pollutant.
- MDO emission is approximately 3,000–6,000 ton which is 1–3 % of global emission of gases [14].

Oil spills are major pollutants during water transportation. As MDO is not easily biodegradable [15], they form a thick, black layer on surface of water body thus poisoning the water and eventually leads to death of aquatic plants and organism. Eventually these oil spills as they are washed to the shore form tar balls thus contaminating coastal side [16].

MDO being fossil fuel has finite number of resources available for future [17]. Given the rapid growth of world economies, the supply of fossil fuels will cease after 30 years [18]. Hence short-term availability of fossil fuel (e.g., MDO) is possible, whereas long-term availability is not assured.

From a safety perspective, MDO is stable and safe when kept in a closed container in normal room temperature and pressure [18]. It is highly flammable with 0.5–5 % [19] but only 24–36 % degradable [20]. Upon decomposition, produces CO_2 and toxic oxides of hydrocarbons [21]. It is also one of the most toxic fuels on combustion as it produces highly toxic gas like hydrogen sulfide that causes gastrointestinal irritation with vomiting and diarrhea. High dose of MDO causes skin irritation and also leads to cancer [21].

Sustainability is questionable as fossil fuels like coal, oil, natural gas, petroleum products are not confined to extinction [22]. Fossil fuels could be made more sustainable in future for long- and short-term benefits by reducing harmful emission of gases like carbon dioxide, carbon monoxide, and sulfur dioxide into atmosphere [23].

7.3.2.3 Pyrolysis Oil

Heating the biomass in the absence of oxygen creates the pyrolysis oil. Pyrolysis oil, also known as bio-oil, is a synthetic fuel under investigation as substitute for petroleum. It is a type of tar and normally contains too high levels of oxygen to be a hydrocarbon, and that's why it is distinctly different from similar petroleum products [24].

Pyrolysis oil has some negative attributes for fuel applications. Those attributes are the high water content, high viscosity, poor ignition characteristics, corrosiveness, instability, and others. However, combustion tests performed in boilers, diesel engines, and turbines have demonstrated the ability of the existing or slightly modified equipment to operate using bio-oil/pyrolysis oil produced in existing pilot plants [25].

Pyrolysis oil can be used as a fuel in combustion applications. However, due to its heterogeneous chemical composition, pyrolysis oil is unsuitable for direct use in diesel engines regardless, in spite of the fact that trials in direct injection engines have been conducted. In addition, pyrolysis oil has achieved greater success in some industrial furnaces such as kilns and boilers [26].

Pyrolysis oil is considered to be expensive especially when compared to cheap fuels such as natural gas. The negative attributes of pyrolysis oil at present include pumping, storage, and transportation challenges that raise the price dramatically. This is one of the major reasons why pyrolysis oil has found little use until now.

We used the average plant size that produces 400 ton/day in order to estimate the price per unit (BTU) and the implementation cost of pyrolysis oil (see Table 7.1). After doing the calculations, the price per unit (BTU) was found to be $6,000.63/1000CFT, and the implementation cost was $999,000.95/1000CFT. Apparently, pyrolysis oil is not an affordable fuel option. However, it is still under research in order to make it available for public at affordable prices [24].

When combusted, pyrolysis oil results in a null net contribution to CO_2 emissions due to the fact that the carbon oxidized in the process is the carbon fixed by the vegetable biomass during its life [27]. The fuel quality of biomass pyrolysis oil is inferior to that of petroleum-based fuels. Pyrolysis oil has the advantage of being generated from renewable resources; hence it does not contribute to the net release of greenhouse gases [25].

On the other hand, there are some environmental issues that relate to the design and operation of the pyrolysis plant/biorefinery producing the pyrolysis oil. There might be some fugitive emissions from collecting, transferring, and processing the oils. Therefore, the oils must be controlled carefully to minimize the emissions. Fugitive emissions from drying the biomass feed also need to be carefully controlled [28].

In terms of degradability, pyrolysis oil is found to be biodegradable with 41–50 % biodegradation rate over a period of 28 days. Since pyrolysis oil has a degradability rate of over 20 %, it is classified as inherently biodegradable [29]. Pyrolysis oil contains substantial amounts of organic acids, which are mostly acetic and formic acid. This results in pH of 2–3 and an acid number of 50–100 mg KOH/g. Thus, pyrolysis oil is corrosive to common construction materials like carbon steel and aluminum; however, it is noncorrosive to stainless steel [25].

As mentioned earlier in this report, the negative attributes of pyrolysis oil present transportation, pumping, and storage challenges. Up until this point in time, pyrolysis oil has found little practical use. However, the development of technologies to effectively stabilize pyrolysis oil and to overcome the storage and aging issues could substantially reduce barriers to use [24]. Canada, Ukraine, Brazil, South Africa, and Baltic have already started the production of pyrolysis oil. The roughly estimated supply of pyrolysis oil from these countries for 2012 was approximately 5 million tons [30]. Therefore, the short-term availability of pyrolysis oil is dependent on those countries since the production of pyrolysis oil for heating purposes is still under research in the United States. It is believed that it will take about 3 years to complete R&D and within 5 years pyrolysis oil fuel substitutes could be certified for public use [24].

Pyrolysis oil has been under research for a long period of time. Based on the research, pyrolysis oil was found to be highly flammable fuel with 0.9–5.9 % flammability rate [31]. As mentioned earlier in this report, pyrolysis oil is biodegradable with 41–50 % biodegradability rate over 28 days, which is considered to be high. In terms of toxicity, pyrolysis oil was found to be irritating to the eyes, respiratory system, and skin. It is also harmful if swallowed, inhaled, or exposed to direct eye contact. The pyrolysis oil produced at temperatures higher than 600 °C can have mutagenic effects. However, environmentally pyrolysis oil is much less harmful than petroleum fuels [25].

Pyrolysis oil is considered to be a highly sustainable fuel source due to the fact that it is generated from renewable resources [25]. Since pyrolysis oil is still under research, it cannot be a reliable fuel option in the short term. However, as mentioned earlier, it is believed that it will take about 3 years to complete R&D and within 5 years pyrolysis oil fuel substitutes could be certified for public use [24]. Thus, pyrolysis oil can be considered as a reliable fuel option that would substitute fossil fuels in the long term.

7.3.3 Decision Model

Figure 7.1 illustrates the HDM for the selection of an alternative fuel for PSU's district heating system. The model is structured with the objective (alternative heating fuel selection), criteria (cost, environment, availability, safety, sustainability), the subcriteria associated with each criterion, and finally all the fuel.

7.3.4 Expert Responses

The inputs for analysis process were performed in three different stages. The first stage included pairwise input sheet 1 for the criteria level. Three utility experts from the facilities department in PSU volunteered to provide us with the inputs. Among the three experts one of them is a manager for campus planning and design, the other one is a utilities manager at PSU, and the third one is the facilities mechanical engineer. Their years of experience at PSU helped us to understand the importance of resource allocation and the opportunity toward being a part of PSU's process for identifying alternate district heating fuel.

The second stage included the pairwise input sheet 2 for the subcriteria/factors level. Four industry experts with over 20 years of experience in the field of oil and gas were contacted for the purpose of evaluating the 12 subcriteria. This stage involves the comparison of the subcriteria within each criterion. The three subcriteria are compared in pairs with each other based on the percentage inputs provided by the industry experts using the PCM software.

The third and the final input stage were for the desirability values for each of the 12 subcriteria. The complete list is available in Appendix 3. The desirability values were rated by one expert between a scale of 10 and 100 % for the factor values ranging between the worst and the best scale provided to them for each of the 12 subcriteria. These values were later projected in the form of a line graph called the desirability graph to identify the metric values for the final TV calculation.

Besides providing us with the values necessary for our analysis purposes, the industry experts also provided us some valuable inputs. These inputs were provided to us based on the study and analysis performed by them as an expert and also based on their valuable experience in the field of oil and gas. One expert who is a Ph.D. in petrochemicals mentioned that "Natural Gas is a gas, hence the question of bio degradation does not arise. It will take substantial time for pyrolysis oil to be adapted regularly as an alternate fuel. Hence considering the cost, environment, availability, safety and sustainability, I feel you should as a group recommend natural gas as alternative heating fuel." The second expert who is a senior consultant in oil and gas for a leading corporation had a very interesting view on the use of pyrolysis oil for PSU. He mentioned that "Natural Gas can be supplied though pipelines while the oils have to be stored in tankers locally. One option for PSU is to have a combination of Natural Gas and Pyrolysis Oil working together. This will slowly reduce the dependency on Natural Gas as a fossil Fuel." The third and the fourth expert who are a senior executive at Bechtel Corp. and retired scientist from DNV energy, respectively, also seem to agree with the first expert on natural gas being the winner. Expert three said that "Countries like Japan are completely moving on to Natural Gas as their primary fuel source as this is both safer and abundantly available. Its only major negative aspect is being a fossil fuel, this is the best fuel available by far.

The next best alternative can be Pyrolysis oil also called bio-fuel as it is renewable and can be produced locally." While expert four's input was "Considering all the factors, natural gas is the best energy alternative for today. More than 50 % of American houses and buildings are heated by natural gas today."

7.4 Results

7.4.1 *Criteria Weights*

See Fig. 7.2.

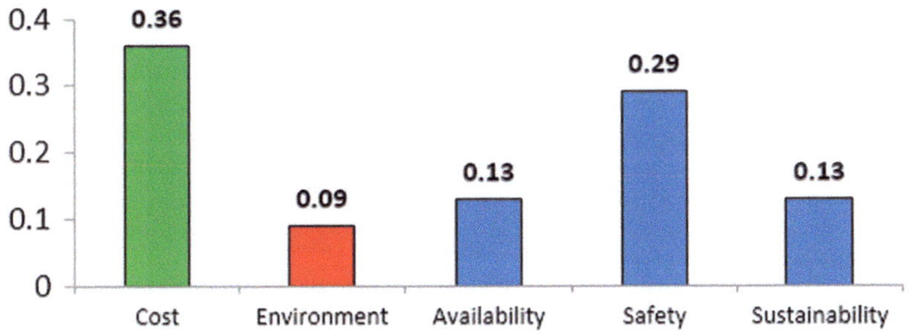

Fig. 7.2 Weights of the criteria

7.4.2 *Factor Weights*

7.4.2.1 Subcriteria Weight Under Criteria

See Table 7.1.

Table 7.1 Highest and lowest factors for each criterion

Criteria	Highest factor	Lowest factor
Cost	Cost per unit (BTU)	Implementation cost
Environment	Degradability	Emission
Availability	Accessibility	Long term
Safety	Flammability	Toxicity
Sustainability	Short term	Long term

7.4.2.2 Subcriteria Weights to Objective

See Fig. 7.3.

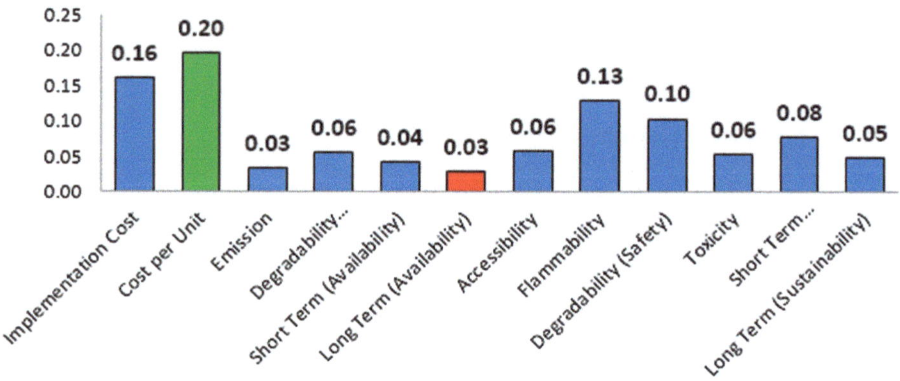

Fig. 7.3 Weights of the subcriteria

7.4.3 Technology Value

See Fig. 7.4.

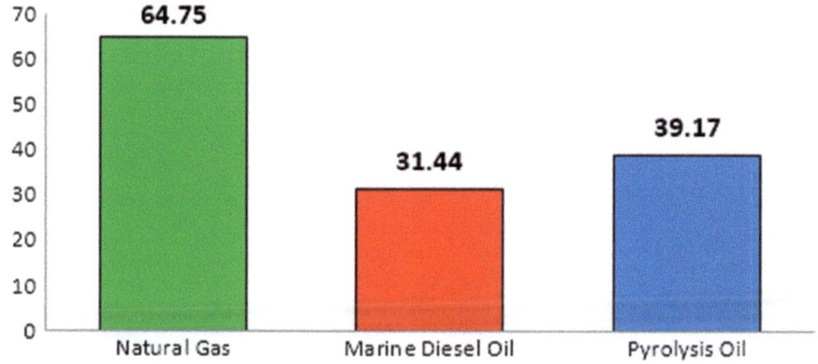

Fig. 7.4 Evaluation scores of the alternatives

7.5 Discussion

Based on the pairwise comparisons from the three experts in the facilities and planning department at PSU, the weights for each criterion in the criteria level were found.

Looking at the inconsistency column, it is plausible to say that the three experts from PSU were consistent when doing the pairwise comparisons as the average inconsistency rate is 0.056, which is considered to be very consistent. The weights for the five criteria are shown Fig. 7.2. Apparently, cost and safety are the most important criteria for PSU. This project is focused mainly on finding the best fuel option for PSU; hence, it makes sense that cost is the most important criterion. Safety is inherently a big concern/issue in the oil and gas industry and that's why it is the second most important criterion. The other three criteria have relatively smaller weights; therefore at PSU the attention is mostly focused on cost and safety.

After obtaining the results for the criteria level from PSU's experts, the four experts in the oil and gas industry were contacted in order to do the pairwise comparisons for the subcriteria/factors level. Those pairwise comparisons were inserted into the PCM software in order to calculate the weights for the each subcriterion. The PCM software can calculate the weights when there are at least three items to compare. In the HDM model in this study, there are three criteria (cost, environment, and sustainability) which have only two subcriteria. Therefore, the weights of their subcriteria are calculated manually using MS Excel. The weights of the other subcriteria under availability and safety are calculated using PCM software because there are three items under each criterion. Therefore, the weights for the subcriteria under availability and safety were calculated using the PCM software. The weights for the other subcriteria were calculated manually using MS Excel.

Based on the pairwise comparisons, the weights for subcriteria under cost, environment, and sustainability were calculated manually using MS Excel. The averages of the pairwise comparisons were calculated in order to find weights for each subcriterion. Figure 7.3 shows the weights for the subcriteria under cost, environment, and sustainability.

Inconsistency rates define rationality of experts' judgement when there are more than two items to compare. Therefore, inconsistencies of the evaluation on the criteria (availability and safety) which consist of three subcriteria in the HDM model were calculated. The ideal inconsistency rate is 0.1.

In terms of cost, the weights for the subcriteria are very close to each other. Therefore, price per unit (BTU) and implementation cost are both important when it comes to making a decision regarding the best fuel option. For environment, the degradability subcriterion is the most important one with a weight of 0.63, which is almost double the weight of the emission subcriterion. Degradability refers to the impact on the environment and the amount of time the fuel takes to naturally degrade. For sustainability, short term is the most important subcriterion when choosing the best fuel option, and it refers to the degree to which system operation may deplete the fuel resources in the short term.

On the other hand, the weights of the subcriteria under availability and safety were calculated using the PCM software. For availability, the subcriteria are numbered as follows in the PCM software: short term (1), long term (2), and accessibility (3). It is obvious that accessibility is the most important subcriterion under availability followed by short term as the second most important criterion. In terms of inconsistency, even though the four industry experts were very consistent when doing the pairwise comparisons for the availability subcriteria, the overall inconsistency rate was in the reasonable range but somehow relatively high (0.093).

In terms of safety, the subcriteria are numbered as follows in the PCM software: flammability (1), degradability (2), and toxicity (3). Based on the PCM results, flammability is the most important subcriterion under safety followed by degradability as the second most important subcriterion. In terms of inconsistency, experts 1, 2, and 3 were very consistent when they did the pairwise comparisons for the subcriteria under safety.

In order to arrive at the final technology value for each of the three fuel options, the desirability percentage values were potted on a graph for each of the 12 subcriteria (factors) as shown under evaluation procedure earlier. The team then identified the actual values for the subcriteria for the three fuel options as presented in Table 3. Please note that the complete table is available in Appendix 2.

For the purpose of this project, the cost values for each of the fuel options have been normalized using the sum for the cost values. For implementation cost, the values were normalized by 1/1,061,320 and price per unit was normalized by 1/28,229. The original cost values are as shown in Appendix 1.

In order for the experts to provide us with the desirability values, the team put together a desirability metric table by choosing the worst and the best limits for each subcriteria/factors. Some of these metrics were quantitative while some of the metrics were qualitatively defined on a 5-point scale ranging from excellent to poor. Table 5 presents the 5-point scale for short-term availability based on various sources. A complete list of the 5-point scale is available in Appendix 6.

Based on the desirability graph the final metric values were identified as percentage of desirability to be used in the TV equation. For the purpose of calculating the TV value, the least among the three desirability percentages were used in the case of costs, emission, and toxicity, while the most among the three desirability percentages were used for all the other subcriteria/factors. Appendices 2, 3 and 4 present our calculation of the TV value for each of the three fuel options.

In the above tables, the technology value is a product of columns 1, 2, and 4 that are the normalized pairwise weights for the criteria, the normalized pairwise weights for the subcriteria/factors, and the desirability value for the subcriteria/factors, respectively. The desirability value is the *Y*-axis value from the graph and the metric value is the *X*-axis on the graph for each subcriteria. Refer to Appendix 5 for a list of desirability graphs.

7.6 Conclusion

HDM model and TV model were used as a part this study to determine the best alternative district heating fuel among natural gas, MDO, and pyrolysis oil for PSU. This approach used in the study has provided a way to quantify both quantitative (e.g., cost) and qualitative (e.g., sustainability). Hence making it easy to add and compare factors and subfactors affecting the selection criteria for fuels.

Considering factors like cost, safety, availability, sustainability, and environment along with their subfactors helped in the analysis to evaluate the technology values of each fuels alternative. The technology values for natural gas have highest score with 91 % followed by pyrolysis with 51 % and MDO with 42 %. The team also considered the fuel options with cost as a constant. This helped us analyze the three alternatives primarily based on its fuel properties in the broad sense without considering the economic impact on the decision. The technology value for the three fuel options with cost values constant was 65 % for natural gas, 31 % for MDO, and 39 % for pyrolysis oil (Refer to Appendices 2, 3 and 4 for the details on the TV value calculations). If PSU were to make the decision on fuel today, we recommend natural gas as the best alternative among the three for use today but for future purposes we recommend pyrolysis oil as the alternative fuel option considering the fact that pyrolysis oil is both renewable and has the least impact on the environment in terms of degradability. According to research natural gas resources available in U.S. would last till 118 years or 2,247 trillion cubic feet, and pyrolysis oil is an alternative provided enough research is done to reduce the acidity levels [9]. Our results were consistent when considering cost as a contributing factor as well as with cost factors constant as mentioned earlier.

The whole purpose of implementing HDM and TV model along with expert assistance was to provide a systematic and scientific approach to identify the best alternative fuel not only for PSU but also for any university in U.S. for today or any day in future also.

Implementation cost for all the three alternatives was assumed to be the same, whereas, in practice, they will differ depending on the available technology. Although sufficient research was done in identifying the factors used in this study, there is an opportunity to consider other factors that will make the study exhaustive. For example, ongoing maintenance costs, tolerance to impurities, and energy output of the fuel. Considering renewable energy sources like solar are also recommended as one of alternate heating fuel option in future.

Appendix 1: Measurement Value of the Three Alternatives

			Measure of effectiveness (limiting values)				
		Measurement unit	Worst	Best	Natural gas	Marine diesel oil	Paralysis oil
Criteria 1: cost							
Factor 11	Implementation cost	[$/1000CFT][a]	1	0	0.00	0.06	0.94
Factor 21	Price per unit (BTU)	[$/1000CFT][a]	1	0	0.00	0.79	0.21
Criteria 2: environment							
Factor 12	Emission	Constant	20	0	12.81	18.30	0.00
Factor 22	Degradability	%	0	100	100 %	30 %	50 %
Criteria 3: availability							
Factor 13	Short term	5-point scale	P	E	Excellent	Acceptable	Acceptable
Factor 23	Long term	5-point scale	P	E	Very good	Acceptable	Good
Factor 33	Accessibility	5-point scale	P	E	Excellent	Acceptable[b]	Acceptable[c]
Criteria 4: safety							
Factor 14	Flammability	%	0	15	5–15 %	0.5–5 %	0.9–5.9 %
Factor 24	Degradability	%	0	100	100 %	30 %	50 %
Factor 34	Toxicity	5-point scale	P	E	Excellent[d]	Acceptable[e]	Poor[f]
Criteria 5: sustainability							
Factor 15	Short term	5-point scale	P	E	Very good	Acceptable	Acceptable
Factor 25	Long term	5-point scale	P	E	Acceptable	Poor	Good

[a]Normalized
[b]Import
[c]Still under lab tests for US
[d]Nontoxic
[e]Medium toxicity
[f]Highly toxic (under research to reduce)

Appendix 2: Criteria, Subcriteria Weights, Desirability Values, and Technology Values

Natural Gas

Criteria	Wt. (1)	Subcriteria	Wt. (2)	Metric (3)	Desirability value (4)	Technology value =(1)×(2)×(4)
Cost	0.36	Implementation	0.45	5.98	30	4.86
		Cost per unit	0.55	6.52	20	3.96
Environment	0.09	Emission	0.38	12.81	59	1.99
		Degradability	0.63	100	100	5.63
Availability	0.13	Short term	0.33	E	100	4.29
		Long term	0.23	VG	80	2.39
		Accessibility	0.45	E	100	5.85
Safety	0.29	Flammability	0.45	12.5	80	10.44
		Degradability	0.36	100	100	10.44
		Toxicity	0.19	E	100	5.51
Sustainability	0.13	Short term	0.61	VG	80	6.37
		Long term	0.39	A	60	3.02
						64.75

Appendix 3: Marine Diesel Oil

Criteria	Wt. (1)	Subcriteria	Wt. (2)	Metric (3)	Desirability value (4)	Technology value =(1)×(2)×(4)
Cost	0.36	Implementation	0.45	5.98	30	4.86
		Cost per unit	0.55	6.52	20	3.96
Environment	0.09	Emission	0.38	18.3	20	0.68
		Degradability	0.63	30	25	1.41
Availability	0.13	Short term	0.33	A	50	2.15
		Long term	0.23	A	50	1.50
		Accessibility	0.45	A	60	3.51
Safety	0.29	Flammability	0.45	3	30	3.92
		Degradability	0.36	30	17	1.77
		Toxicity	0.19	A	40	2.20
Sustainability	0.13	Short term	0.61	A	50	3.98
		Long term	0.39	P	30	1.51
						31.44

Appendix 4: Pyrolysis Oil

Criteria	Wt. (1)	Subcriteria	Wt. (2)	Metric (3)	Desirability value (4)	Technology value =(1)×(2)×(4)
Cost	0.36	Implementation	0.45	5.98	30	4.86
		Cost per unit	0.55	6.52	20	3.96
Environment	0.09	Emission	0.38	0	100	3.38
		Degradability	0.63	50	49	2.76
Availability	0.13	Short term	0.33	A	50	2.15
		Long term	0.23	G	70	2.09
		Accessibility	0.45	A	60	3.51
Safety	0.29	Flammability	0.45	3.85	35	4.57
		Degradability	0.36	50	32	3.34
		Toxicity	0.19	P	10	0.55
Sustainability	0.13	Short term	0.61	A	50	3.98
		Long term	0.39	G	80	4.03
						39.17

Appendix 5: DesirabilityCurves

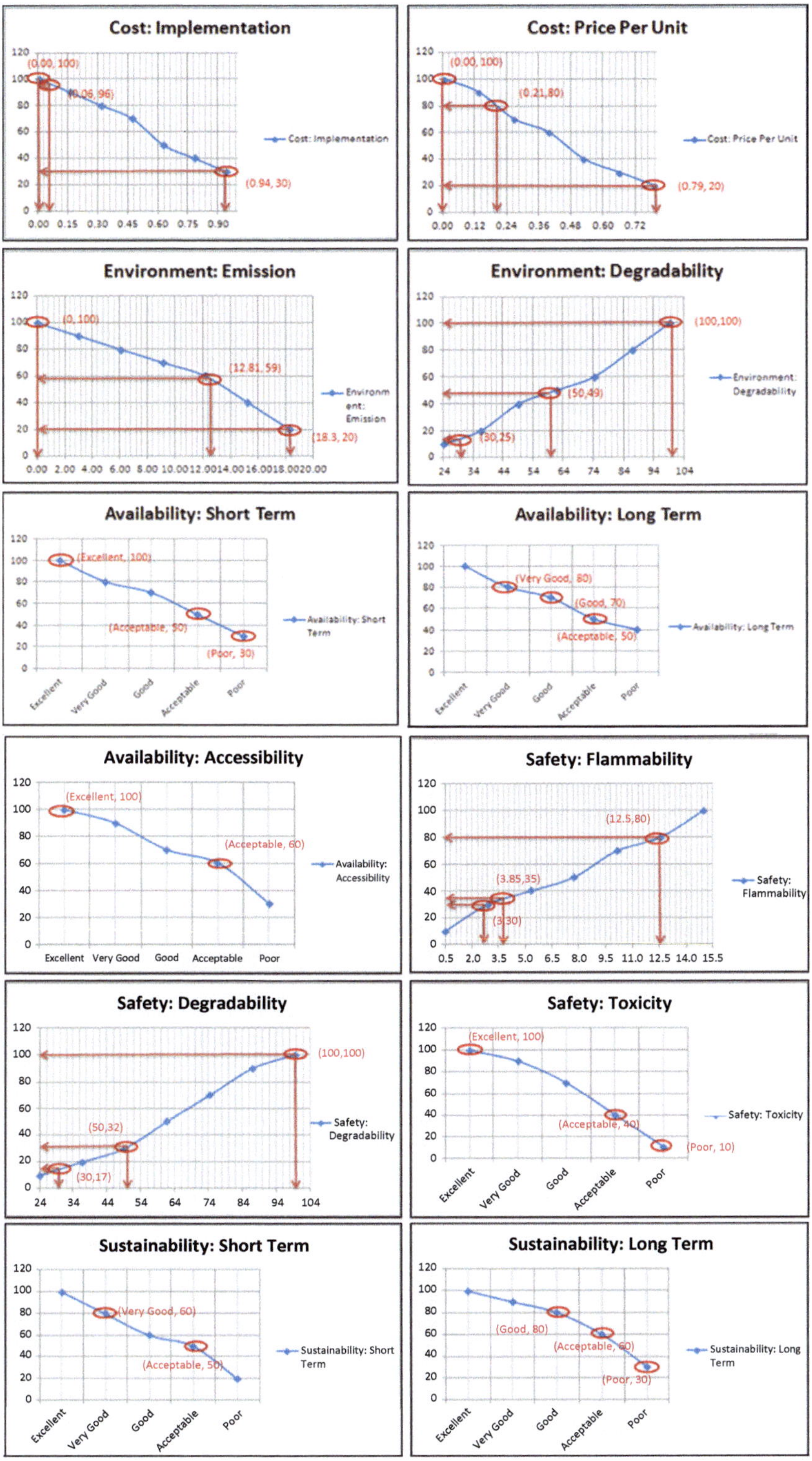

Appendix 6: Description of 5-Point Scale for Factors

5 Point Scale Measurement Description		
Factors	**5 Point Scale**	**Description**
Short Term Availability	Excellent	Fuel available in surplus locally to fulfill short term demand in the United States
	Very Good	Adequate availability of fuel to fulfill majority of the short term demand in the United States
	Good	Adequate availability of fuel to fulfill partial demand in short term in the United States
	Acceptable	Adequate availability to fulfill peak scenarios of demand in the United States for the short term
	Poor	Inadequate availability to fulfill current demand in the United States
Long Term Availability	Excellent	Fuel available in surplus locally to fulfill long term demand in the United States
	Very Good	Adequate availability of fuel to fulfill majority of the long term demand in the United States
	Good	Adequate availability of fuel to fulfill partial demand in long term in the United States
	Acceptable	Adequate availability to fulfill peak scenarios of demand in the United States for the long term
	Poor	Inadequate availability to fulfill future demand in the United States
Accessability	Excellent	The local suppliers can deliver and transport the fuel easily and on time. No wait time necessary as the fuel is locally available
	Very Good	The local suppliers can deliver and transport the fuel easily but might include some amount of transportation time.
	Good	The local suppliers can deliver and transport the fuel with added import cost and minimilistic time delays.
	Acceptable	The local suppliers can deliver and transport the fuel with added import cost and on a need basis. The fuel is made available based on the projected need though imports.
	Poor	The fuel is non accessable or requires considerable amount of time and efforts in the form of research and delivery. One example is the Pyrolysis oil although it looks promising for the future.
Toxicity	Excellent	Totally clean. No toxic treatment needed.
	Very Good	Toxicity within the safety allowance but may cause irritability and skin problems when exposed to the fuel on a large scale over an extended period of time. It is not life threatening.
	Good	Positive measures like reduction in acidity level and processing are needed that can reduce the impact of the fuel exposure to humans and animals.
	Acceptable	Toxic if a spill happens or comes directly in contact with humans or animals over an extended period of time. Adequate care is necessary with storage and transportation. The is long term impact to the environment.
	Poor	Highly levels of toxic composition. Regarded unstable for usage as a fuel for general use.

References

1. D. Kocaoglu, N. Gerdsri, A quantitative model for the strategic evaluation of emerging technologies, Engineering and Technology Management Department, Portland State University
2. N. Gerdsri, An analytical approach to building a technology development envelope (TDE) for roadmapping of emerging technologies, Portland State University, 2005
3. NaturalGas.org [Online], http://www.naturalgas.org/overview/background.asp. Accessed 17 June 2012
4. Natural Gas Annual Report, Energy Information Administration, 2010

5. Natural Gas Monthly, Energy Information Administration, May 2012
6. Natural gas pipelines in the western region, U.S. Energy Information Administration (eia) [Online], http://www.eia.gov/pub/oil_gas/natural_gas/analysis_publications/ngpipeline/western.html. Accessed 31 May 2012
7. S. Alban, How safe is natural gas? HowStuffWorks.com, August 2012 [Online], http://science.howstuffworks.com/environmental/energy/safe-natural-gas.htm. Accessed August 2012
8. F.-Y. Liang, M. Ryvak, S. Sayeed, N. Zhao, The role of natural gas as a primary fuel in the near future, including comparisons of acquisition, transmission and waste handling costs of as with competitive alternatives. Chem Cent J **6**(Suppl 1), S4 (2012)
9. M.B.V. Ghent, *Everything you need to know about marine fuels* (Chevron Marine Products, Houston, TX, 2008)
10. T. Argina, Marine and rail lubricants, (2012), p. 3
11. J.J. Corbett, J.J. Winebrake, Emissions tradeoffs among alternative marine fuels: total fuel cycle analysis of residual oil, marine gas oil, and marine diesel oil. J Air Waste Manag Assoc **58**(4), 538–542 (2008)
12. Bunker, Bunker price, Bunker World, 8 June 2012 [Online], http://www.bunkerworld.com/prices. Accessed 31 May 2012
13. Oil cost hits ship studies: nature news [Online], http://www.nature.com/news/2008/080723/full/454372a.html. Accessed 18 June 2012
14. Raunekk, How emission from marine engines increase air pollution, Bright Hub, 17 June 2010 [Online], http://www.brighthub.com/engineering/marine/articles/24522.aspx. Accessed 31 May 2012
15. Chevron Marine Products, Material safety data sheet, read and understand material safety data sheet before handling or disposing of product, p. 5
16. K. Sleight, Effects of marine pollution on the sea, Bright Hub, 4 June 2010 [Online], http://www.brighthub.com/engineering/marine/articles/37397.aspx. Accessed 31 May 2012
17. G. Robertshaw, Realizing finite availability of fossil fuel is fundamental to climate talk, The Good Human, 5 Jan 2009 [Online], http://thegoodhuman.com/2009/01/05/realizing-the-finite-availability-of-fossil-fuels-adds-a-new-twist-to-climate-change-talk/. Accessed 31 May 2012
18. U. S. Tetra Tech Systems, Final report on low sulphur marine fuel availability, (Apr. 2008), p. 34
19. E. S. and Technology D. Environment Canada, Marine diesel fuel oil, p. 1
20. G. Soares, Y. Garbatov, N. Fonseca, Maritime transportation and exploitation of ocean and coastal resources. in *Proceedings of the 11th International Congress of the International Maritime Association of the Mediterranean*, Lisbon, Portugal, 26–30 September 2005
21. Valero Marketing and Supply, *Material safety data sheet—valero*, vol. 1 (2010), p. 2
22. wiki, Fossil fuels with examples, Wiki Answers, 2012 [Online], http://wiki.answers.com/Q/What_are_fossil_fuels_Also_list_examples. Accessed 31 May 2012
23. K.S. Lackner, Sustainability of fossil fuels (American Geophysical Union, Spring Meeting, 2002)
24. A. A. Boateng, Pyrolysis oil, SunGrant BioWeb, 16 June 2011 [Online], http://bioweb.sungrant.org/Technical/Biopower/Technologies/Pyrolysis/Pyrolysis+Oil/Pyrolysis+Oil.htm. Accessed 13 June 2012
25. A. Oasmaa, S. Czernik, Fuel oil quality of biomass pyrolysis oils-state of the art for the end users. Energy Fuels **13**(4), 914–921 (1999)
26. D. Chiaramonti, A. Oasmaa, Y. Solantausta, Power generation using fast pyrolysis liquids from biomass, Science Direct, July 2005
27. R. Calabria, F. Chiariello, P. Massoli, Combustion fundamentals of pyrolysis oil based fuels, Science Direct, Nov 2006
28. M. Ringer, V. Putsche, J. Scahill, Large-scale pyrolysis oil production: a technology assessment and economic analysis (National Renewable Energy Laboratory, 2006)
29. J. Blin, G. Volle, P. Girard, T. Bridgwater, D. Meier, Biodegradability of biomass pyrolysis oils: comparison to conventional petroleum fuels and alternatives fuels in current use, Science Direct, Apr 2007
30. D. Bradley, European market study for biooil (pyrolysis oil) (Climate Change Solutions, 2006)
31. Joffre, Materials safety data sheet, Nova Chemicals, 15 Sep 2011

Chapter 8
Technology Adoption: Residential Solar Electric Systems

Kevin C. van Blommestein and Tugrul U. Daim

Abstract The purpose of this study is to evaluate the adoption and aggregated diffusion of solar electric systems in the residential sector. The goal of this paper is to try answer the following questions using an Agent-Based Model (ABM):

1. Is there evidence of a delay in the aggregate adoption of solar electric systems? If so, how can the adoption be improved?
2. What is the relationship between increasing electricity prices, price preference, and rate of adoption?
3. What impact does changing the incentive structure have on the overall electricity savings?

The model could be used by electric utility companies, energy program administrators, and government and state agencies for planning purposes.

8.1 Aggregated Bass Diffusion Model

One approach followed in determining the impact of incentives is to predict the diffusion of a technology using an s-curve (e.g., Bass model), and use a single parameter to model the impact of these incentives. The problem with this model is that the same characteristics are assumed for all consumers. For the Bass model [1], it is assumed that the effect of advertising and the effect of word of mouth is the same for all consumers, and all consumers are well-mixed and homogeneous (Fig. 8.1). This is obviously not the case, since consumers all behave differently, each have their own characteristics, and interact differently with other consumers. The Bass model therefore does not take into account heterogeneous market dynamics.

K.C. van Blommestein (✉)
Portland State University, 1900 SW 4th Avenue 50-02, Portland, OR 97207, USA
e-mail: kevin9@pdx.edu

T.U. Daim
Department of Engineering and Technology Management, Portland State University, 1900 SW 4th Avenue 50-02, Portland, OR 97201, USA
e-mail: tugrul@etm.pdx.edu

T.U. Daim et al. (eds.), *Policies and Programs for Sustainable Energy Innovations*, Innovation, Technology, and Knowledge Management,
DOI 10.1007/978-3-319-16033-7_8

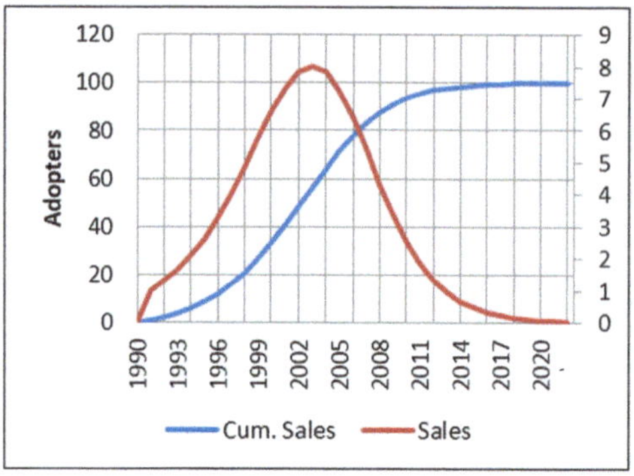

$$S(t) = \left[\left(p + \frac{q}{m}\right) N(t-1)\right] [m - N(t-1)]$$

where:

$S(t)$: *the number of adopters at time t*

$N(t)$: *the number of cumulative adopters*

m : *the total market potential*

p : *the coefficient of innovation (advertising effect)*

q : *the coefficient of imitation (word of mouth effect)*

Fig. 8.1 Bass diffusion model

8.2 Agent-Based Modeling and Diffusion

Rahmandad and Sterman [2] compare Agent-Based Modeling (ABM) and Differential Equation Modeling (DEM) by creating a contagious disease diffusion model with varying network structures and levels of heterogeneity. These types of models have also been used extensively in evaluating technology diffusion. The authors mention that both models have their own strengths and weaknesses, which should be aligned with the purpose of the model. The granularity of ABM increases computational requirements and it is difficult to link the behavior of a model with its structure as its complexity increases. However ABM can capture diverse relationships among individuals with heterogeneous attributes. For DEM, individuals in each stock (compartment) are assumed to be homogeneous and well mixed, which is not always an accurate assumption; however, it is more computationally efficient. According to the authors, a trade-off needs to be made between keeping a model simple that can be analyzed thoroughly, or disaggregated to capture heterogeneous attributes and relationships of individuals.

The authors created an AB model of the deterministic Susceptible, Exposed, Infected, Removed (SEIR) model, with the same parameters. The AB model was created using five network structures (fully connected, random, small world, scale-free, and lattice), and evaluated with both homogeneous and heterogeneous agents. It was determined that for a highly connected network (fully connected) diffusion for ABM was the closest to DEM, while networks with high clustering (lattice) had a slower rate of diffusion. This is understandable since connected networks allow diseases to spread easily to all other individuals, while for clustered networks the disease can only spread easily to local individuals that are part of the same cluster. For heterogeneous individuals, the initial growth accelerated faster, however the total number of infected individuals reduced. This is because high-contact individuals become infected sooner but are also removed sooner, therefore the number of potential infected high-contact individuals reduce. This is an important concept for technology diffusion that will be discussed under the analysis section.

Bohlmann et al. [3] made use of ABM to examine the effect of different network structures and interpersonal and intersegment communication on the diffusion of a technology. They considered two different types of agents, namely innovators and followers, and used four different network structures (random, cellular automata, small-world, and power-law networks). First, they examined the effect of different network structures and different adoption thresholds by using a fixed amount of initial adopters (innovators), the same adoption thresholds for all agents per simulation, and a fixed external influence by initial adopters. The second part of the study examined relational heterogeneity by using a two-segment model, consisting of an innovators' segment and followers' segment. The heterogeneous relationships among innovators, innovators and followers, and followers and followers, were represented by weighted network links. From these two studies they managed to answer nine research questions affecting technology diffusion. First, they established that the initial location of innovators varied the rate of diffusion for different network structures. Second, the diffusion for high-adoption thresholds was more likely to occur with clustered networks, and varied significantly between network structures. Finally and most interestingly, focusing on innovator adoption rather than innovator-follower communication can speed up diffusion.

Laciana and Rovere [4] utilized the Ising model to simulate technology adoption. The Ising model is suitable for social processes since it deals with the interaction between an object and its neighbors. The Isling model can be adapted for adoption, by taking into account the social influences from an agent's social network and the contribution of an agent's individual preference towards a technology. The model also allows for individual preferences between two alternative options (e.g., adoption and disadoption, competing products, etc.). The authors used the model to evaluate two competing products and determined that a new product with a clear advantage diffuses quicker with a disperse distribution of early adopters, while when a new product is only slightly better, diffusion does not saturate. The authors also determined that when there was a hub agent with a large number of ties in the network, the rate of adoption accelerated. This type of person is commonly referred to as an opinion leader.

Zhang and Nuttall [5] used ABM to evaluate the effectiveness of the United Kingdom (UK) government policies for encouraging the adoption of smart metering devices. They used two types of agents, namely consumers and electricity suppliers. The consumers interacted with their direct neighbors and a random amount of remote consumers. Three indicators were examined as an output of the model, namely the impact of multiple scenarios on the diffusion of the technology, electricity suppliers' market share, and the switching of consumers among electricity suppliers. Using real market data to determine initial parameters, the model produced the expected results of a stable market share, the s-curve pattern of typical technology diffusion, and the dynamic switching of consumers among electricity suppliers. Based on the results of the model, the UK was pursuing the least effective scenario.

All the agents, agent behaviors, and agent–agent relationships are summarized in Table 1 under appendix.

The articles discussed above all had different objectives in mind; however, they have a lot of similarities. Most of the reasoning behind using ABM is the requirement for heterogeneous populations and complex interaction among individuals. ABM was seen as a good representation of the complexities that exist within current markets. The main areas of focus in these articles were on the social network structures and the position of initial adopters. The outcome of using different network structures varied significantly, which was represented well in these articles, and was not captured in previous aggregated approaches of evaluating technology diffusion. Additionally, the impact of the location of early adopters in the network was crucial to the rate of diffusion, which cannot be captured by an aggregated model. From all these findings, it is understandable why ABM would be the preferred approach to researching technology adoption and diffusion.

Bohlmann et al. [3] mention that a phenomenon is emergent when it "arises from the heterogeneous and complex interactions among agents in a social system." Based on this definition, all the above articles demonstrated emergent behavior. However, some of the outcomes from these studies were relatively self-explanatory. As an example, highly connected networks had the quickest rate of diffusion, which is expected. Only once the complexity of these systems was increased by adding additional parameters, or when two alternatives closely resembled one another, were the results more surprising. As an example, Bohlmann et al. varied the adoption threshold for multiple network structures and determined that for high-adoption thresholds, diffusion was more likely to occur for clustered networks. Rahmandad and Sterman [2] did not vary the threshold and determined that the rate of diffusion was the slowest for highly clustered networks. This coincides with what Bohlmann et al. determined using a low adoption threshold; however, Rahmandad and Sterman failed to capture the effect of higher thresholds. Therefore would the outcome of their study be different if they adjusted the threshold they used for the AB and DE models? Additionally, Laciana and Rovere [4] and Bohlmann et al. determined the effect of different initial positions of early adopters. If this was also taken into account by Rahmandad and Sterman, would their results still be the same?

Laciana and Rovere [4] were the only authors to include a hub in the social network, to resemble an important agent with many connections. This increased the rate of diffusion substantially, and could be a good representation of advertising through online social networks. Their unique approach of using a physics model, Ising model, to representation of social interaction, added another dimension to the previous two studies. Their study was the only one to allow for disadoption of a technology and could represent more than one technology in a market. Even though this was the most unique approach, We did not find the results the most compelling. As an example, an increase in the dispersion of initial adopters decreased the time required for market saturation, which is expected when adopters are interacting more with other non-adopters, instead of between themselves. Additionally, when a hub agent was introduced, diffusion increased, which is again expected. There was no real surprising emergent behavior.

Zhang and Nuttall [5] were the only authors to actually apply ABM to a real world situation. They focused on one network structure and used a more complicated but established model for technology adoption, known as the Theory of Planned Behavior (TPB) consumer acceptance model. This model included additional consumer parameters such as price sensitivity, motivation to comply with other agents, and enthusiasm towards the technology. They did not take into account different possible network structures; however, their study was based on the location of households in a 2d environment. Their assumption of households communicating only with neighbors and a few random remote households therefore seemed plausible. As Laciana and Rovere [4] mentioned, when physical proximity among nodes is important, a regular lattice network provides a good approach. Additionally, Beinhocker [6] mentions that people in social networks generally have a cluster of friends and some random friends, which is clearly demonstrated in this article. This was the only study where a surprising emergent behavior from the model was backed up by actual market observations. In the model, consumers moved from one electricity supplier to another, but the electricity suppliers' market share remained constant, which was actually observed in the market. Also, this was the only study where there were central figures (electricity suppliers) that could influence multiple consumers by adjusting prices.

Of the eight attributes of agents introduced by Gilbert and Troitzsch in the "Simulation for Social Scientist" [7], all models only covered four of these attributes, namely "knowledge and belief," goals, "knowledge representation," and inference. Agents in all models based their decision on the current status of the agents they were connected to, and inferred that if a certain amount of agents had adopted the technology, and they preferred the technology, then they should adopt it. This was clearly illustrated by Laciana's and Rovere's model that used social influence and individual preference to represent the relative utility of adopting the technology for a specific agent. The goal of agents in all models was to adopt a technology, and the knowledge representation was seen as each agent storing the current status of their connected agents and their own preferences. Based on the description by Gilbert and Troitzsch, none of the agents from the studies had social model attributes, since they did not build the social models while the simulation ran. There was also no representation of language between agents, since agents only determined the current status of their neighbors and there was no back-and-forth communication between them. Finally, there was no planning or emotions by any of the agents.

Epstein and Axtell [8] referred to how simple rules create surprising behavior in the Sugarscape model. Based on all the articles discussed, the rules and goals for each agent were relatively simple. The goal for each agent was simply to reach a certain threshold in order to adopt a technology or disease. The rules for each agent were to assess their neighbors' states and in most cases, their own individual preferences towards the technology, before adopting. However, unlike the Sugarscape model, there were predefined structures specifying which agents could interact with one another. For the Sugarscape model, agents move around the landscape and interact with one another and the environment. It is therefore possible for an agent to interact with any other agent, similar to the fully connected network, however with a delay requirement for movement. From the articles discussed, there was no movement by the agents or any interactions with the environment. The models are purely based on agent behavior and agent–agent interaction for any changes in the overall behavior of the model. These models can therefore be extended to evaluate the influence of the environment on the behavior and interaction of agents. A good example of this would be the adoption of renewable energy devices, where the solar intensity and wind speeds of an agent's environment would influence the end decision.

Based on what has been identified throughout these articles, and the fact that individuals in all types of markets have heterogeneous behaviors and interactions, ABM is an obvious approach to understanding technology adoption and diffusion. The only problem noticed from the articles is that the outcomes of the models are not used by themselves for decision making purposes, but mainly as a supplementary tool. As an example, Zhang and Nuttall [5] were the only authors to use ABM for a real market. These authors illustrate the concept of their model as an approach to evaluating policies, but do not go into detail when discussing the outcome. The model was presented more as a concept than an actual forecasting tool for evaluating future adoption. This coincides with Rahmandad and Sterman [2], where they state that a trade-off needs to be made between keeping a model simple that can be analyzed thoroughly, or disaggregated to capture heterogeneous attributes and relationships of individuals.

8.3 Methodology

The methodology for this paper is as follows:

1. Design the model by specifying the agents, environment, relationships among agents, and the decision-making process of each agent
2. Build the model according to the variables and procedures specific under appendix, in Table 2–4
3. Test and debug the code, and verify the operation of the model using boundary values
4. Determine the maximum interaction radius and number of random interactions, using BehaviorSpace, in order to determine the minimum number of links that create ≈100 % awareness
5. Use actual values for electricity price and incentives in Oregon, and installed solar electric system costs in the United States
6. Answer the questions previously mentioned under the paper purpose

8.4 Model Development

8.4.1 Agents, Environment, and Relationships

Figure 8.2 is a simple diagram depicting the agents, patches, and relationships in the solar electric system adoption model. A complete list with descriptions for the agents and the patches can be found under appendix, in Table 3 and 4. The basic idea is that the patches represent the land on which the houses are built. Patches are represented by sun hours, land cost, and land area. For this specific paper only sun hours for each patch is used, which is the amount of solar energy in kWh that can be delivered per day (kWh/day). This value is used to determine what size solar electric system would be required by a household to supply 100 % of the electricity consumption.

Households can influence their neighbors who are within a randomly assigned interaction radius, and influence remote households, also by a randomly assigned value. The interactions are directed, and when each household builds its own network, undirected links form when influences can take place in both directions. When a household influences another household, the id of the influencer is added to a list of influencers of the influenced household. The length of this list is used to determine whether a household is above a certain threshold and should move to the next adoption stage. This will be described in more detail in the following section.

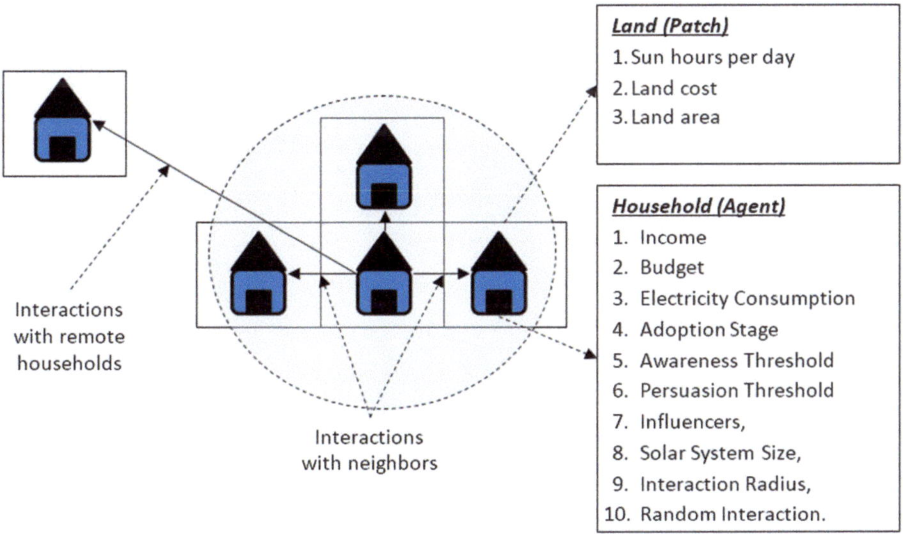

Fig. 8.2 Agents, environment, and relationships

8.4.2 Agent's Decision Making Process

Each household can be in four possible states; unaware of the technology, aware but not persuaded, persuaded but not decided, and finally decided. A flow diagram showing the decision points for transitioning between stages is illustrated in Fig. 8.3. All agents start off in the unaware stage, except for a few initial adopters who start off in the aware stage. For each tick, all agents not in the unaware stage select a randomly connected neighbor and add their id to the selected agent influencers list, with a probability of adoption stage × 10. The adoption stage is represented by 1 for unaware up until 4 for decided. The idea behind this probability is that the higher the influencing agent's adoption stage is, the more influential they will be. In other words, if the agent has already decided on the technology, then the agent is going be more influential than an agent who is only aware of the technology.

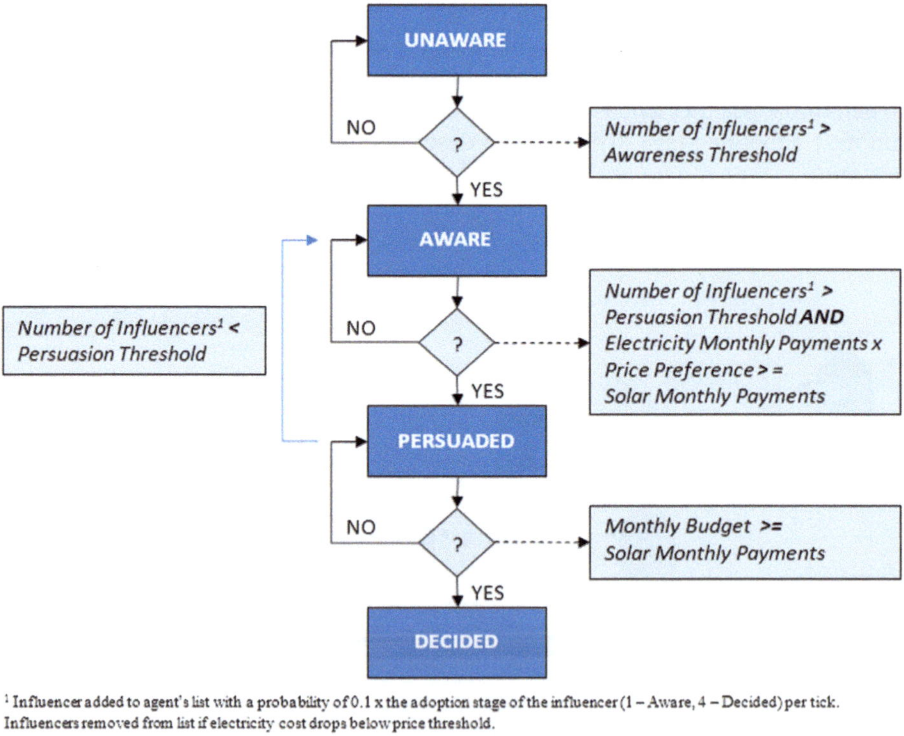

Fig. 8.3 Agent's decision-making process

In order for an agent to move from unaware of the technology to aware, the number of influencers needs to be greater than an awareness threshold. The awareness threshold is a randomly assigned value between 1 and 3 for each agent. Each agent needs at least 1 other agent in its influencers list to become aware. Once the agent becomes aware, all ids are removed from the influencers list.

For an agent to move from aware to persuaded, the number of influencers in the new list must be greater than a persuasion threshold. The persuasion threshold is a randomly assigned value between 0 and 10. Influencers are added to the list in the same manner as described above. In addition to this threshold, the monthly electricity payments by the agent multiplied by a price preference must be greater than or equal to the monthly payments required to purchase a solar electric system. Since the lifespan of a solar electric system is generally assumed to be around 20 years, the total purchase cost of the system is divided by 20 years × 12 months = 240 months, to determine the equivalent monthly payments (not taking into account inflation and present value). The price preference was included to take into account the preference of the agent when making the decision, and is a randomly assigned value between 0 and a max price preference assigned by the user. If the price preference is greater than 1, then the agent will not wait for the solar payments to drop below the electricity payments, and vice versa.

As an example of the above mentioned process, if an agent consumes 1,200 kWh of electricity per month at $0.1 per kWh, the monthly electricity cost would be 1,200 × 0.1 = $120 per month. The size of the solar system required to supply 1,200 kWh per month is calculated to be 8.54 kW using the equation specific under appendix, in Table 3. At $5/W for solar, an incentive of $1/W, and a 20-year payment duration, the monthly solar cost would be 8,540 × (5 − 1)/240 ≈ $142 per month. Therefore if the price preference was 1, then $120 × 1 is not greater than or equal to $142 and the agent will not become persuaded. If the price preference was 2, then $120 × 2 is greater than $142, and the agent would become persuaded. The idea behind using a price preference is that not all agents would react the same to increasing or decreasing prices of electricity and/or solar electric systems. Additionally, the price preference can also take into account agents who are not basing their decision purely off price, but also other factors such as environmental concerns, etc.

Once an agent is persuaded, a decision is made to determine whether the agent has enough budget to pay for the monthly solar payments. If the agent's monthly budget is greater than the solar payments, then the agent will change to decided and purchase the solar electric system. If the agent remains in the persuaded stage and the electricity monthly payments × price preference becomes less than the solar monthly payments, then an influencer is removed from the agent's list for each tick that this is true. When the number of influencers becomes less than the persuasion threshold, then the agent moves back to the aware stage.

8.4.3 Communities

The Netlogo model world is divided into four neighborhoods, each with different levels of income. The quadrants represent low-income, medium-income, high-income, and a distribution of income-level households. The intention of creating multiple neighborhoods was to determine whether households joined together to purchase a larger shared solar electric system. However, due to time constraints, sharing of solar electric systems was not included as part of this paper, but can easily be extended for future work.

8.4.4 Assumptions

The following assumptions were taken into account when creating the model:

1. Electricity consumption, income, budget, and price preference for each agent remains constant throughout the simulation
2. No new households are added to the model throughout the simulation
3. Direct links are used to connect the households, assuming that the possibility of influencing another household would take place in one direction
4. A household will only buy a solar electric system that will replace 100 % of its electricity consumption
5. The solar price and electricity price will remain the same unless the user adjusts the sliders while the simulation is running
6. All agents will compare the monthly solar costs against the monthly electricity costs when deciding to purchase a solar electric system
7. The world does not wrap horizontally or vertically

8.4.5 Netlogo Model

The Netlogo model for this paper is illustrated in Fig. 8.4. The colors of the houses represent the current adoption stage of the household (red—unaware, blue—aware, orange—persuaded, green—decided). The color of the patches is scaled according to sun hours, with lighter yellow representing higher sun hours and vice versa. The user adjustable sliders are described in detail under appendix, in Table 2. The graph represents the number of households in each stage of adoption, which is also shown in the monitors below the graph. Additionally, the total electricity consumed, total electricity saved, and total incentives paid out are shown in the monitors below the sliders. Finally, the two switches next to the go button allow the user to hide the links representing the connections between households, and hide the patch colors representing the sun hours for each patch (Fig. 8.4).

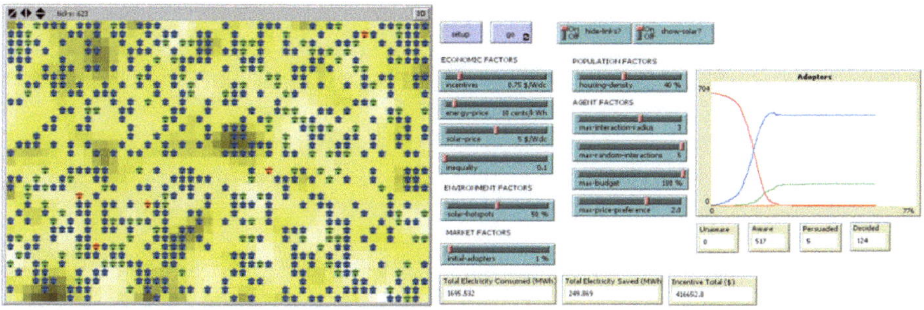

Fig. 8.4 Netlogo model

8.5 Model Verification

In order to conduct a basic verification of the model, beyond optimizing the code and debugging potential errors, a boundary test for each user adjustable variable was done. The idea of the boundary test was to set the variables to their extreme, and follow the agent's decision-making process. The boundary test started by setting all variables to 0 except for the housing density. As a result there were no connections between households and no initial adopters, with all households remaining unaware of the technology.

The next test was to introduce a small amount of initial adopters and create a network connecting the households. This allowed the initial adopters to communicate, resulting in all households becoming aware of the technology. None of the households could become persuaded since the maximum budget was set to 0 % and therefore no households could afford the solar system. The next test was to increase the maximum budget to 100 %, keep the solar hotspots at 0 %, and set the solar price to $0/Wdc. Households all became persuaded; however, because there were no solar hotspots (i.e., sun hours), no electricity could be produced by these solar systems and therefore no household ended up purchasing them. The next test set solar hotspots to 100 % resulting in all previously persuaded households to become decided.

Finally, by setting all variables to their maximum value, except housing density and initial adopters, households ended up in three adoption stages, namely aware, persuaded, or decided. This was expected since some households are not influenced by the increased electricity prices and remain aware, some households are influenced but cannot afford a solar electric system and remain persuaded, while the rest have decided to purchase the solar electric system.

BehaviorSpace was finally used to determine the minimum number of links required for ≈100 % awareness by adjusting the maximum interaction radius and maximum random interactions. The values obtained were, 3 for maximum interaction radius, and 5 for maximum random interactions. These values were used for the remainder of this paper (Table 8.1).

Table 8.1 Boundary tests

Test	Housing density (%)	Initial adopters (%)	Solar hotspots (%)	Incentives ($/Wdc)	Electricity price ($/kWh)	Solar price ($/Wdc)	Max interaction radius	Max random interaction	Max budget (%)	Max price preference	All adoption stages
1	40	0	0	0	0	0	0	0	0	0	Unaware (UA)
2	40	1	0/100	0/5	0/100	10	5	5	0	0	Aware (A)
3	40	1	0	0/5	0/100	0	5	5	0/100	0/2	Persuaded (P)
4	40	1	100	0/5	0/100	0	5	5	0/100	0/2	Decided (D)
5	40	1	100	5	100	10	5	5	100	2	A, P, D

8.6 Results and Discussion

8.6.1 *Delay in Aggregated Adoption*

Is there evidence of a delay in the aggregate adoption of solar electric systems? If so, how can the adoption be improved?

It is common knowledge in marketing literature that a "Chasm" exists in the technology adoption life cycle of disruptive technologies [9] (Fig. 8.5). This is the difficult step of making the transition from early adopters to early majority. The idea for crossing this Chasm is to focus on each of the five main segments of the adoption life cycle, one at a time, and use each segment as a foundation for marketing for the next segment. By allowing agents to be in four different adoption stages in the model, it was possible to understand potential reasons why the early majority were not adopting the technology.

By using actual values for electricity price [10] and incentives in Oregon [11], and installed solar electric system costs for the United States [12], the output shown in Fig. 8.5 was obtained. The maximum price preference was set to 2 and the initial adopters were set to 2.5 %, as shown in the technology adoption life cycle as innovators. The total number of adopters (green) is approximately 20 % of the total number of households, which is slightly more than the 16 % of adopters before the chasm in the technology adoption life cycle; however, 16 % is only an approximation.

It was possible to identify reasons why majority of households were still in the aware stage (blue) and very few households in the persuaded stage (orange). All households had more influencers than the persuasion threshold, therefore this was not restricting agents from moving to the following stage in the adoption process. The restriction was only based on the price of the solar electric system. Agents could not move from the aware to persuaded stage because the monthly payments

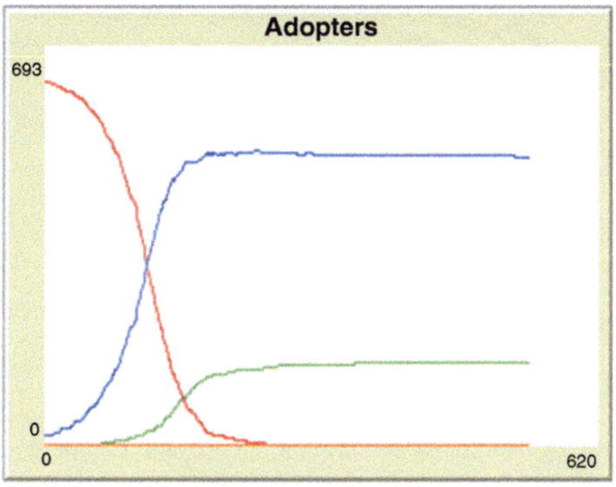

Fig. 8.5 Technology adoption life cycle and result (*red*—unaware, *blue*—aware, *orange*—persuaded, *green*—decided)

were too high, and if some agents did move to persuaded, they could not afford the solar electric system due to their budget.

There are two main methods of increasing the adoption, either by increasing the incentives offered to reduce the purchase cost of the system, or by increasing the price of electricity, thereby making the solar electric system the preferred alternative. These two methods are examined in the following two questions.

8.6.2 Impact of Increasing Electricity Prices

What is the relationship between increasing electricity prices, price preference, and rate of adoption?

As was mentioned in the previous section, one way of increasing adoption is by increasing the price of electricity. Figure 8.6 illustrates the impact of increasing electricity prices on the adoption of solar electric systems, using the same parameters as the previous question. As expected, adoption increases with increasing electricity prices, for all maximum price preferences except 0. As was previously stated, each household is randomly assigned a price preference between 0 and the maximum price preference. The equation used in the agent's decision making process is: electricity monthly payments × price preference ≥ solar monthly payments. By selecting a price preference of 2, there is a 50 % chance that a household will wait after the solar monthly payment falls below the electricity monthly payment, before adopting the technology. There is also a 50 % chance that households will not wait for the cost to be less. The higher the price preference, the less the number

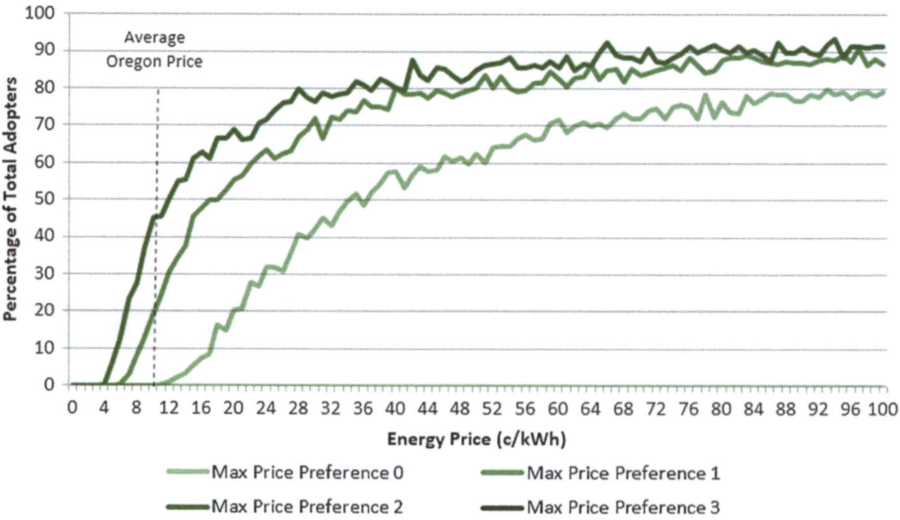

Fig. 8.6 Impact of increasing electricity prices

of households delaying their purchases, thereby increasing the number of adopters at lower electricity prices. It can be seen that the price preference has a large impact on adoption and it is therefore important to try determining what price preference most accurately represents the community of interest during a study.

8.6.3 *Impact of Incentives*

What impact does changing the incentive structure have on the overall electricity savings?

Figure 8.7 illustrates the impact of increasing incentives on the overall electricity savings and the total investment made by the organization supplying the incentives. The total electricity savings was simply calculated by adding up all households' electricity consumption who had already adopted the solar electric system. The total investment was simply calculated by adding up all incentives given to the households who had adopted the system.

As part of the current incentives offered by Energy Trust of Oregon for solar electric systems [11], \$5,000 is the maximum incentive that can be offered. Therefore, for large consumers of electricity they will need to pay the remaining amount that is not covered by the incentive. At \$5/W for a solar system [12], the average household would need a 8.5 kW system (based on the average monthly consumption in Oregon of 991 kWh and estimated sun hours of 5 kWh/day), which relates to \$42,500 without incentives, and \$37,500 with the maximum incentive. Even with the maximum incentive, this is still a substantial amount to pay for a solar electric system. As a result, only the small consumers of electricity see a benefit in adopting with incentives. This is the reason why the electricity savings remain relatively constant in the left graph in Fig. 8.7. The right graph shows the result if the maximum limit is removed. The total investment is substantially more without a limit but so is the total savings.

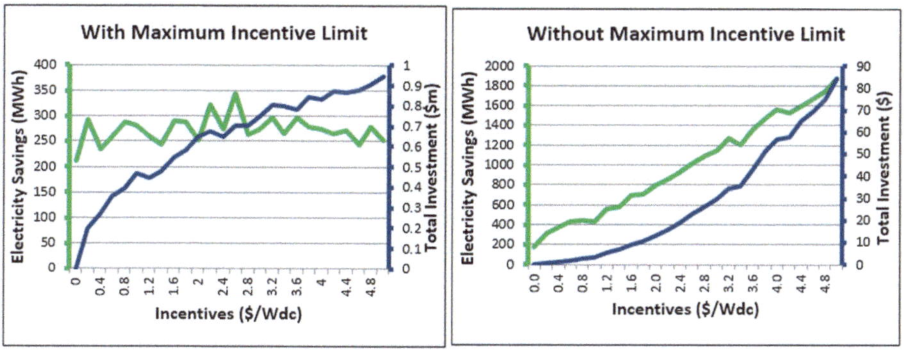

Fig. 8.7 Impact of incentives of the overall adoption

8.7 Conclusion

Even though the model could not be validated using market data, patterns commonly referred to in technology adoption literature were identified in the model. By separating the adoption process into stages allowed for the potential identification of the "Chasm." Also by observing which households were aware of the technology but did not adopt, can aid marketers in understanding how they can increase the adoption and overcome the "Chasm." The two main methods of increasing adoption were analyzed by increasing the electricity price and increasing incentives.

The behavior of the model by increasing electricity prices was not really surprising. The higher the price preference, the quicker the number of adopters increased. Since the behavior of the model was easily described by the decision making of the individuals in the model, this does seem to me as an emergent behavior. In other words, we think the aggregated behavior can simply be broken down into the behavior of the individuals. However, the outcome did show us that the price preference had to be heavily weighted towards the solar electric system in order to get households to adopt the technology. High prices for solar electric systems and very low prices for electricity make this decision very easy for the households.

The behavior of the model for increasing incentives was more interesting than the increasing electricity. At first, we thought there was a problem with the model when the electricity savings remained relatively constant as incentives increased. You would think that increasing incentives would increase the adoption and thereby the electricity savings. The problem is, there is just too big of a gap between the cost of electricity and the cost of solar electric systems. Unless the upper limit is removed, it is difficult to see how adoption can increase. The model however does not take into account other motives why households would adopt these systems other than financial reasons.

There are too many unknowns for the model to be used for decision making purposes; however, going through the process of creating the model definitely aids in obtaining a better understanding of the system. Even though this model seemed too "engineered," it did give some valuable insight that possibly would have been overseen if looking at the system from the top–down. The best example of this was the effect of the incentive limit on the actual adoption. ABM is definitely a good match for understanding adoption, and with the right market research and understanding of the consumers, we think it can be used effectively for successful decision-making purposes.

The following is a list of improvements and extensions that could be done to this paper:

- A sun hours per day map [13] can be imported into Netlogo if real world conditions are required
- Market research can be conducted to understand the decision-making process of the households in more detail
- Actual sales data for solar electric systems can be obtained to determine whether the model represents what is actually happening in the market

- Data from the Residential Building Stock Assessment [14] can be used to represent the households more accurately
- Different social network structures [2] can be modeled to determine the impact of each on the adoption process

Appendix: Model Parameters

Agents and Environment

The global variables, agents, and patches for the solar electric system adoption model.

Global variables (user adjustable variables)	Housing density: The percentage of total patches that are occupied by houses
	Incentives: The incentives offered by utility companies and others to encourage the adoption of the technology
	Energy price: The price of electricity (kWh)
	Solar price: The price of a solar electric system per kWh
	Inequality: The distribution of wealth for one of the communities under evaluation. This distribution of wealth is defined by the equation min-income × exp(random − exponential(1/inequality)). The intention of this equation is to try creating the Pareto distribution
	Solar hotspots: The percentage of patches that will have the highest solar intensity level. Random patches are chosen and the solar intensity diffuses to the neighborhood patches
	Initial adopters: The percentage of households that will already have a solar electric system before the simulation starts
	Maximum interaction radius: Each household can influence other households within a random radius between 0 and maximum interaction radius
	Maximum random interactions: Each household can influence a random number of other households between 0 and maximum random interactions
	Maximum budget: Each household has a random percentage of their income, between 0 and 100 %, which they can spend
	Maximum price preference: A household will decide to move from aware to persuade if electricity monthly payments × price preference ≥ solar monthly payments. This is a random value between 0 and 3
	Hide-links: Hide the links that are connecting the households
	Show-solar: Show the sun hours by using a scaled yellow color for each patch

Procedures	Setup global: Initializes all the global variables
	Update plot: Update the plots displaying the number of households in each stage of adoption, the distribution of in and out links, and the distribution of income
	Update display: Observes the status of show-solar? and hide-links? to determine whether the display should be updated, even when the simulation is running

Agent	Households: Houses are randomly placed on patches. The number of households is controlled by the global housing-density variable

Characteristics	Income: The total monthly income from all members of the household
	Budget: Percentage of income which the households can spend per month
	Electricity consumption: The amount of electricity consumed by the household per month (kWh). This is calculated by multiplying the average electricity consumption in Oregon with the ratio of the households income to the median income
	Adoption stage: A household can be unaware, aware, persuaded, or decided on the technology
	Price preference: A household will decide to move from aware to persuaded if electricity monthly payments × price preference ≥ solar monthly payments
	Awareness threshold: How many households need to mention the technology to this household before they change their adoption stage from unaware to aware
	Persuasion threshold: How many households need to mention the technology to this household before they change their adoption stage from aware to persuaded
	Interaction radius: The radius of the circle in which a household can influence other households, or be influenced
	Random interaction: How many random households outside the radius can be influenced? This resembles the random friends discussed by Beinhocker
	Solar size-required: The size of the solar electric system required. This is calculated by electricity consumption × 12 × 1,000/(365 × [sun-hours] of patch-here × derate-factor). The derate factor for solar electric systems is generally assumed to be 0.77
	Own incentives: The amount of incentives the household can obtain for their solar electric system. The maximum amount defined by Energy Trust of Oregon is $5,000

References

1. N. Meade, T. Islam, Modelling and forecasting the diffusion of innovation—a 25-year review. Int. J. Forecasting **22**(3), 519–545 (2006)
2. H. Rahmandad, J. Sterman, Heterogeneity and network structure in the dynamics of diffusion: comparing agent-based and differential equation models. Manag. Sci. **54**(5), 998–1014 (2008)
3. J.D. Bohlmann, R.J. Calantone, M. Zhao, The effects of market network heterogeneity on innovation diffusion: an agent-based modeling approach. J. Prod. Innov. Manag. **27**(5), 741–760 (2010)
4. C.E. Laciana, S.L. Rovere, Ising-like agent-based technology diffusion model: adoption patterns vs. seeding strategies. Phys. A: Stat. Mech. Appl. **390**(6), 1139–1149 (2011)
5. T. Zhang, W.J. Nuttall, Evaluating government's policies on promoting smart metering diffusion in retail electricity markets via agent-based simulation. J. Prod. Innov. Manag. **28**(2), 169–186 (2011)
6. E. Beinhocker, *Origin of Wealth* (Harvard Business Press, Boston, 2006)
7. N. Gilbert, K. Troitzsch, *Simulation for the Social Scientist*, 2nd edn. (Open University Press, New York, 2005)

8. J. Epstein, R. Axtell, *Growing Artificial Societies: Social Science from the Bottom Up* (Brookings Institution Press, Washington, DC, 1996)
9. G.A. Moore, *Crossing the Chasm: Marketing and Selling High-Tech Products to Mainstream Customer* (HarperCollins, New York, 1999)
10. U.S. Energy Information Administration, Electric Price Monthly [Online] (2013), http://www.eia.gov/electricity/monthly/epm_table_grapher.cfm?t=epmt_5_6_a. Accessed 1 Dec 2013
11. Energy Trust of Oregon, Solar Electric for Homes: Your Home Resource for Clean Energy [Online] (2013), http://energytrust.org/residential/incentives/solar-electric/SolarElectric/. Accessed 1 Dec 2013
12. G. Barbose, N. Darghouth, S. Weaver, R. Wiser, *Tracking the Sun VI: An Historical Summary of the Installed Price of Photovoltaics in the United States from 1998 to 2012* (Lawrence Berkeley National Laboratory, Berkeley, 2013)
13. National Renewable Energy Lab, Dynamic Maps, GIS Data & Analysis Tools: Solar Maps [Online] (2013), http://www.nrel.gov/gis/solar.html. Accessed 1 Dec 2013
14. D. Baylon, P. Storm, K. Geraghty, B. Davis, *Residential Building Stock Assessment: Single-Family Characteristics and Energy Use* (Northwest Energy Efficiency Alliance, Seattle, 2012)

Chapter 9
Portfolio Optimization in the Electricity Market-Investor's Perspective

Umit Kilic and Gulgun Kayakutlu

Abstract International energy investors are interested in investing new markets where there is a possibility of constructing a portfolio composed of fossil and renewable energies. Capital allocation decision is to be made considering the socio-political effects caused. Hence the market is analyzed in terms of production possibilities, price volatility, and social acceptance. The portfolio with maximum possible revenue is to be created with the least environmental effect and the smallest technological risks. This study offers a weighted goal programming (WGP) model, where the weights are calculated using Analytical Network Planning (ANP). Case study is realized in Turkey, because the Turkish electricity market is experiencing significant structural changes and a rapid transformation process. Liberalization and constantly increasing electricity demand in the country have drawn a lot of interest. The same model can be applied in any country by changing the energy resources and country-based criteria.

9.1 Introduction

Strategic planning for the medium- to long-term expansion of the electricity generating capacity of a specific country has been an important issue in the past, when electricity markets were regulated. The major concerns in regulated markets were mainly the dependence from imported fuels, stability and reliability of the transmission grid, as well as quality and security of supply [1].

The electricity sectors of many countries have faced numerous changes in their structure and their business environment during the last years. First of all, the electricity markets have gone through a deregulation process, which has introduced competition in a formerly state-regulated economic sector. Therefore, the planning for new power plant additions and existing plant replacements has shifted its focus

U. Kilic • G. Kayakutlu (✉)
Istanbul Technical University, İstanbul, Turkey
e-mail: gkayakutlu@gmail.com

T.U. Daim et al. (eds.), *Policies and Programs for Sustainable Energy Innovations*, Innovation, Technology, and Knowledge Management,
DOI 10.1007/978-3-319-16033-7_9

from strategic fuel selection to economic considerations, such as the minimization of the production cost. This shift has also amplified the effects of uncertainty in fuel prices, since now its effects are even more crucial for an investment decision [2, 3].

Enactment of Kyoto Protocol is another factor that has a significant effect on investment decisions, since electricity generation based on conventional sources lead to additional costs in the developed countries. Although developing countries currently have no restrictions on greenhouse gas emissions, future uncertainties on the scope of the agreements and volatility of greenhouse gas allowances make decision-makers of developing countries to take into consideration this factor as well.

The scope of this work is to incorporate all above-mentioned criteria in a mathematical model to help investors factor in different goals while establishing an optimum electricity generating portfolio. The remainder of this paper is organized as follows: in Sect. 9.2, a literature review is presented. In Sect. 9.3, an overview of methodology is given while in Sects. 9.4 and 9.5, details of mathematical model are provided through a case study. Section 9.6 contains the results of the implementation. Finally, in Sect. 9.7 a detailed analysis is presented with suggestions for further studies.

9.2 Literature

The issue of the optimum electricity generating portfolio has long troubled researchers. Ref. [4] is among the first to introduce the portfolio analysis in the power sector. More recent research [5–7] has extended the analysis to various power expansion mixes. Mean-variance portfolio techniques have been applied in various instances [8, 9], presenting also various risk measures [10].

Along with mean-variance approach, large linear programming models have been used extensively over several decades to address Electricity Supply Industry (ESI) modeling [11–13]. Modeling with single objective functions has been a powerful tool in optimizing power station expansion under specific environmental constraints, as well as for examining the economic feasibility of new options in the energy market. This type of analysis, done in partial equilibrium frameworks, has provided policy makers with the "perfect market" response to future scenarios that are valid for both regulated, centrally planned power markets, as well as for efficient fully deregulated markets. Although this type of modeling has enjoyed some success for integrated resource planning in the past, resource planning today has become a far more complex task [12]. What such an approach fails to deliver is explicit consideration of trade-offs between different objectives and the need to address uncertainty in the modeling process [14]. Multiple-objective programming models are developed to handle this problem and analyze the trade-offs between different objectives [14, 15].

9.3 Methodology

9.3.1 General Approach

A methodology is proposed for decision-making about establishing electricity generating portfolio when taking into account different technologies, profitability of portfolio, fuel price volatility, social acceptance, greenhouse gas emissions, and capital expenditures. A Weighted Goal Programming (WGP) model has been developed and weights of the goals calculated through Analytical Network Process (ANP). Profitability calculations are done based on Levelized Cost of Electricity (LCOE). The techniques embedded in the methodology are briefly explained in this section of the paper.

9.3.2 Weighted Goal Programming

Goal Programming is a multiobjective programming technique. The ethos of GP lies in the Simonan [16] concept of satisfying of objectives. Simon conjectures that in today's complex organizations the decision-makers do not try to maximize a well-defined utility function. In fact the conflicts of interest and the incompleteness of available information make it almost impossible to build a reliable mathematical representation of the decision-makers' preferences. On the contrary, within this kind of decision environment the decision-makers try and achieve a set of goals (or targets) as closely as possible. Although Goal Programming was not originally conceived within a satisfying philosophy it still provides a good framework in which to implement this kind of philosophy [17].

Goal Programming models can be classified into two major subsets. In the first type, the unwanted deviations are assigned weights according to their relative importance to the decision-maker and minimized as an Archimedean sum. This is known as WGP. The algebraic formulation of a WGP is given as:

$$\min \quad z = \sum_{i=1}^{k} \left(u_i n_i + v_i p_i \right) \tag{9.1}$$

$$s.t. \quad f_i\left(x\right) + n_i - p_i = b_i, \quad i = 1 \ldots Q, \quad x \in C_s \tag{9.2}$$

where $fi(x)$ is a linear function(objective) and bi the target value for that objective. ni and pi represent the negative and positive deviations from this target value. ui and vi are the respective positive weights attached to these deviations in the achievement function z. These weights take the value zero if the minimization of the corresponding deviation variable is unimportant to the decision-maker. Cs is an optional set of hard constraints as found in linear programming (LP) [17].

9.3.3 Analytical Network Process

The ANP was first introduced by Saaty to provide a framework for dealing with decision-making problems. Since the introduction, it has been applied to a large variety of decision-making and forecasting problems. The ANP is a general form of the well-known decision theory, Analytic Hierarchy Process (AHP). Similar to the AHP, the ANP is based on deriving ratio-scale measurement to be used to allocate resources according to their ratio-scale priorities. Whereas AHP models assume an unidirectional hierarchical relationship among decision levels, ANP does not require this strictly hierarchical structure and allows for more complex interrelationships among the decision levels. ANP generalizes the pairwise comparison process, so that decision models can be built as complex networks of decision objectives, criteria, stakeholders, alternatives, scenarios, and other environmental factors that all influence one another's priorities. The key concept of the ANP is that influence does not necessarily have to flow only downwards, as is the case with the hierarchy in the AHP. Influence can flow between any two factors in the network causing nonlinear results of priorities of alternative choices. The ANP can be described as a system of N components (which may be part of a cluster of components) that form a network, where every component (Cn) can interact with or have an influence on itself or some or all of the other components of the system. The network, N, equals $\{C_a, C_b, C_c, \ldots, Cn\}$ and $\{\{C_a, C_a\}, \{C_a, C_b\}, \{C_a, C_c\}, \ldots, \{Cn, Cn\}\}$ represents the set of pairwise linkage within or between components of the network. This multicriteria decision-making model derives priorities or weights for each of the "n" criteria or components, Cn, of the model based on their judged relative importance to the overall goal. The derivation of the ANP priority weights, which use pairwise assessment based on statistical or judgmental relevance, is quite different from more traditional methods [18].

9.3.4 Levelized Cost of Generating Electricity

The notion of levelized costs of electricity (LCOE) is a handy tool for comparing the unit costs of different technologies over their economic life. It would correspond to the cost of an investor assuming the certainty of production costs and the stability of electricity prices. In other words, the discount rate used in LCOE calculations reflects the return on capital for an investor in the absence of specific market or technology risks. Given that such specific market and technology risks frequently exist, a gap between the LCOE and true financial costs of an investor operating in real electricity markets with their specific uncertainties is usually verified. For the same reason, LCOE is also closer to the real cost of investment in electricity production in regulated monopoly electricity markets with loan guarantees and regulated prices rather than to the real costs of investments in competitive markets with variable prices.

Despite these shortcomings, LCOE remains the most transparent consensus measure of generating costs and remains a widely used tool for comparing the costs of different power generation technologies in modeling and policy discussions. The calculation of the LCOE is based on the equivalence of the present value of the sum of discounted revenues and the present value of the sum of discounted costs. The LCOE is, in fact, equal to the present value of the sum of discounted costs divided by total production adjusted for its economic time value [19]. The formulation given below represents LCOE calculation:

$$\frac{\sum_t \left(Inv_t + O\,\&\,M_t + Fuel_t + Car_t + Decomm_t \right) * \left(1+r\right)^{-t} \Big)}{\sum_t \left(Electricity_t * \left(1+r\right)^{-t} \right)} \tag{9.3}$$

In this study, LCOE is only used for profitability calculations. Though it is an important factor affecting the results that can be get from our model, the decision-making process doesn't merely rely on the usage of this tool. The above-mentioned shortcomings of LCOE are compensated through incorporating environmental, fuel volatility, and social acceptance related goals in the model.

9.4 Case Study

9.4.1 Turkish Electricity Sector Status

Turkish electricity sector is chosen for the implementation of mathematical model developed because the Turkish electricity market has been considerably restructured in recent years. In order to open it further for private sector participation, the formerly vertically integrated companies have been unbundled, leaving only the transmission grid as a natural monopoly. During the last two decades, the Turkish electricity market has been rapidly growing, with an average annual growth rate of more than 6 % [20]. Rapidly growing demand and deregulation draw interests of international and local investors. Before and after entering the market, capital allocation and electricity generating portfolio construction is one of the most important challenges that all players face, since there are lots of criteria that need to be taken into consideration and inherent uncertainties related to the market.

All data fed into the model is provided in this section of the study. The same model can be applied in any country by changing the energy resources, country-based criteria, and relevant data.

9.4.2 Levelized Cost of Electricity for Turkey

In 2010, International Energy Agency (IEA) and Organisation for Economic Co-operation and Development (OECD) published a detailed report on levelized costs of electricity on country and source basis. The study contains data on electricity

Table 9.1 Source-by-source data on electricity generating costs for Turkey at 5 % discount rate—USD/MWh [19]

Technology	Investment costs	O&M	Fuel and carbon	LCOE	CF (%)
Coal—PCC	15.24	11.29	20.19	46.72	85
Natural gas	10.46	4.7	38.3	53.46	85
On-shore wind	44.6	17.96	0	62.56	35
Off-shore wind	77.63	43.3	0	120.93	45
Hydro	18.08	3.93	0	22.01	50
Solar—PV	133.67	19.52	0	153.19	20
Biomass	32.26	26.25	19.13	77.73	85
Geothermal	34.02	5.47	0	39.48	85

generating costs for almost 200 power plants in 17 OECD member countries and 4 non-OECD countries.

The levelized costs and the relative competitiveness of different power generation technologies in each country are highly sensitive to the discount rate and slightly less, but still significantly sensitive, to the projected prices for CO_2, natural gas, and coal. For renewable energy technologies, country- and site-specific load factors also play an important role [19]. In the report, there is no data provided for the electricity generating costs in Turkey. Since carbon prices for nonmember countries are zero as in Turkey and there are lots of similarities in terms of economy, gross domestic product per capita, and economic growth rate between Turkey and non-OECD countries (China, Russia, Brazil, South Africa) analyzed in report, LCOE for Turkey is assumed to be the arithmetic average costs of non-OECD countries. In many studies, China, Russia, Brazil, South Africa, and Turkey are classified in the same group as emerging countries and upper middle income economies [21, 22]. The above-mentioned facts about Turkey and non-OECD countries justify the assumption made for the LCOE calculations. Table 9.1 presents the results of LCOE analysis for Turkey.

9.4.3 Renewable Energy Policies and Wholesale Electricity Prices in Turkey

Turkey aims to utilize its energy potential, including from renewable sources in a cost-effective manner. In its efforts to promote renewable energy, the government has focused on electricity. The renewable energy-related legislation has been intensified. The cornerstone of Turkey's legislation on electricity from renewable sources is the Law on the Utilization of Renewable Energy Resources for the Purpose of Generating Electricity, enacted in May 2005, and its subsequent amendments. Also relevant are the 2001 Electricity Market Law, 2007 Energy Efficiency Law, and

Table 9.2 Electricity Prices based on different sources—USD/MWh [24, 26]

Energy resources	Prices
Hydro	70.3
Wind	70.3
Geothermal	105
Biomass	133
Solar	133
Fossil fuel—Wholesale price	79.7

2007 Law on Geothermal Resources and Natural Mineral Waters. Together laws include the following instruments:

- Feed-in tariffs and purchase obligation
- Connection priority
- Reduced license fees
- Exemptions from license obligation for small-scale generators
- Reduced fees for project preparation and land acquisition [23, 24].

The electricity generated from a fossil fuel source can either be sold through day-ahead market or to retail and wholesale electricity companies. In line with Electricity Market Law, day-ahead electricity prices are decided to be taken as the basis for the average electricity wholesale prices [25]. Source-by-source electricity prices, which are fed in the model as well, are presented in Table 9.2.

9.4.4 Gaseous Pollutants Emissions

During the last 20 years, half of all increases in energy-related CO_2 emissions were from electricity generation [27]. In fact, no power source is entirely impact free. All energy sources require energy and give rise to some degree of pollution from manufacture of the technology. The environmental impacts can depend greatly on how energy is produced and used, the fuel mix, the structure of the energy systems, and related energy regulatory actions and pricing structures.

Measured gaseous pollutants emissions for various fuel types such as CO_2, CH_4, NO_x, and SO_2 are presented in Table 9.3. The figures shown in Table 9.3 are based on the life cycle assessment technique and indicate gaseous emissions emitted during the whole process.

9.4.5 Social Acceptance

Citizens' preferences for different energy sources have come to play an increasingly central role in decisions about energy investments to be undertaken in the regions or the countries they live in even sometimes in their neighboring countries [31]. The

Table 9.3 Main gaseous pollutants—g/kWh [28–30]

Fuel type	CO_2	CH_4	NO_x	SO_2
Natural gas	386	1.076	0.351	0.125
Hydropower	32	0.135	0.056	0.055
Coal	838	4.716	0.696	0.351
Oil	760	4.216	0.622	0.314
Nuclear	17	–	0.047	0.072
Biomass	–	–	0.350	0.087
Geothermal	21	0.059	–	–
Wind	38	0.169	0.055	0.071
Solar	319	0.083	0.408	0.494

Table 9.4 Opposition to and endorsement of energy investment alternatives [35]

Opposition			Endorsement	
Number of respondents (%)			Number of respondents (%)	
Coal	1,855	82.9	86	3.6
Natural gas	394	17.6	881	37.3
Dams	135	6	1,539	65.2
Renewables	90	4	1,414	60.4
Nuclear	1,399	62.5	170	7.2

public resistance that escalated dramatically after the recent Fukushima accident, for instance, means that governments in a number of developed countries are no longer free to easily opt for nuclear energy. Germany is one such case where strong public pressure in favor of a nuclear phase-out, backed by the political pressure of the antinuclear Green Party, has forced the current government to reverse traditional German national policies on nuclear energy [32, 33]. The impact of citizens' preferences on energy policies extends, albeit to a lesser degree, to developing countries as well. Strong local resistance in India, for example, led to the withdrawal of the World Bank from funding the Sardar Sarovar Dam project [34].

Given the above-mentioned concerns related to energy investments and the importance of the issue, social acceptance is taken into consideration while constructing our model too. A study based on data from a face-to-face survey of 2,422 residents randomly drawn from urban Turkey is used for the social acceptance calculations in this study. Table 9.4 shows the results reached through the survey.

9.4.6 Fuel Price Volatility

High volatility in price returns often appears in deregulated energy markets. The market participants such as energy producers and distributors always face such high volatility risks from energy markets. Lots of volatility models both in continuous

Table 9.5 Average volatility for fossil fuels [37]

Fuel	1986–1990	1991–1995	1996–2000	2001–2005
Coal	0.0176	0,0182	0,0189	0,0158
Gas	0.1034	0,0702	0,1020	0,1251

and discrete time were developed in financial markets, and they are directly applied to the volatility models in energy markets without any adjustment for the energy characteristics [36]. Both continuous and discrete time models can be found in abundance in the literature. In our model, a linear programming approach has been embraced for the minimization of fuel price volatility in the portfolio. The characteristics of each energy source are defined through a volatility matrix and weights used in the model are given in Table 9.5.

9.5 Mathematical Model

Nine different electricity generation methods have been included in the analysis, almost all of them with different fuel source. For each one of them, the best available technology has been selected. Goals that should be taken into consideration are defined in order to reach the optimum electricity generating portfolio. After defining the goals, weights attained to each goal are calculated and factored in the model.

9.5.1 Goals

In order to determine the electricity generating portfolio capable of meeting the multiple-objectives market participant defined, the model will be developed with the following goals:

Goal 1: Maximize profit
Goal 2: Minimize carbon emissions
Goal 3: Minimize fuel volatility in portfolio
Goal 4: Maximize social acceptance or minimize social opposition

Mathematical expression of each goals is given as follows:

$$Max.\ G_1 = \sum_{n=1}^{n} (Sp_n - LCOE_n).X_n.Cf_n.T \tag{9.4}$$

$$Min.\ G_2 = \sum_{n=1}^{n} Ef_n.X_n.Cf_n.\mathrm{T} \tag{9.5}$$

$$Min.\ G_3 = \sum_{n=1}^{n} Fv_n.X_n \tag{9.6}$$

$$Min.\ G_4 = \sum_{n=1}^{n} So_n.X_n \tag{9.7}$$

9.5.2 Constraints

Hard constraints that no tolerance is given for deviations are commonly encountered in decision-making problems. Two constraints are decided as utmost important factors to be considered from an investor's point of view and incorporated in the model.

Constraint 1: Installed capacity limitations for each energy source
Constraint 2: Installed capacity limitation of portfolio

Turkey's technical potential in renewable energy sources is high; however, once the feasibility and various technical constraints related to developing renewable energy projects are taken into consideration, the economic potential of renewable investments seems limited [38]. As for the fossil fuel sources, the country possesses rich lignite deposit with generally low in calorific values. In addition, there are also hard coal and asphaltite basins that most of them are currently in use. The aforementioned facts about the energy sources of country force decision-makers to apply constraints while making capital allocation. On the other hand, portfolio diversification is another factor making capacity limitations significant in terms of source-based risk aversion while constructing a portfolio.

In general, private investors have a maximum capacity horizon in a certain market. Therefore, constraint 2 is applied to the model to simulate the capacity cap.

9.5.3 Weights

ANP technique is used for defining the weights of the goals in the model. Hierarchical structure of the model is comprised of four subnetworks representing the goals of the model and subnetworks are composed of aforementioned criteria for choosing eight different alternatives.

Pairwise comparisons of the alternatives for criteria in each subnetworks are performed through questionnaires in order to take the judgments of 7 experts. The fundamental scale for pairwise comparison is given in Table 9.6. By taking into account the 1–9 Saaty scale, Tables 9.1, 9.2, 9.3, 9.4, 9.5 are used to construct the pairwise comparison matrix of the alternatives for criteria of all subnetworks. Super8 Decisions software was used for the analysis. Weights calculated through pairwise comparison is presented in Table 9.7.

Table 9.6 Numerical ratings associated with pairwise comparison

Intensity of importance	Definition
1	Equally important
3	Moderately important
5	Strongly important
7	Very strongly important
9	Extremely important
2,4,6,8	Intermediate values
Reciprocals of above	If activity *i* has one of the above nonzero numbers assigned to it when compared with activity *j*, then *j* has the reciprocal value when compared with *i*

Table 9.7 Weights of the goals

Goal	Weights (0–100)
1	48.3
2	24.4
3	16.1
4	11.2

9.6 Results and Discussions

The WGP is run in LINGO to apply Eqs. 9.4, 9.5, 9.6, 9.7. A hypothetical 2000 MW portfolio is constructed and the results are presented in Fig. 9.1. Although the most important criterion for an investor was chosen to be profitability factor by our ANP model, renewable energy sources hold the majority of the portfolio with 63.2 %. Contrary to general belief in the industry, renewable energy sources prove their competitiveness once appropriate incentive framework is provided.

Biomass, hydro and geothermal energy sources seem to get maximum possible capital allocation within constraints applied in the model. Current power generation mix of the country verifies that the importance and benefits of hydroelectricity is well understood by the investors. Since hydroelectricity is a well-developed technology that has a longer history comparing to other renewable power sources, the status of this source in the mix is not surprising. However, it seems that biomass and geothermal energy sources were unable to attract investors' attention as it was indicated by the results of the model. Geothermal energy has lots of advantages in terms of profitability, carbon emissions, and public acceptance; the main obstacles in the way of project development are high upfront investment and irrevocable capital expenditures at site development stage. Different kind of incentive mechanisms can be developed to overcome aforementioned challenges through modifications in the regulatory framework.

The share of biomass in the electricity generation of Turkey is less than even 1 %. The results of our model show that an investor should exploit from all available potential that can be got from this energy source. The reason behind the lack

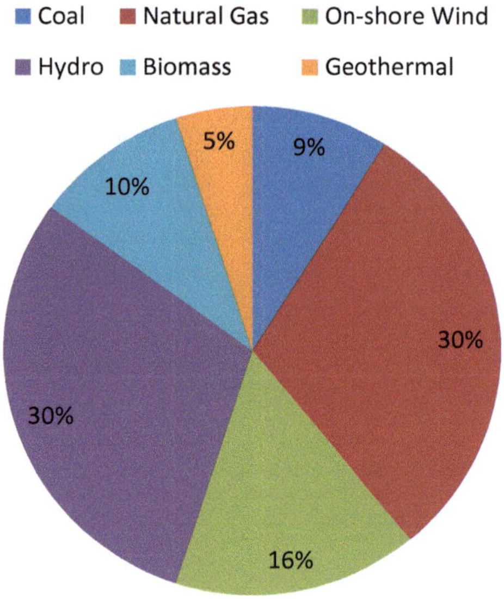

Fig. 9.1 Results generated by the model

of usage of various types of biomass is technical, supply, and quality problems. Given the fact that the problems related to developing biomass energy projects can be easily handled, a significant increase in the share of this source is expected in the near future.

The developed model proposes no capital allocation to off-shore wind and solar power technologies. This is not an anticipated result, since the electricity sale prices for these technologies are considerably lower than their levelized costs. Feed-in tariffs for these technologies should be revised, if private investment is expected.

Interestingly, the share of on-shore wind power plants in the portfolio is low unlike the recent developments on wind energy in the country. Especially in the last 5 years, there has been an unprecedented increase in the wind energy investments due to the popularity of renewable energy in the world; however, the results generated by the model show that investment climate for wind is not better than many energy sources analyzed in this study. The electricity prices defined for wind energy in the feed-in tariff is very close to the levelized cost of this technology, therefore the investors should analyze on-shore wind energy consciously in a more quantitative way.

Fossil fuel power stations' share in the portfolio is about 37 %. Although coal-fired power plants are very competitive in terms of profitability, its high carbon emissions and low social acceptance drop its share lower in the portfolio. Usage of recently developed clean firing technologies can make coal-fired power station investments attractive from investor's point of view. The increase in the investment costs once clean firing technologies used should be subsidized to offset the economic losses of investors.

Almost 50 % of Turkey's electricity generation comes from natural gas-fired power stations. There has been a long debate about the installation of natural gas-fired power plants in the country due to the security of supply and rise in the natural gas prices. Our model suggests to allocate maximum share possible for the natural gas-fired power plants in the portfolio. Despite the aforementioned shortcomings of natural gas for policy makers, natural gas-fired power stations are still important alternatives from an investor's point of view.

9.7 Conclusions and Suggestions

This paper has developed a multiple criteria decision-making model to construct an electricity generating portfolio. Weights of the model are calculated according to the different perspectives by means of the ANP.

The mathematical model developed is implemented to Turkish electricity market as a case study for demonstration purposes, but it is applicable to any market by changing the energy resources and country-based criteria.

One of the major limitations of this model is that all data fed into the model represents today's investment climate. It is widely known that the construction and production costs of all renewable energy technologies are decreasing. It would also be useful to compare future cost projections of renewable energy with cost projections for fossil fuels. However, predicting the price of oil and natural gas has proven even more difficult in the past few years than predicting the cost of renewable energy. Similarly, as technologies mature, they may improve on other critical performance dimensions [39]. Prices of fossil fuels are country specific and for renewable energy technologies, country- and site-specific load factors play an important role, a more precise LCOE calculation can be done in the future for the country used in the case study. Moreover, model can be further developed through incorporation of uncertainties in carbon prices, country risks, exchange risks, security of supply risk, supply and demand projections of relevant market.

The model gives decision-makers a tool to use in making strategic decisions on matters related to energy investments. Clearly, private investors do not have to abide strictly by the results given by this, or any other model, but the present model provides quantitative results that can help improve the decision-making process.

References

1. T. McGovern, C. Hicks, Deregulation and restructuring of the global electricity supply industry and its impact upon power plant suppliers. Int. J. Prod. Econ. **89**, 321–337 (2004)
2. A. Rentizelas, G. Tziralis, K. Kirytopoulos, Incorporating uncertainty in optimal investment decisions. World Rev. Entrep. Manag. Sustain. Dev. **3**(3/4), 273–283 (2007)
3. G. Kumbaroglu, R. Madlener, M. Demirel, A real options evaluation model for the diffusion prospects of new renewable power generation technologies. Energy Econ. **30**, 1882–1908 (2008)
4. D. Bar-Lev, S. Katz, A portfolio approach to fossil fuel procurement in the electric utility industry. J. Finance **31**, 933–947 (1976)

5. S. Awerbuch, M. Berger, *Energy Security and Diversity in the EU: A Mean-variance Portfolio Approach* (International Energy Agency, Paris, 2003)
6. S. Awerbuch, Portfolio-based electricity generation planning: policy implications for renewables and energy security. Mitig. Adapt. Strat. Glob. Change **11**, 693–710 (2006)
7. M. Bazilian, F. Roques, *Analytic Methods for Energy Diversity and Security*. Applications of Mean Variance Portfolio Theory (A Tribute to Shimon Awerbuch, London, 2008)
8. I. Fortin, S. Fuss, J. Hlouskova, N. Khabarov, M. Obersteiner, J. Szolgayova, An integrated CVaR and real options approach to investments in the energy sector. J. Energy Markets. **1**, 61–85 (2008)
9. F.A. Roques, D.M. Newberry, W.J. Nuttall, Fuel mix diversification incentivesin liberalized electricity markets: a mean-variance portfolio theory approach. Energy Econ. **30**, 1831–1849 (2008)
10. A.A. Rentizales, Investment planning in electricity production under CO_2 price uncertainty. Int. J. Prod. Econ. **140** (2), 622–629 (2012)
11. C. Cormio, M. Dicorato, A. Minoia, M. Trovato, A regional energy planning methodology including renewable energy sources and environmental constraints. Renew Sust. Energy **7**, 99–130 (2003)
12. B.F. Hobbs, Optimization methods for electric utility resource planning. Eur. J. Oper. Res. **83**, 1–20 (1995)
13. B.F. Hobbs, P.M. Meier, *Energy Decisions and the Environment—A Guide to the Use of Multicriteria Methods* (Kluwer Academic, Boston, 2000)
14. G. Heinricha, M. Howellsb, L. Bassona, J. Petrie, Electricity supply industry modelling for multiple objectives underdemand growth uncertainty. Energy **32**, 2210–2229 (2006)
15. E. Svensson, T. Berntsson, A.B. Strömberg, M. Patriksson, An optimisation methodology for identifying robust process integration investments under uncertainty. Energy Policy **37**, 680–685 (2009)
16. H.A. Simon, *Models of Man* (Wiley, New York, 1955)
17. M. Tamiza, D. Jonesa, C. Romero, Goal programming for decision making: an overview of the current state-of-the-art. Eur. J. Oper. Res. **111**, 569–581 (1998)
18. M.P. Niemira, T.L. Saaty, An analytical network process model for financial crisis forecasting. Int. J. Forecast. **20**, 573–587 (2004)
19. IEA, OECD, *Projected Costs of Generating Electricity* (IEA, OECD, Paris, 2010)
20. TEIAS: Electricity Generation - Transmission Statistics of Turkey, http://www.teias.gov.tr/istatistik2009/index.htm
21. C. Kearney, Emerging markets research: trends, issues and future directions. Emerg. Mark. Rev. **13**, 159–183 (2012)
22. World Bank: Country classification http://data.worldbank.org/about/country-classifications/country-and-lending-groups#Upper_middle_income
23. M. Tukenmez, E. Demireli, Renewable energy policy in Turkey with the new legal regulations. Renew. Energy **39**, 1–9 (2012)
24. *Law on Utilization of Renewable Energy Sources for the Purpose of Generating Electrical Energy*, http://www.mevzuat.adalet.gov.tr/html/1477.html
25. *Electricity Market Law*, http://www.mevzuat.adalet.gov.tr/html/1130.html
26. TEIAS, *Day-ahead Market Prices*, http://www.teias.gov.tr/
27. S.M. Rashad, F.H. Hammad, Nuclear power and the environment: comparative assessment of environmental and health impacts of electricity-generating systems. Appl. Energy **65**, 211–229 (2000)
28. TAEA, *Sustainable Development and Nuclear Energy* (Turkish Atomic Energy Authority, Ankara, 2000)
29. A. Strupczewski, Accident risks in nuclear-power plants. Appl. Energy **75**(1–2), 79–86 (2003)
30. EURELECTRIC, *Efficiency in Electricity Generation* (Union of the Electricity Industry-EURELECTRIC, Brussels, 2003)

31. N.F. Pidgeon, I. Lorenzoni, W. Poortinga, Climate change or nuclear power—No thanks! A quantitative study of public perceptions and risk framing in Britain. Glob. Environ. Chang. **18**, 69–85 (2008)
32. D. Adam, Parochial energy policy. Nat. Clim. Change (Commentary) **1**, 230–232 (2011)
33. M. Poumadere, R. Bertoldo, J. Samadi, Public perceptions and governance of controversial technologies to tackle climate change: nuclear power, carbon capture and storage, wind, and geoengineering. WIREs Clim. Change **2**, 712–727 (2011)
34. R. Dwivedi, Displacement, risks and resistance: local perceptions and actions in the Sardar Sarowar. Dev. Change **30**, 43–78 (1999)
35. P.E. Akyazı, F. Adaman, B. Ozkaynak, U. Zenginobuz, Citizens' preferences on nuclear and renewable energy sources: evidence from Turkey. Energy Policy **47**, 309–320 (2012)
36. T. Kanamura, A supply and demand based volatility model for energy prices. Energy Econ. **31**, 736–747 (2009)
37. E. Regnier, Oil and energy price volatility. Energy Econ. **29**, 405–427 (2007)
38. MENR, *Ministry of Energy and Natural Resources*, www.etbk.gov.tr
39. T.U. Daim, G. Kayakutlu, K. Cowan, Developing Oregon's renewable energy portfolio using fuzzy goal programming model. Comput. Ind. Eng. **59**, 786–793 (2010)

Chapter 10
Energy Efficiency Technologies: Pacific NW US Case

Ibrahim Iskin and Tugrul U. Daim

Abstract This chapter presents a number of energy efficiency technologies that currently make up energy efficiency technology inventory of Bonneville power Administration (BPA). In Sect. 10.2, you will be presented with regional efforts toward uncovering the next generation of energy efficiency technologies along with a technology management framework that is currently being used in the region. In Sect. 10.3, you will also be presented with a list of prior energy efficiency technologies that has come out as an output of aforesaid framework and the full list of technologies that are currently being assessed.

10.1 Energy Efficiency Technologies Under Deployment in the Pacific Northwest

Energy efficiency programs pursued by Bonneville Power Administration support diffusion of wide range of energy efficiency technologies in the Pacific Northwest. These technologies can be grouped into eight service areas which are lighting, water heating, heating, ventilating and air conditioning (HVAC), commercial kitchen and food service, envelope, new home construction, industrial and agriculture, others. Each service area has multiple technologies or technology areas that are proven to be more efficient than baseline alternatives. Please refer to Table 10.1 to see full list of technologies/technology areas that will be presented in this section.

I. Iskin (✉)
Portland State University, Portland, OR USA
e-mail: ibrahimiskin@gmail.com

T.U. Daim
Department of Engineering and Technology Management, Portland State University, 1900 SW 4th Avenue 50-02, Portland, OR 97201, USA
e-mail: tugrul@etm.pdx.edu

T.U. Daim et al. (eds.), *Policies and Programs for Sustainable Energy Innovations*, Innovation, Technology, and Knowledge Management,
DOI 10.1007/978-3-319-16033-7_10

Table 10.1 Existing energy efficiency technologies under respective focus areas

Focus area	Technology/technology area
Lighting	Compact fluorescent lamps
	Light emitting diodes
Water heating	Heat pump water heaters
	Electric storage water heaters
	Gravity film heat exchangers
	Pipe insulation materials
HVAC	Ductless heat pumps
	Air source heat pumps
	Geothermal heat pumps
	Line voltage thermostats
	Web-enabled thermostats
	Duct sealing
Commercial kitchen and food service	Commercial refrigerators
	Commercial freezers
	Ice makers
	Steam cookers
	Hot food holding cabinets
	Combination ovens
	Convection ovens
	Electric fryers
	Dishwashers
	Prerinse spray valves
	Clothes washers
Envelope	Insulation materials
	Air sealing materials
	Window materials
New home construction	
Industrial and agriculture	Variable frequency drives
	Irrigation hardware
Other energy efficient technologies	Low-flow shower heads
	Smart power strips

10.1.1 Energy Efficient Lighting Technologies

Energy efficiency programs pursued by Bonneville Power Administration has been supporting two lighting energy efficiency technologies which are compact fluorescent lamps (CFLs) and light emitting diode lamps.

BPA has been supporting the diffusion of CFLs in a variety of programs dedicated for various market segments. For instance, BPA Change a Light program allows utilities to make bulk purchases and directly ship high-quality CFLs to residential customers. Specialty CFL and hard to reach CFL program supports distribution of specialty CFL bulbs by providing reimbursements whereas Direct Install

Energy Star CFLs program provides reimbursements to Energy Star qualified CFL bulbs and fixtures. Industrial, commercial, and agricultural lighting efforts are provided through Commercial and Industrial lighting program which provide reimbursements for standard T8 fluorescent lamps and custom projects. BPA also supports the diffusion of LED lighting in a number of market segments such as exit signs and retrofit kits and traffic lights. Moreover, there are several demonstration projects aiming to uncover potential of light emitting diodes (LEDs) in different market segments such as street lighting project by Seattle City Light, decorative lighting by Jordan Schnitzer Museum of Arts and Bonneville Power Administration Headquarters [1].

10.1.1.1 Compact Fluorescent Lamps

CFLs have been accounting for significant part of energy savings accomplishments in the Pacific Northwest region, and there still remains considerable amount of savings potential.

Fluorescent lamps produce light through fluorescence using a mixture of argon and mercury gas stimulated by an electric current. Ultraviolet rays produced by the interaction strike a coat of fluorescent phosphor coated on the interior surface of the bulb and produce visible light [2]. CFLs are smaller (size of a standard incandescent light bulb) versions of conventional fluorescent lamps that have been used in offices, factories, stores, and schools for over 60 years. CFLs typically have 3–4 times more luminous efficacy, saving around 75 % energy and approximately six times more lifetime compared to incandescent light bulbs. CFLs can function with both AC and DC electrical power and can easily be used with incandescent light bulb fixtures, making it easier to install in old type fixtures. Due to use of mercury in CFLs, environmental concerns have been raised by the public. As a response, programs for disposal of CFLs have been established for environmental safety purposes.

10.1.1.2 Light Emitting Diodes

LED applications have widely spread in electronic devices such as electronics signs, clocks, televisions, monitors, laptops, flashlights, etc. LED technology has recently started diffusing into other markets especially in the areas of street lighting, traffic lights, automobiles, and decorative lighting.

LED technology in definition refers to a semiconductor diode that emits visible light when voltage is applied to it [3]. Typical LEDs can produce several colors which are red, green, and blue; however, recent developments have enabled them to produce white light as well. R&D efforts are still underway to exploit full potential of LED technology. A typical LED lamp has longer life span and lighting efficacy than its substitutes. For instance, LED lamps have approximately six times longer lifetime and two times better lighting efficacy than standard CFLs. LED lamps only work with DC electrical power and need special fixtures of its own for operation, making it harder to replace older lighting technologies.

10.1.2 Energy Efficient Electric Water Heating Technologies

There are four energy efficient electric water heating technologies that Bonneville Power Administration has been providing incentives for. For instance, these technologies are heat pump water heaters, electric storage water heaters, gravity film heat exchangers, and pipe insulation materials.

Deployment of heat pump water heaters is currently being supported through Energy Smart labeling programs; however, efficiency gains through heat pump water heater technology have been falling behind the expectations due to lack of major US manufacturers. As of 2009, supply chain problems have been resolved and diffusion of heat pump water technology is expected to uptake [4]. Moreover, field tests and demonstration projects are also underway to enable BPA reimbursements. Since electric storage water heating units with low initial costs are more preferable by the end users, BPA has been providing reimbursements for more efficient electric storage water heaters whose cost of ownership is cheaper in the long run. BPA has been providing reimbursements for qualified electric water storage units with various tank sizes in both residential and commercial sectors. BPA provides reimbursements for installation of gravity film heat exchangers in multifamily and single family end use in the residential sector. BPA also supports diffusion of insulation materials that have a minimum of R-3 rating in residential sector.

10.1.2.1 Heat Pump Water Heaters

According to Energy Star, replacing standard water heaters with heat pump water heaters can save average households $300 per year on electric bills, having potential to avoid 19.6 million pounds of carbon dioxide (emission equivalent of 1.6 million cars) in case of full diffusion.

Heat pump water heaters work opposite way to the refrigerators do. Heat pump water heaters take heat from the surrounding air and use it to heat up water. Heat pump water heaters use liquid refrigerant that is vaporized by heat pump's vaporizer and passed through the compressor. As the pressure is increased in the compressor chamber, refrigerant temperature is increased; heated refrigerant is sent through the condenser coil that circles around the storage tank. As the hot refrigerant completes its cycle, heat is transferred to the water and refrigerant cools down and condenses into liquid form again. By running the condensed refrigerant through the expansion valve, the pressure is reduced and the cycle is restarted again [5].

10.1.2.2 Electric Storage Water Heaters

According to Energy Star [6], depending on the family size, heater location, and the size, high efficiency water heaters can operate with 10–50 % less energy than baseline models in the market.

Storage water heaters can work on various fuel resources such as natural gas, propane, fuel oil, and electricity. Natural gas and propane-fueled water heaters operate similarly by using a gas burner that heats the water tank from below whereas oil-fueled water heaters use a mix of oil and air in a vaporizing mist and ignites the mixture with an electric spark. Electric water heaters use two electric resistance units. One of the units is kept at the bottom of the tank, maintaining the minimum temperature requirement, whereas the upper unit is used to provide hot water whenever demand occurs. Storage water heaters usually have tank size of 20–80 gal, operating by releasing hot water at the times of demand and refilling the tank with cold water to reheat for the next use. Electric storage water heaters use thermostats to keep the water within the specified temperatures. Since water is constantly heated in the tank ready for use, energy can be wasted in case of long duration hours. In order to eliminate the standby loss, insulation of the water storage tank becomes significant. For instance, DOE recommends water tanks with thermal resistance of R-12 to R-25 [7].

10.1.2.3 Gravity Film Heat Exchangers

Showers, tubs, sinks, dishwashers, and clothes washers are some of the use areas where hot water is required. According to DOE, every year 350 billion kWh worth of hot water is wasted through the drains whereas large portion of this energy could be reused. Gravity film heat exchangers have been developed under a grant provided by the DOE Inventions Program, aiming to reclaim heat from waste water. Although savings differ depending on the hot water consumption patterns, installation specifications, and other variables, it is expected that gravity film heat exchangers can save 30–50 % energy from heat required from shower use [8].

In simple terms, gravity film heat exchanger is a vertical heat exchanger design that extracts heat from the drained water and applies the reclaimed heat to preheat cold water that is entering the building. The design consists of a central pipe and thinner copper pipes that are circling around it. As the hot waste water travels through the central pipe, cold water is circled through the copper coil and heated as a result of heat exchange. Due to inability to store waste water, gravity film heat exchangers are not suitable for heating water that is to be used for later. This technology is the most efficient where there is production of hot waste water and need for hot water simultaneously such as showers [8].

10.1.2.4 Pipe Insulation

According to DOE [9], insulated pipes can reduce heat loss and increase water temperature by 2–4° compared to an uninsulated pipes. Benefits of pipe insulation can be observed through a number of ways such as longer pipework lifetime due to improved condensation control, avoided pipe freezing, less noise, reduced health hazards, reduced water consumption, and energy loss.

Pipes may operate at temperatures that are very different than ambient temperature. Because of the temperature differences between the ambient air and the pipe, amount of heat loss might be significantly high. Thermal pipe insulations are used to cover pipework in order to increase thermal resistance and reduce heat flow between the environment and the pipe. Depending on the environment the pipework is located, type and thickness of the insulation material need to be chosen carefully. For instance, there are a number of insulation methods that can be used such as spiral-wrap fiberglass, self-sealing foam tubing, fiberglass shell pipe covers, and asbestos insulation [10]. Performance of insulation materials is characterized with R values associated with them. For instance, the greater the R value the better efficiency an insulation material can provide.

10.1.3 Energy Efficient Heating, Ventilating, and Air Conditioning Technologies

In the area of HVAC, there are five technologies that BPA has been developing programs for. These technologies are ductless heat pumps (DHPs), air source heat pumps, geothermal heat pumps, duct sealing materials, electronic line voltage thermostats, and web-enabled programmable thermostats.

BPA has been providing reimbursements for ductless heat pump units in both residential and commercial sectors. BPA has completed a couple of research efforts focusing on field testing of DHPs in single family homes and laboratory test of some commercial ductless heat pump products. Current research efforts have been focused on field tests of DHPs in residences and small commercial buildings, and evaluation of pilot incentives program for DHPs in single family homes [11]. BPA also provides incentives for a number of different air source heat pump applications in residential sector. BPA's completed research efforts consist of two field tests for cold climate heat pumps with staged compressor in an attempt to understand potential of this technology in the Pacific Northwest [12]. Another HVAC technology in promotion is geothermal heat pumps whose one of the drawbacks is its high initial costs which can be several times more expensive than air source heat pump systems. In order to mediate the high initial cost barrier, BPA has been providing reimbursements for installation of geothermal heat pump units in the residential sector single family homes end use. BPA has completed a field test assessing potential of web-enabled thermostats in classrooms and started providing reimbursements. Accordingly, BPA is providing reimbursements for increasing the diffusion web-enabled programmable thermostats in modular classroom end use. In addition to that, BPA also provides reimbursements for use of line voltage electronic thermostats in residential sector end use as well. Duct sealing is also an area of BPA's energy efficiency efforts. BPA provides reimbursements for duct sealing efforts in residential sector manufactured homes, existing and new single family homes end uses.

10.1.3.1 Ductless (Mini Split) Heat Pumps

DHPs have been installed in existing commercial buildings for over 20 years in different parts of the world, and recently diffusing into the US market. In simple terms, a ductless heat pump system is a heating and cooling unit that functions without using conventional ducts. Thus, DHPs are reported to be 25–50 % more efficient than baseline technologies due to eliminated heat loss from ducts and on/off cycling losses [13].

Conventional heat pumps usually have a single indoor unit consisting of a refrigerant coil and an air handler, and a single outdoor unit consisting of a condenser and a compressor. Conventional heat pumps serve an entire house by cooling the air at the evaporator coil and distribute it using ductwork. Ductless heat pump systems usually have one outdoor unit serving multiple indoor units, each containing a refrigerant coil and blower. Refrigerant is distributed around the place using a small diameter insulated refrigerant lines which are linked to each zone. Each individual evaporator unit blows the cooled or heated air through a fan [14]. Moreover, DHPs use variable speed (frequency) drives which align the pump speed with the demand, optimizing the energy use. Variable frequency drive technology is discussed more in detail in the following sections due to its various applications.

One of the biggest nonenergy benefits of DHPs is its ability to provide simultaneous heating and cooling at the same time for multiple zones, increasing occupant comfort levels. DHPs also do not require as invasive ductwork as the baseline technologies, making it cheaper and easier to install in both existing and new constructions. DHPs also provide more flexibility in interior design considerations through its modern looking design and ability to be mounted on different parts of the houses such as ceilings, walls, etc. Furthermore, DHPs operate at lower noise levels than the baseline technologies and can be operated using a remote controller [15].

10.1.3.2 Air Source Heat Pumps

Technological developments such as more precise refrigerant control, variable speed blowers, improved coil design, electric motor and dual speed compressors, and better copper tubing design have enabled new generation air source heat pump units to operate one-and-a-half to two times more efficient than those technologies available 30 years ago. For instance, when installed properly, air source heat pumps can provide one-and-a-half to three times more heat energy than the electricity it consumes [16]. (Please note that, aforementioned statement is scientifically correct since heat pumps do not convert energy, but they move energy from one place to another.)

There are two common types of heat pumps which are air source heat pumps and geothermal heat pumps. Air source heat pumps work using the same principle as the refrigerators do. An air source heat pump system consists of a compressor and two copper coil units enhanced with aluminum fins for greater heat transfer. One of the coils remains inside and the other one remains outside. Air source heat pumps force liquid refrigerant to outside coils in order to extract heat from the air. As the liquid temperature increases it turns into gas which is forced toward the indoor coil where heated refrigerant gas releases heat and condenses into liquid form again. Using a reversing valve the refrigerant is sent back to outside coil and the cycle is restarted. In the cooling mode; refrigerant fluid is evaporated in the indoor coil, extracting heat from the indoor environment. Refrigerant gas is compressed to increase its pressure in order to allow the gas to condense at high outside temperatures. Pressurized gas is passed through the outside coil where it releases heat to outside air and condenses into liquid form again. Similar to heating mode, the cycle is repeated. Using the same principles there are a number of advanced technologies which are under development. These technologies are named as reverse cycle chillers, cold climate heat pumps, and all climate heat pumps [16].

Technically, there is always heat available in the space as long as the temperature is above absolute 0 K. Air source heat pump efficiency has been observed to decrease in cold climates where extracting heat from the environment starts requiring more energy. For instance, as the outside temperature goes below 40 °F, less efficient electric resistance coils are set functional to provide indoor heating. Although air source heat pump technology can be used in most part of the United States, it is the most efficient in milder climates where summers are not extremely hot and winters are not extremely cold [17].

10.1.3.3 Geothermal Heat Pumps

Geothermal heat pumps have been available in the market since 1940s. Working principle of geothermal heat pumps is very similar to that of air source heat pumps with a slight difference in heat exchange medium. Geothermal heat pumps use earth as the heat exchange resource whose temperature fluctuations are relatively negligible compared to outside air. For instance, depending on the latitude ground temperatures range between 45 and 75 °F; this is usually warmer than a regular winter day and colder than a summer day. This feature allows geothermal heat pump efficiencies to be significantly high in cold environments. For instance, geothermal heat pumps can reach 300–600 % efficiency levels whereas the efficiency of air source heat pumps ranges between 175 and 250 % [18].

Geothermal heat pumps are also referred as geoexchange, earth coupled, ground source, or water source. It should be noted that geothermal heat pumps are different than conventional geothermal power which uses high temperature as heat source to generate electricity. Similar to air source heat pumps, geothermal heat pumps extract heat from the outside environment and transfer it to a specified place, giving it both cooling and heating capability. Depending on the climate and soil conditions different types of geothermal heat pump technologies are available. It is also possible to combine different types of geothermal heat pumps as well as air source heat pumps for improved efficiency.

Benefits of geothermal heat pumps are that they operate quieter and more efficiently along with having longer lifetime (20+ years) and need for little maintenance compared to air source heat pumps. Furthermore, less dependency on outside air also makes them more reliable in terms of performance and energy savings. Ground source heat pumps are also effective in maintaining humidity levels of places, making it a very effective choice in humid areas [19].

10.1.3.4 Line Voltage (Electronic) and Web-Enabled Programmable Thermostats

Rooftop HVAC units, mostly operated by single zone thermostats, are currently operational in around 40 % of the commercial building spaces. Field tests have shown that most of the existing thermostats provide minimum energy control features which waste energy due to poorly executed fan scheduling, and cooling and heating setpoints. Web-enabled smart thermostats are believed to have 10–60 % savings potential in the Pacific Northwest.

A thermostat is a device that controls heating and cooling units in order to maintain a constant temperature specified by the end users. BPA is currently supporting diffusion of two different types of thermostats which are line voltage electronic and web-enabled programmable thermostats.

Conventional thermostats were simple electromechanical devices intended to capture heating and cooling orders manually by the users. As a result of technological developments in communication technologies and electronics, manufacturers have started integrating better control and communication features. For instance, web-enabled thermostats use internet, mobile, land lines, and similar communication protocols to establish remote access and allow multiple zones to be monitored and controlled from a single controller unit. This feature is believed to be very useful in small commercial buildings with split systems where heating and cooling needs of multiple zones can be managed from a remote place leading to reduced onsite maintenance staff cost. Web-enabled thermostats are also proven to be useful in supporting utility demand-side management programs by enabling utilities to reduce cooling loads of participating buildings at peak load times [20].

Line voltage thermostats are the most commonly used thermostat types with electric baseboard heating units. Line voltage thermostats are wired to the heating and cooling units giving them ability to switch the devices on and off. While conventional line voltage systems were mostly mechanically operated, recently electronically controlled thermostats became widely available in the market. Different than mechanical substitutes, electronically controlled line voltage thermostats can be programmed by the users to match variable heating and cooling needs throughout the day.

10.1.3.5 Duct Sealing

According to National Association of Home Builders, repairing poorly functioning ductwork may save 15–30 % of total energy spent on heating and cooling houses, having savings potential of five billion dollars annually in the United States [21]. Furthermore, on the user level, well insulated ductwork can save an average house several hundred dollars.

Duct systems are branching network of tubes that carry air from furnaces and central air conditioners to each room. Duct systems are made of sheets of metal, fiberglass, and other materials [22]. Due to aging and environmental conditions, ducts may start functioning poorly in various ways such as leaking, broken parts, misaligned joints, etc. Depending on the problem with the ducts system, there are different ways of sealing ducts such as tapes, mastic, and a recently developed aerosol sealants. Tapes are efficient in repairs where mechanical strength is required whereas mastic sealants are useful in situations where flexibility is critical such as joint points.

Invented by Lawrence Berkeley National Lab in 1987, aerosol sealants are relatively newer than the previous alternatives. Aerosol sealant is in the form of gas which is pumped through the duct systems, fixing open spots up to an inch in diameter. Principle of aerosol sealing technology is that the gas mixture contains both air and sealant particles. As the mixture is pushed through the ductwork, air speeds up and turns sharply from the leaky points where it is forced out; however, bigger sealant particles cannot change direction as fast as air and end up clumping around these points. As the process is continued for enough time, duct system is sealed without leaving leaky points. Aerosol sealants are found to be cost effective than the other methods due to its ease of application, reducing labor-intensive nature of the ductwork repairs [23].

Apart from energy savings and indirect impacts on reduced carbon dioxide emission, duct sealing also provide a number of benefits. Well insulated ductworks can improve comfort problems such that some of the rooms cannot be heated or cooled easily. Duct sealing can also improve indoor air quality and reduce health hazards by preventing pollutants such as insulation particles, chemicals from entering duct system. Furthermore, poorly functioning ducts may cause backdrafting problem which is referred to the situation when combustion gases such as carbon monoxide, carbon dioxide, etc., are drawn back to the living space rather than outside. Sealing ducks can prevent such incidents from happening and reduce health risks [24].

10.1.4 Commercial Kitchen and Food Service Energy Efficient Technologies

Bonneville Power Administration supports a number of commercial kitchen and food service measures. These measures include commercial refrigerators, freezers, ice makers, steam cookers, hot food holding cabinets, combination and convection ovens, electric fryers, dishwashers, prerinse valves, and clothes washers.

It should be noted that it is difficult to classify these energy efficiency measures using technologies, but end-use areas, performance levels, and metrics. Thus, in this section there will be brief explanation of what each end use means and what performance levels are qualified for reimbursements. For instance, BPA uses Energy Star and Consortium for Energy Efficiency (CEE) ratings for defining measure specifications.

10.1.4.1 Commercial Refrigerators

Commercial refrigerator is defined as "a refrigeration cabinet designed for storing food products at temperatures above 32 °F but no greater than 40 °F and intended for commercial use" [25]. Commercial refrigerators are distinguished into three main types which are vertical solid door refrigerators, vertical glass door refrigerators, and chest refrigerators. BPA reimburses commercial refrigerators that have performance levels equal or greater than Energy Star 2.0 performance level.

10.1.4.2 Commercial Freezers

According to definition of ASHRAE, commercial freezer is defined as "a refrigeration cabinet designed for storing food products at temperatures of 0 ° F and intended for commercial use" [25]. Commercial freezers have three main types which are vertical solid door freezers, vertical glass door freezers, and chest freezers. BPA reimburses for commercial freezers that have performance levels equal or greater than Energy Star 2.0 performance level.

10.1.4.3 Ice Makers

According to definition of AHRI, commercial ice maker is defined as "factory-made assembly, including a condensing unit and ice-making section operating as an integrated unit, with means for making and harvesting ice. It is an assembly that makes no less than 50.0 lb of ice per day and up to 4,000.0 lb of ice per day" [26]. Technology basically turns water into ice by removing heat through condenser units. Air cooled and water cooled freezers are the two main types of ice makers available in the market. BPA reimburses air cooled commercial ice makers that have performance levels equal or greater than Energy Star 2.0 or CEE Tier 3 performance levels.

10.1.4.4 Steam Cookers

A commercial steamer is defined as "a device with one or more food steaming compartments in which the energy in the steam is transferred to the food by direct contact. Models may include countertop models, wall-mounted models and floor-models mounted on a stand, pedestal or cabinet-style base" [27]. Steam cookers are split into two main categories which are, namely, pressureless/convection steamers and pressure steamers. BPA reimburses commercial steamers that have performance levels equal or greater than CEE Tier 1A and 1B performance levels.

10.1.4.5 Hot Food Holding Cabinets

Defined by CEE, commercial hot food holding cabinet is "an appliance that is designed to hold hot food at a specified temperature, which has been cooked using a separate appliance" [28]. There is a number of hot food holding cabinets available in the market with different sizes and mobility features. BPA reimburses commercial hot food holding cabinets that have performance levels equal or greater than CEE Tier 2 performance level.

10.1.4.6 Combination Ovens

A commercial combination oven is defined as "an oven that combines the function of hot air convection (oven mode) and saturated/superheated steam heating (steam mode), or both (combi mode), to perform steaming, baking, roasting, rethermalizing, and proofing of various food products" [29]. BPA reimburses electric combination ovens that have cooking efficiency of 70 % or greater and an idle energy rate of 3.5 kW or less.

10.1.4.7 Convection Ovens

According to definition of CEE, convection oven is defined as "a general-purpose oven that cooks food by forcing hot dry air over the surface of the food product. The rapidly moving hot air strips away the layer of cooler air next to the food and enables the food to absorb the heat energy" [29]. BPA reimburses commercial electric convection ovens that have performance levels equal or greater than Energy Star performance level.

10.1.4.8 Electric Fryers

An electric fryer is defined as "an appliance, including a cooking vessel, in which oil is placed to such a depth that the cooking food is essentially supported by displacement of the cooking fluid rather than by the bottom of the vessel. Heat is delivered to the cooking fluid by means of an immersed electric element or band-wrapped vessel (electric fryers), or by heat transfer from gas burners through either the walls of the fryer or through tubes passing through the cooking fluid" [30]. There are four main types of fryers available in the market. These are open pot fryers, flat bottom fryers, pressure/kettle fryers, and large vat fryers. BPA reimburses commercial electric fryers that use less than 10,000 kWh per year and have been approved by Regional Technical Forum (RTF).

10.1.4.9 Dishwashers

A dishwasher is defined as “A machine designed to clean and sanitize plates, glasses, cups, bowls, utensils, and trays by applying sprays of detergent solution (with or without blasting media granules) and a sanitizing final rinse” [31]. There are many different types of commercial dishwashing machines; however, four main types are undercounter, door type, conveyor, and flight type dishwashers. BPA reimburses commercial dishwashers that meet or exceed RTF efficiency standards.

10.1.4.10 Prerinse Spray Valves

Defined by CEE, a prerinse spray valve is “a handheld device that uses a spray of water to remove food waste from dishes prior to cleaning in a commercial dishwasher. Pre-rinse spray valves consist of a spray nozzle, a squeeze lever that controls the water flow, and a dish guard bumper” [32]. BPA supports diffusion of prerinse spray valves with flow rate of 0.65 gal per minute or lower in electrically heated dishwashing water facilities where there are ten or more meal shifts per week.

10.1.4.11 Clothes Washers

A washing machine is defined as a machine designed to wash laundry using water as the primary cleaning solution. Washing machines are primarily categorized in two types which are front loaded and top loaded. Conventional washing machines which are also referred as top loaded use a large vertical drum to soak and wash the laundry; however, it has been reported that horizontal axis (front loaded) washing machines use 50 % less energy and 18 gal less water [33]. BPA reimburses commercial clothes washers that have performance levels equal or greater than Energy Star performance level and are used in multifamily or commercial laundry areas.

10.1.5 Energy Efficient Envelope Technologies

Similar to commercial kitchen and food service measures, envelope measures are referred with the application areas and performance levels. Envelope measures supported by BPA can primarily be split into three categories which are insulation, air sealing, and windows measures.

BPA provides reimbursements for insulation projects in low income, single family, multifamily, manufactured homes, small office, and retail end uses. Projects can include attic, wall, and floor insulation. Air sealing projects are also qualified for BPA reimbursements. Qualified applications need to be in standard and low income housing end uses. For residential end use, BPA provides reimbursements for replacement of metal framed, single or double pane prime windows with substitutes rated by National Fenestration Rating Council. BPA also provides reimbursements for windows replacement projects in small commercial end uses with electric heating. Replacing windows need to be rated with U value of 0.30 or lower, and patio doors need to be rated with 0.35 or lower.

10.1.5.1 Insulation Materials

Heat naturally moves from warmer to cooler spaces. For instance, in winter, heat flows from heated spaces to adjacent unheated spaces such as attics, garages, basements, ducts, walls, floors, etc. Similarly, in summers, outside heat flows inside the cooler living spaces where cooling units are in operation. Both of the aforementioned situations cause decrease in efficiency and increase in energy consumption. Insulation materials increase resistance to heat flow in places, lowering heating and cooling costs [34]. Performance levels of insulation materials are referred with R values; the higher the R value an insulation material has, the greater the heat resistance it shows. Depending on what part of the house to be insulated, there are various forms of insulation materials available such as blanket, concrete block, foam board, insulating concrete forms, loose fill, reflective, rigid fiber board, sprayed foam, and structured insulated panels.

10.1.5.1.1 Blanket Insulation

Blanket insulation is the most common insulation type that is in the form of batts and rolls. Blanket batts and rolls may be made of various materials such as fiberglass, mineral wool, plastic, and natural fibers. They are the most useful in applications such as unfinished walls, foundation walls, floors, and ceilings [35].

10.1.5.1.2 Concrete Block Insulation

Insulated concrete blocks consist of two attached concrete blocks which have insulation material in between. Insulation material in between the concrete walls may be made of polystyrene, polyisocyanurate, and polyurethane. Concrete block insulation is suitable for insulating unfinished walls, foundation walls, new construction projects, and major renovations [35].

10.1.5.1.3 Foam Board Insulation

Foam board/rigid panel type insulation is a block of insulation material that can be used to insulate any part of the house. Foam board insulation materials have good thermal resistance that can reduce heat conduction through structural elements. Foam board insulation may be made of polystyrene, polyisocyanurate, and polyurethane. Foam board insulation materials are the most efficient in applications such as unfinished walls, foundation walls, floors and ceilings, and unvented low slope roofs [36].

10.1.5.1.4 Insulating Concrete Forms

Insulating concrete forms are made of insulation foam structure with a hallow inside which is filled with concrete. Insulating concrete forms have high thermal resistance and are easy to interconnect with other panels. Insulating concrete forms can be in the form of foam board and blocks. Best application areas for this type of insulation are unfinished walls, foundation walls, and new construction projects [37].

10.1.5.1.5 Loose Fill Insulation

Loose fill insulation is a material that is a mix of cellulose, fiberglass, mineral wool, and other materials. Loose fill insulation's advantage is that it can be applied to any space without disturbing the structural unity, making it a good alternative for retrofit applications. This type of insulation material is the best in filling cavities of enclosed or open new walls, unfinished attic floors, and hard to reach places [38].

10.1.5.1.6 Reflective Insulation

Reflective insulation systems consist of aluminum foils which are supported by a variety of materials such as kraft paper, plastic film, polyethylene bubbles, and cardboard. Reflective insulation systems' heat resistance depends on the direction of the heat flow. For instance, they are the most effective in preventing downward heat flows. They are most useful in applications such as unfinished walls, ceilings, and floors [39].

10.1.5.1.7 Rigid Fiber Board Insulation

Rigid fiber board insulation materials are made of resin bonded inorganic glass fibers and mineral wool. Rigid fiber board insulation is mostly used in insulating air ducts and areas where there is exposure to excess heat [40].

10.1.5.1.8 Sprayed Foam Insulation

Sprayed foam insulation can be sprayed, injected, or poured in place when the application takes place. This type of insulation has ability to fill in smaller cavities and open areas with better insulation resistance than batt insulation types. Sprayed foam insulation can be made of variety of materials such as cementitious, phenolic, polyisocyanurate, and polyurethane. They are the most efficient when applied to cavities of enclosed or open new walls and unfinished attic floors [41].

10.1.5.1.9 Structural Insulated Panels

Structural insulated panels are made of two oriented strand boards or similar structural materials which have thick foam boarding in between. Most common types of insulation materials are made from expanded polystyrene or polyisocyanurate. Structural insulated panels are reported to be better than most of the conventional insulation methods and also reduce outside noise as well. Unfinished walls, ceilings, floors, and new construction applications are some of the best areas for use of this insulation material [42].

10.1.5.2 Air Sealing Materials

Air sealing methods and materials aim to prevent air leaks around the house such as windows, cracks, doors, etc. Air sealing can provide significant savings as well as increasing comfort and durability of homes. Materials for air sealing can be divided into two main categories such as caulking and weatherstripping. Below, you will be provided brief information about each of the methods.

10.1.5.1.1 Caulking

Caulking materials and methods are suitable for places where there is need for flexibility. For instance, these methods are suitable for sealing cracks, gaps, or joints that are smaller than one quarter inch in diameter. If applied around the faucets, ceiling fixtures, water pipes, drains, bathtubs, etc., caulking can also prevent water damage as well. Depending on the application area, material type, strength tolerance, etc., there are many types of caulking materials available. For instance, there are silicone, polyurethane spray foam, water-based foam, butyl rubber, latex and oil, or resin-based sealants [43].

10.1.5.1.2 Weatherstripping

Weatherstripping materials and methods are suitable for places where there are movable joints such as windows and doors. Depending on the application area, weather, friction, temperature changes, mechanical resistance, visibility, etc., there are different types of weatherstripping materials available. For instance, these are tension seal, felt, tape, rolled or reinforced vinyl, door sweep, magnetic, tubular rubber and vinyl, reinforced silicone, door shoe, bulb threshold, frost brake threshold, and interlocking metal channels [44].

10.1.5.3 Window Insulation

Energy flow through windows can happen in three different ways which are nonsolar heat losses and gains (convection and conduction), solar heat gains in the form of radiation and airflow. Nonsolar heat flow occurs due to temperature differences between outside and inside of the buildings. Resulting heat flow causes energy loss and inefficient use of heating or cooling resources depending on the season. Energy efficiency performance of windows is rated with U factor which is a measure of the rate of nonsolar heat flow through a window. For instance, the lower U factor (higher R value) a window has the more heat flow resistant it is [45].

There are fundamentally three ways to increase energy efficiency of window products. For instance, first method involves in changing the chemical and physical composition of the glass material or use of coating materials on the surface of the windows; second approach is to use less conductive gases between layers of window panes; whereas the third option is to reduce heat conductivity properties of the window frames [46].

10.1.5.1.1 Low-E Coatings

Low emittance technology is relatively a new coating technology which contains invisible, very small metal or metallic oxide deposited on a glazing surface. Low emittance technology reduces the U factor by reflecting radiative heat flow. Depending on the amount of light to be led through the windows surface and energy efficiency performance, intensity of the low-e coatings can be determined. For instance, there are three main types of low-e coatings which are named as low, medium, and high solar gain windows in an order of increasing R values [47].

10.1.5.1.2 Gas Fills

Another improvement that can be done to increase energy performance of the windows is the gas filling technology. Principle of this technology depends on the characteristics of the gas between the panes. For instance, by filling the space with gases which are less conductive, more viscous, or slow moving, conduction through the gas is minimized and heat transfer is reduced. For instance, there are two main types of gasses which are argon and krypton as well as mix of two. In applications where the thickness of the windows is important krypton is preferred, whereas argon proves inexpensive solutions in conventional uses of windows [48].

10.1.5.1.3 Frame Types

Frames account for 10–30 % of the total surface area of the windows, thus they constitute significant potential in improving energy efficiency. Material used to construct the structure of the window not only impacts the physical characteristics and look of the windows, but also thermal resistance of the windows as well. Due to structural properties of aluminum, it has become an important frame material since 1970s; however, its high conductive feature reduces the energy efficiency of windows. In order to address similar issues associated with windows frames, a number of technologies have been developed. One technological approach is to replace aluminum or similar highly conductive materials with nonmetal materials such as wood, wood clad, vinyl, and hybrid. In order to further enhance energy efficiency properties of the nonmetal window frames, fiberglass, thermoplastics, composite materials, thermally improved wood, and vinyl are used as well [49].

10.1.6 Energy Efficient New Home Construction Technologies

There is no specific technology associated with energy efficient homes, but combination of individual technologies. Majority of these technologies have widely been mentioned earlier. According to Energy Star website, Energy star qualified houses are at least 15 % more energy efficient than homes built in compliance with 2004 International Residential Codes [50]. Efficiency areas associated with energy efficient homes can be grouped under five categories which are effective insulation, high performance windows, tight construction and ducts, efficient heating and cooling equipment, and energy efficient lighting and appliances [51]. Energy efficient homes are promoted through labeling programs such as Energy Star. In order for an energy design to be labeled as energy efficient, Energy Star conducts energy tests on homes that are submitted for accreditation.

BPA supports a number of energy efficient new home construction designs. For instance, BPA provides reimbursements for electrically heated manufactured homes which are certified by Northwest Energy Efficient Manufactured Homes Program (NEEM), Eco-Rated Homes, or Energy Star. BPA also provides reimbursements for site built homes which are certified by Northwest Energy Star Home Standards, electrically heated homes compliant with Montana House specifications, as well as new residential multifamily constructions compliant with RTF specifications.

10.1.7 Energy Efficient Industrial and Agriculture Technologies

Most of the industrial energy efficiency efforts at BPA are carried out through custom projects approach due to difficulty of developing deemed or calculated measures. Custom project approach requires data collection before and after energy efficient equipment implementation, and base savings between the two. Furthermore, depending on the industry end use, there are many types of technologies which can be applied. As a result, it has been observed that energy efficiency efforts are most of the time not based on the technologies, but the savings and specific projects which involve implementation of multiple technologies. In this section, two of the most significant industrial energy efficiency technologies, variable frequency drives (VFD) and irrigation hardware will be presented.

10.1.7.1 Variable Frequency Drives

Conventional motor driven systems are most of the time designed to handle peak work demands in order to ensure safety and reliability. This situation often leads to inefficient use of energy during the operations of low work demand [52]. VFD is an electronic controller unit that adjusts the power that is being supplied to a motor based on the work demand [53], optimizing the energy consumption based on the nature of the work. VFD are also known as adjustable frequency drives, variable speed drives, microdrives, or inverter drives. VFD have wide range of applications areas such as pumps, fans, hoists, conveyors, etc.

VFD also have nonenergy benefits which are, namely, extended equipment life and reduced maintenance requirement due to elimination of abrupt motor starts. This is achieved by reduced voltage soft starters which gradually increase the speed of the motor, decreasing deteriorating mechanical forces; whereas single speed motors would start with high constant speeds, creating greater mechanical stress [54].

10.1.7.2 Irrigation Hardware

BPA supports diffusion of a number of irrigation-related mechanical improvements that are in compliance with the specifications. These improvements may not be considered as technologically highly complicated, however they are proven to be more water efficient than baseline systems. For instance, BPA provides reimbursements for water sprinklers, nozzles, wheel lines, wheel line hubs, wheel line levelers, drain gaskets, pressure regulators, drop tubes, gooseneck elbows, and pivot bases. Apart from irrigation hardware, BPA also provides reimbursements for freeze-resistant livestock water tanks as well.

10.1.8 Other Energy Efficient Technologies

Some of the energy efficiency measures do not bundle up to address a particular focus area, thus those measures will be presented under this section. For instance, these technologies are low-flow showerheads and smart power strips.

10.1.8.1 Low-Flow Showerheads

Since Bonneville Power Administration is significantly dependent on its hydro system along the Columbia River System, low-flow showerheads have significant energy savings potential both in saving water and electricity through reduced water heating need. BPA have been supporting diffusion of Energy Star low-flow showerheads in both residential and commercial sector.

Showers are the largest water consuming fixtures in most of the residential sector and have significant potential to save energy through reduced water heating requirements. In order to capture this potential, U.S. Department of Energy has passed a standard in 1994 forcing residential showerheads to have maximum flow rate of 2.5 gal per minute. There are two types of low-flow showerheads which are distinguished by whether air is mixed with the water stream or not. For instance, aerating showerheads achieve efficient use of water by mixing air with the water stream to alter nozzles through advanced flow principles whereas the nonaerating type showerheads use less water through high oscillation of the spray stream. According to standards, a low-flow showerhead is a showerhead that has flow rate of equal or less than 2.5 gal per minute compared to standard showerheads with 4.5 gal per minute [55].

10.1.8.2 Smart Power Strips

Consumer electronics sector is significantly growing in the recent years, due to addition of new and more electricity demanding appliances such as flat TVs, monitors, smart phones, etc., increase energy savings potential from plug load. It is expected that each qualified power strip has potential 100 kWh annual savings [56].

Smart power strips are improved version of conventional power strips, achieving energy savings through switching off the appliances that are in standby mode. Smart power strips are able to detect whether an appliance go in standby mode by constantly monitoring level of current in each socket. Smart strips also allow users to define time duration after standby mode until switching off feature becomes functional. More advanced models have an arrangement of master and slave sockets, cutting down the current to whole strip when the appliance on the master socket goes into standby mode. There is also another type of smart power strip which detects existence of occupants through motion sensors and trigger energy saving feature whether occupant is in place or not.

Bonneville Power Administration has been providing reimbursements for power strips in commercial sector for office applications. Due to end of measure life, BPA is currently running a pilot program to gather data for redeeming this measure. BPA is also working with NEEA toward providing reimbursements for energy efficient monitors and TVs in the coming years. More information regarding the program considerations can be found in the Question 1C.

10.2 Regional Perspective on Importance of Emerging Technologies

Please note that the term "Emerging technology" in the context of energy efficiency may be used to refer two difference cases. First case may occur when a technology is not commercialized yet meaning that it is still in the R&D development phase. Second case may occur when a technology is currently available in the market; however, it has never been promoted through an energy efficiency program, meaning that there is not much known about its energy savings potential. The sixth power plan requires regional energy efficiency actors to meet aggressive energy conservation targets. In order to meet these targets, importance of emerging technologies has been highly emphasized. It is noted that many of the current successfully diffused energy efficiency technologies were resulted from research projects completed in the 1980s and 1990s.

These technologies are CFLs, resource-efficient clothes washers, super efficient windows, and premium efficiency motors. Due to deregulations taken place in the mid-1990s, emerging technologies efforts have halted significantly. Impacts of deregulations are felt today in a way that there is no technology that has a promising potential as CFLs. In order to fill this gap, public power, investor-owned utilities, and other energy efficiency organizations have decided to create a consortium which will guarantee the region's energy efficiency technology pipeline will be filled and maintained. Accordingly, Northwest Energy Efficiency Alliance (NEEA) and BPA were given the responsibility to identify and track wide range of energy efficient technologies that could have potential for the region [57]. An initiative named E3T Emerging Technologies group within the BPA energy efficiency group has been started in 2008. Collaborating with NEEA, BPA has taken the lead to collaborate with universities, national labs, and other experts from the utilities in developing a portfolio of technologies which have potential to address regional needs. Inventory of emerging technologies is created and maintained through extensive technology roadmapping efforts that cover wide range of energy end uses in the Pacific Northwest.

10.2.1 Northwest Energy Efficiency Technology Roadmap Portfolio

As mentioned before, to address the challenges of increasingly ambitious energy savings and climate change mitigation goals, BPA and partner organizations have been working toward developing a technology inventory that will enable reaching medium- and long-term energy efficiency goals. Accordingly, 30+ region wide technology, market and utility program specialists from 20 organizations have participated in a number of expert panels to provide input for creation of a regional technology roadmap. Role of this technology roadmap has been set to define a research agenda for the Pacific Northwest in order to both drive technological development toward addressing regional needs and eliminate duplication of similar efforts underway elsewhere. First roadmap has been developed in 2010 and is being updated yearly considering the evolving emerging technologies and drivers in the region. For instance, initial roadmap has been developed by focusing on the energy efficiency technologies in residential and commercial sectors. Recent efforts in 2012 have included industrial energy efficiency technologies in food processing industry as well as combined heat and power technologies [58].

With the addition of industrial section, regional energy efficiency technology roadmap portfolio has diversified into eight sections, with 45 product and service level roadmaps within each section. For instance, these sections are:

Retrofit Building Design/Envelope
New Construction Building Design/Envelope
Lighting
Electronics
Heating, Ventilation, and Air Conditioning
Sensors, Meters, and Energy Management Systems
Food Processing Industry
Combined Heat and Power

Service level roadmaps within each section consist of several layers of information which are interconnected within each other. These layers provide information with respect to the following contexts:

Driver layer: This layer consists of list of drivers associated with environmental, global, market, policy and regulatory, and technology innovation-related issues. Key drivers set the strategic directions of energy efficiency efforts in the Pacific Northwest US.

Gap layer: This layer converts identified drivers into product and service performance gaps which are needed to be addressed.

Goal layer: This layer sets the threshold performance levels required for the gaps identified.

Technology layer: This layer includes a number of technological solutions that have potential to address the gaps and goals identified. Technologies may either be existing in the market or need to be developed in order to bring nonexisting products to market

R&D program gaps layer: This layer attempts to identify what the required R&D programs are in order to match technology performance levels with required performance levels in the goal layer.

Currently, Northwest Energy Efficiency Technology Roadmap Portfolio contains hundreds of energy efficiency technologies as well as R&D program needs. Fifteen out of identified R&D program needs have been selected as high priority due to their potential in enabling next generation of high efficiency technologies.

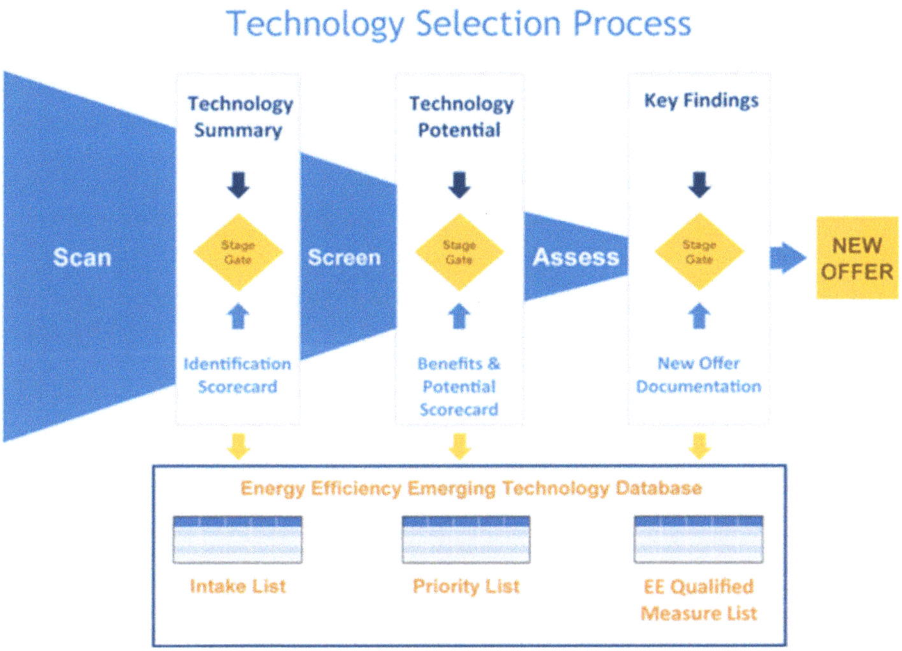

Fig. 10.1 E3T energy efficiency technology management framework

10.2.2 Energy Efficiency Technology Management Framework at BPA

E3T program operates within BPA and leads regional efforts to identify and assess new energy efficiency technologies for measuring development. The goal of E3T is to provide a robust pipeline of energy efficiency technologies for BPA and its customers, aiming to contribute to the region's medium- and long-term energy savings targets. In order to successfully manage the energy efficiency technology portfolio, emerging technologies group uses a framework that consists of several phases of assessment efforts. For instance, the framework consists of five stages with decision points for efficiently moving potential technologies from detection to program development. Framework stages include technology identification, technology benefits, technology potential, action plan and assessment, and new offer. You can see the aforementioned framework in Fig. 10.1.

E3T program identifies emerging energy efficiency technologies through a number of channels which are technology considerations from technical advisory groups, Northwest energy efficiency technology portfolio, and energy efficiency programs within BPA. E3T emerging technologies group has assessed a number of energy efficiency technologies, each of them being in different stages along the funnel. For instance, E3T group has:

Identified over 371 emerging energy efficiency technologies
Reviewed 150 technologies
Produced 55 technology summary reports
Produced reports on potential of seven technologies
Selected nine measures, eight of which have moved to field assessments or directly into programs

In the following section, you will be provided further information about the energy efficiency technology inventory of the Pacific Northwest along with a number of high priority technologies.

10.3 Emerging Energy Efficiency Technologies in the Pacific Northwest

As mentioned earlier, E3T group has developed an electronic database to where qualified experts can make technology suggestions. Accordingly, E3T team currently has 371 technology alternatives under review in its portfolio. These technologies spread across multiple focus areas such as HVAC, lighting, energy management, industrial agriculture, consumer electronics, envelope, water heating, commercial kitchen, and integrated design. Breakdown of these technologies under focus areas can be seen from Table 10.2. (Please refer to Appendix 1 for list of all emerging technologies with respect to focus areas.) Accordingly, it is easily observed that there is big emphasis on HVAC, lighting, and energy management technology areas. One of the biggest reasons for this situation is that sixth Power Plan emphasizes these areas as having the biggest energy savings potential. This topic will be discussed in greater detail in Question 1C.

Since the number of all emerging technologies is too large to handle, author has decided to present only those technologies which have been identified as high priority and moved to offer stage. There are 13 high priority technologies, falling under

Table 10.2 Emerging energy efficiency technologies under respective focus areas

Focus area	Number of technologies
HVAC	147
Lighting	84
Energy management	75
Industrial/agriculture	29
Consumer electronics	19
Envelope	11
Water heating	4
Commercial kitchen	1
Integrated design	1

Table 10.3 High priority emerging energy efficiency technologies under respective focus areas

Focus area	Technologies
HVAC	Demand-controlled ventilation systems
	Variable capacity compressor
	Air-side economizers
	Variable refrigerant flow systems with heat recovery
	Advanced design rooftop HVAC units
	Evaporative cooling systems
Lighting	Bi-level lighting controls
	Integrated lighting systems
	Wireless control systems
Energy management	Advanced rooftop unit controls with remote access and energy monitoring
	Low-cost building energy management and control systems
	Building energy performance analytics software
	Non intrusive load monitoring

three major focus areas such as HVAC, lighting, and energy management. Please refer to Table 10.3 for breakdown of high priority energy efficiency technologies with respect to each focus area. In the following section, you will be briefly informed about these high priority technologies with respect to corresponding focus areas.

10.3.1 High Priority Technologies in HVAC Focus Area

There are six high priority technologies under HVAC focus area. These technologies are demand-controlled ventilation systems, high occupancy buildings, variable capacity compressors, air side economizers, variable refrigerant flow (VRF) systems with heat recovery, advanced design rooftop HVAC units, and evaporative cooling.

10.3.1.1 Demand-Controlled Ventilation Systems

Demand-controlled ventilation technology has promising energy efficiency potential in two applications which are commercial kitchens and high occupancy buildings. Below, you can find further information about each application.

Commercial kitchen ventilation: BPA currently considers this technology promoting in commercial kitchen ventilation. According to the American Gas Association, commercial kitchen has two billion dollars worth of energy savings potential annually due to excess ventilation that runs at constant speeds for long hours [59]. Demand-controlled ventilation systems, combination of VFD, and optical and thermal sensors, adjust speed of the fans by detecting the amount of heat and smoke in the kitchen environment. Fan energy savings from this technology is expected to range 30–60 %. Apart from energy savings these systems also provide improved comfort by eliminating excess fan noise, heat, and smoke [60].

High occupancy buildings: For over a decade, demand-controlled ventilation units have been preferred in high but varied occupant density places such as conference rooms, auditoriums, etc. Due to improved carbon dioxide censor technology, demand-controlled ventilation units are believed to have potential in other places with less varied occupant density such as classrooms, restaurants, office areas, etc. This technology promises solutions to those places where it is difficult to meet code standards requiring certain amount of outside air to be used in ventilation. Demand-controlled ventilation systems make use of carbon dioxide censors to adjust its ventilation speed and meet code standards in an optimized manner. Depending on the occupant density, ventilation speed is adjusted and energy efficiency savings are realized [61].

10.3.1.2 Variable Capacity Compressor

BPA considers promoting variable capacity compressors in packaged rooftop HVAC units. Manufacturer claimed energy efficiency savings for the technology are around 40 %. Variable capacity compressors use a technology called "Digital Scroll" which achieves variable capacity with constant rate motors by using a pair of loading and unloading scrolls. Savings are stated to occur due to the fact that the motor runs at constant speeds in an optimized manner, but the scrolls are engaged and disengaged to achieve required demand. This technology is stated to be as efficient as and significantly cheaper than VFD which works on the principle of variable speed [62].

10.3.1.3 Air-Side Economizers

Air-side economizers are currently an interest of BPA in for reducing cooling loads in data centers. Although there is no savings data currently available, American Society of Heating, Refrigerating and Air-Conditioning Engineers and Lawrence Berkeley National Laboratory have demonstrated successful results. The technology is based on 100 % use of outside air which is circulated around the units that need cooling whereas the current practices use mechanical cooling without economizers and benefit from outside air at a minimum level. Accordingly, in order to process the outside air suitable for cooling electronics equipment, air-side economizer systems need additional ductwork, dampers, additional filtration, and humidity management equipment [63].

10.3.1.4 Variable Refrigerant Flow Systems with Heat Recovery

VRF systems are capable of adjusting its operation levels to the changing load requirements by using VFD. Compared to baseline systems, VRF systems operate more efficiently due to their part load efficiency and reduced duct loss.

Basic principle of this technology is that it uses a hot or cold refrigerant created by an outside unit and distributes it around the space using refrigerant lines. VRF systems have very wide range of application types, thus energy savings vary greatly depending on the baseline and installed system, occupant and building characteristics, climate zone, etc.; however, modeled savings propose 20–30 % savings compared to air to air heat pump units [64].

Technology can be used as a single unit which is also referred as DHPs or split heat pumps. DHPs are considered as easier to install in homes with electric resistance zonal heating systems. It is expected that this technology can provide approximately 200 average megawatts of savings at a cost less than 60 dollars per megawatt hour by 2030. Furthermore, DHPs have widely been used in other parts of the world such as northern Europe, Asia, Australia, and New Zealand; however, it has just been introduced to US market. NEEA, BPA, and Energy Trust of Oregon (ETO) have developed market transformation programs to support diffusion [4].

A number of units can also be connected with each other via ventilation ducts and heat recovery system can be activated with the help of a control system. System enables transfer of heat from one place to another place depending on the occupant preferences. Ability to provide heating and cooling function enables this technology to provide savings in large buildings with different temperature needs. Variable refrigerant systems also promise nonenergy benefits such as quieter operation, low space requirements, and less or eliminated ductwork [64].

10.3.1.5 Advanced Design Rooftop HVAC Units

Advanced design rooftop HVAC units use the same technology as the standard units; however, the components have been redesigned to enable the systems operate more efficiently. For instance, enhancements can be applied to a number of components such as fans, coils, filters, dampers, compressors, condensers, controls, and airflow path. Advanced design rooftop units are stated to be significantly more efficient than its rooftop substitutes; however, the first cost is considerably high [65].

10.3.1.6 Evaporative Cooling Systems

Evaporative cooling systems work on the principle that water takes heat from the environment as it evaporates. The heat transferred from air to water is called latent heat which is the amount of heat needed to evaporate the water. Heat drawn during the process is basically the source of cooling. Shortly, evaporative cooling systems work in a two-stage manner using the aforementioned principle. In the first stage, warm air is precooled by passing it through a heat exchanger that is cooled by evaporation on the outside. In the second stage, precooled air is sent trough a water soaked pad and picks up humidity as it cools down and reaches it maximum cooling potential. Conditioned air then is ventilated into the specified environment [66].

Evaporative cooling technologies have varying energy savings potential depending on the difference between dry and wet bulb temperatures. This technology has been stated to be promising for the Western Oregon and Washington regions since the weather is considerably dry in the summer season. For instance, the saving potential is estimated to be around 60 % throughout the BPA region. It is believed that this technology may have potential to spread widely, and it has already been proven cost effective in many applications such as industrial plants, commercial kitchens, laundries, dry cleaners, greenhouses, loading docks, warehouses, factories, construction sites, athletic events, workshops, garages, kennels, poultry ranches, hog, and dairy. One of the drawbacks with this technology has been reported to be causing disturbances to occupants in wet climates due to increased humid in the air [66].

10.3.2 High Priority Technologies in Lighting Focus Area

There are three high priority technologies under lighting focus area. These technologies are bi-level lighting controls, integrated lighting systems, and wireless lighting control systems.

10.3.2.1 Bi-Level Lighting Controls

Bi-level light fixtures and controls are used to combine a lighting source with occupancy censors. Typical lighting technologies are T5 and T8 type fluorescent lamps; however, similar control technologies are expected to be developed for LED lighting as well. This combination allows lighting systems to be more energy smart by adjusting illumination to a lower level or higher level depending on the need. This technology is the most suitable for stairwells, emergency exits, and elevators which are mandated by building codes to be illuminated at all times although they are not always occupied. Furthermore, it can also be used in store rooms, restrooms, corridors, and laundry rooms where illumination requirement varies throughout the day. Bi-level lighting controls technology typically sets illumination level to 50 % of full lighting potential if the place is unoccupied and, goes up to 100 % in case of occupation. According to studies conducted in California, bi-level lighting control systems have 40–80 % more efficiency depending on the occupancy characteristics of the place. A nonenergy benefit of this technology is stated to be longer lamp life and reduced maintenance costs. BPA is currently assessing potential of bi-level lighting technology in stairwell, office, and parking space lighting applications [67].

10.3.2.2 Integrated Lighting Systems

Integrated lighting system is a combination of efficient light fixtures, fluorescent light sources, and lighting control which allows users to set the illumination level based on their preferences. BPA is assessing potential of this technology in classroom applications where need for illumination varies during lecture presentations and student tasks. Lighting control system allows teachers to manage light level distribution in any part of the classroom. Case studies have shown that these systems can provide 30–50 % energy efficiency savings. Nonenergy benefits of this technology have been stated to be its contribution to performance of the students due to less glaring and increased attention. Integrated lighting controls can also be combined with daylighting systems which allow them to operate more efficiently by taking daylight levels into consideration [68].

10.3.2.3 Wireless Lighting Control Systems

Wireless controls systems use radiofrequency signals to communicate between light units and controller units. Lighting levels can be controlled at both single and multiple unit level with respect to a number of sensor inputs that would increase occupant comfort. These systems can provide better control options than the wired systems by enabling controlling of multiple zones from a single controller unit. Furthermore, this technology does not require wiring, thus making it easier and less invasive to implement in both existing and new construction projects. Wireless control systems are also more flexible than wired systems in case of structural changes in an environment by enabling reconnection of lighting and controller units on a web-based platform.

Use of wireless technology has some important opportunities for future improvements. For instance, wireless technology allows communication to be made two ways which allow room for integration of lighting systems with energy management systems. Wireless control systems are also promising with regard to meeting constantly improving building codes such as ability to harvest daylighting opportunities more efficiently and ability to integrate with utility demand response system.

Wireless systems are observed to be more flexible, easy to install, and less invasive in both existing and new construction projects. It is also verified that these systems increase occupants' comfort level by offering wide control options [69].

10.3.3 High Priority Technologies in Energy Management Focus Area

There are four high priority technologies under energy management focus area. These technologies are advanced rooftop unit controls with remote access and energy monitoring, low cost building energy management and control systems, building performance analytics software, and nonintrusive load monitoring.

10.3.3.1 Advanced Rooftop Unit Controls with Remote Access and Energy Monitoring

A total of 60–80 % of the economizers in the Pacific Northwest have been found to work improperly. This situation is believed to waste significant amount of electricity. Advanced rooftop unit controls with remote access and energy monitoring technology is a combination of rooftop HVAC unit controls and energy monitoring which is accessible via web interface. Web interface allows users to manage scheduling, setpoint adjustments, energy and performance monitoring, fan control, and demand-controlled ventilation operations. Advanced rooftop unit controls with remote access and energy monitoring systems are projected to provide 50–75 % energy savings in the Pacific Northwest along with side benefits such as improved equipment life, quieter operation, and reduced maintenance costs. BPA is currently considering this technology for small commercial building and big box stores [65].

10.3.3.2 Low Cost Building Energy Management and Control Systems

In terms of functionality, low cost building automation systems are very similar to advanced rooftop unit controls with remote access and energy monitoring systems; however, these systems are also able to provide a whole energy management solution by integrating with fire, safety, lighting, and security systems. Majority of small- and medium-sized commercial buildings do not have building automation systems which would turn off all unnecessary operations during unoccupied hours such as nights, holidays, and weekends. It is projected that such systems can provide 10–30 % energy efficiency savings; however, in order to capture this potential, building energy management and control systems need to be low cost and easy to implement. Currently, a number of manufacturers are working on improving the existing technology for wider adoption [70].

10.3.3.3 Building Energy Performance Analytics Software

Building energy performance analytics software collects energy consumption data of buildings and uses real time analytics, fault diagnosis, and reports to provide suggestions for improved energy performance. These software packages are referred with a variety of names such as building analytics software, energy information systems, building dashboards, fault detection and diagnostics, ongoing commissioning, and smart building systems.

Building performance analytics software has wide range of functions that can be implemented in buildings. For instance, this technology can provide energy saving suggestions by benchmarking energy consumption of similar buildings using hourly utility bill and weather data. Similarly, it is also possible to collect other data types such as temperature and humidity to track and identify incidents where there is excess energy consumption for improved energy management. Currently, this technology is almost not in use in the market although, it has been found to improve energy efficiency 1–30 % depending on the application area [71].

10.3.3.4 Nonintrusive Load Monitoring

Nonintrusive load monitoring technology monitors buildings' energy consumption from a single point and provides consumption information about individual components residing inside. One of the most important benefits of this technology is that it eliminates submetering requirement which is invasive and costly. In simple terms, technology works by capturing signals with special hardware equipment and uploading the data on a platform where the data can be analyzed using pattern matching algorithms. These algorithms help identify and specify individual loads which are later fed into energy management systems for final presentation and analysis. Potential applications of nonintrusive load monitoring are providing data to energy efficiency research organizations, energy auditing tool, and work in parallel with energy management systems.

Although this technology has significant potential in all end-use sectors, BPA is currently interested in assessing the potential of this technology in residential sector. Currently, this technology is still in commercialization phase, development efforts are toward reducing the cost of the applications, increasing accuracy of measurements, ability to integrate with home energy management systems, and decreasing intrusiveness. Energy savings potential of this technology remains unknown as of now [72].

10.4 Energy Efficiency Conservation Achievements in the Pacific Northwest US

Northwest Conservation and Electric Plan is a regional utility resource planning activity that is dedicated for addressing the electricity need of Pacific Northwest within 20 years of time horizon. Plan is updated periodically by the Northwest Power and Conservation Council by taking constantly changing variables into consideration. This section will mostly be based on the analyses and findings that have been included in the most recent conservation plan which is also referred as the sixth Power Plan.

Energy conservation programs have achieved savings of 4,000 average megawatts of electricity, meeting the half of the demand growth in Pacific Northwest between 1980 and 2008. Conserved amount of electricity is expressed as being enough to power state of Idaho, Western Montana, and city of Eugene for 1 year. Conservation programs helped keep electricity rates decrease, avoiding 8–10 new coal or gas fired power plants and saving ratepayers $1.8 billion. Please see Figs. 10.2 and 10.3 for further information about the historical accomplishments [4].

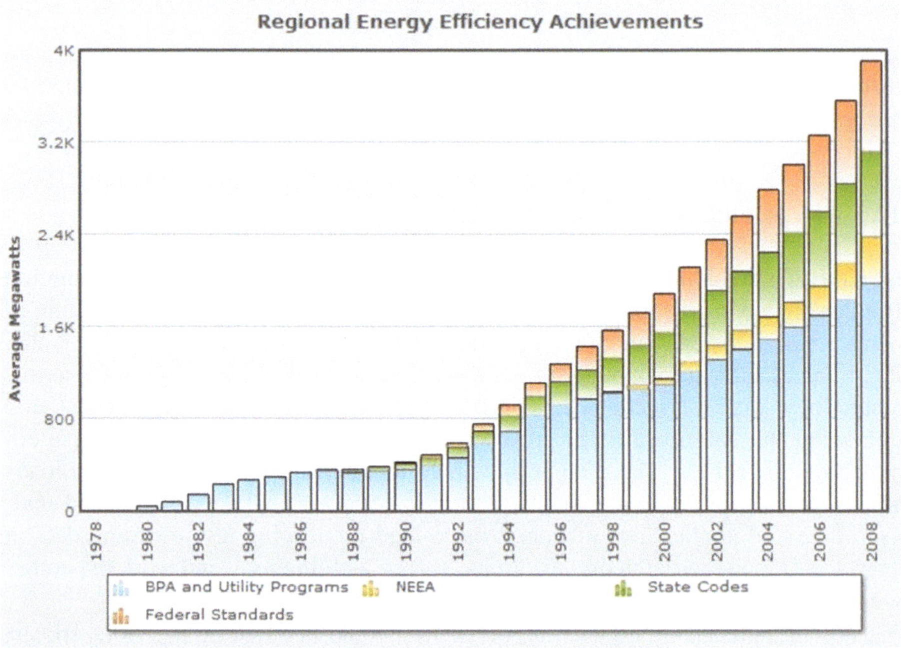

Fig. 10.2 History of cumulative electricity savings Pacific Northwest (1978–2008), Figure adapted from NWPCC [4]

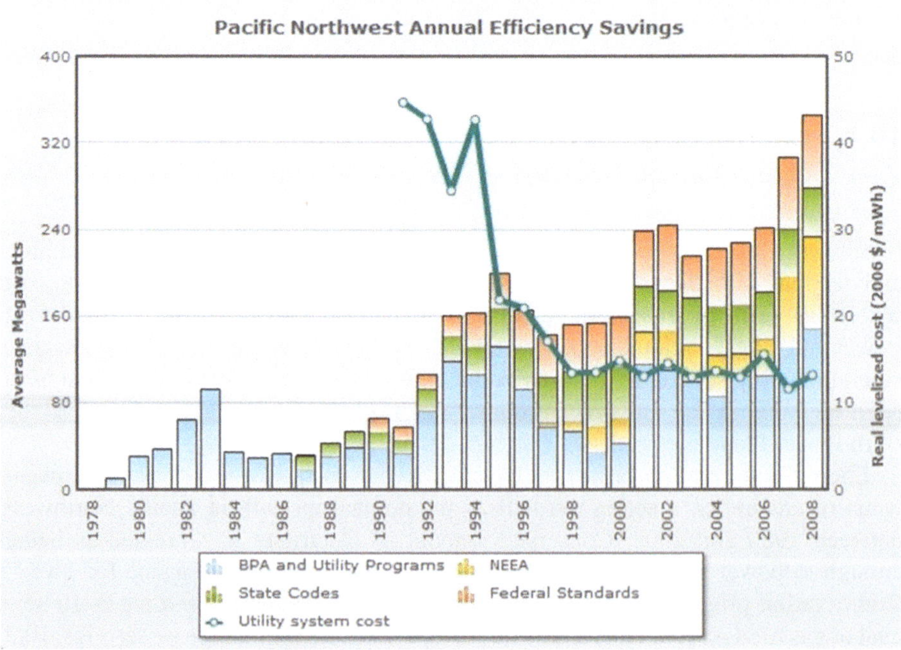

Fig. 10.3 History of electricity savings in the Pacific Northwest (1978–2008), Figure adapted from NWPCC [4]

10.5 Motivations for Energy Efficiency Conservation in the Pacific Northwest US

In the sixth Power Plan, it is stated that increasing cost of energy resources, rapidly developing renewable energy generation technologies, and concerns about the climate change have increased the value of energy conservation. For instance, it is emphasized that meeting most of the electricity demand growth in the next 20 years by nongeneration alternatives will provide the necessary time until risks involved in fuel prices, climate change policies, deployment, and development of emerging technologies are reduced. Apart from mediating the uncertainties stemmed from macro environment, energy conservation is stated to stand out as important piece of energy portfolio due to its potential to address a number of challenges and drivers that electric utilities will face in the coming years. In this section, you will be presented about those drivers and challenges and how they motivate regional actors toward considering energy efficiency as significant part of the solution. For instance, drivers and challenges for deployment of energy efficiency technologies in the Pacific Northwest are as follows [4].

1. Cheapest and least risky resource: energy conservation stands out as the cheapest and the least risky resource option compared to all available generation technologies.
2. Carbon policies: energy conservation helps reducing potential carbon costs by reducing carbon dioxide emission through more efficient use of electricity.
3. Increasing and changing electricity demand: according to analyses, energy efficiency potential in the Pacific Northwest is large enough to meet the increasing energy demand as well as mitigate load shape changes due to various reasons such as impacts of aging population on energy use, diffusion of new technologies, and population growth.
4. Reduced operating capacity, flexibility, and reliability: BPA is hugely dependent on its hydropower system to balance supply and demand load fluctuations. Energy efficiency technologies help BPA maintain its balancing operations without requiring additional generation capabilities. Currently, two major issues are considered to limit operating capacity, flexibility and reliability of hydropower system in the region.
   - Impacts of renewable energy portfolio standards: renewable energy portfolio standards dictate integration of wind farms to the BPA transmission grid; however, intermittent nature of wind resource pushes BPA's operating flexibility to its limits and risk service reliability. Energy efficiency programs can help reducing peak load and improve system reliability.
   - Impacts of fish and wildlife programs: fish and wildlife programs require BPA to maintain specific amount of water flow through the dams for reducing detrimental impacts on local fish fauna. This situation reduces operating flexibility at the times when power generation needs to reserve/flow water for peak demand times. Energy efficiency programs can reduce use of electricity and conserve water through the hydrosystem.

5. Uncertainties in fuel prices and growing cost of energy: uncertainties in fuel prices create an unstable environment for making cost-effective decisions in the long run. For instance, volatility of oil prices have been experienced to be detrimental on keeping electricity rates low. Energy efficiency programs can minimize negative impacts of price volatilities by reducing inefficient energy use.
6. Regional development and nonenergy benefits: some of the nonenergy benefits gained through more efficient technologies are included in cost–benefit ratio calculations since avoided costs are contributing to regions' residents.

In the following sections, you will be provided with greater detail of information about each of the aforesaid motivations for deployment of energy efficiency technologies in the case of Pacific Northwest.

10.5.1 Cheapest and Least Risky Resource

Considering the amount of available energy efficiency potential at reasonable costs in the region, energy conservation has become one of the most important pieces of energy portfolio. For instance, it has been proposed in the sixth Power Plan that energy conservation programs are viewed as the least cost and risky resources compared to generation alternatives. Furthermore, cost-effective energy efficiency programs are reported to have potential to meet 85 % of the region's electricity growth in the next 20 years. This estimated conservation potential equals to almost 6,000 average megawatts, saving 17 million tons of carbon emission per year by 2030, and can be captured by measures which cost under $100 per megawatt hour. Moreover, 4,000 MW out of the total potential is stated to be captured with measures costing under $40 per megawatt [4].

In the sixth power plan it has been stated that existing generation technologies are mature and cost effective in short to medium terms. For instance, wind resource is expected to provide considerably cheap electricity in the short term along with some small-scale local renewable generation alternatives such as geothermal. It is also considered that natural gas fired generation alternatives are also cost competitive and may be utilized as a resource in the coming years. It is further emphasized that new coal fired generation units are difficult to site due to new plant emission standards forced by many states and do not stand as strong alternatives in the future unless there are significant technological developments on carbon separation and sequestration technologies. According to the analysis by Northwest Energy Conservation and Power Council, if carbon dioxide emission costs are included, cost of the majority of the generation resources in the Pacific Northwest would range between $70 and $105 per megawatt hour (levelized to $2,006). Generation cost for each alternative is presented below [73] (Fig. 10.4).

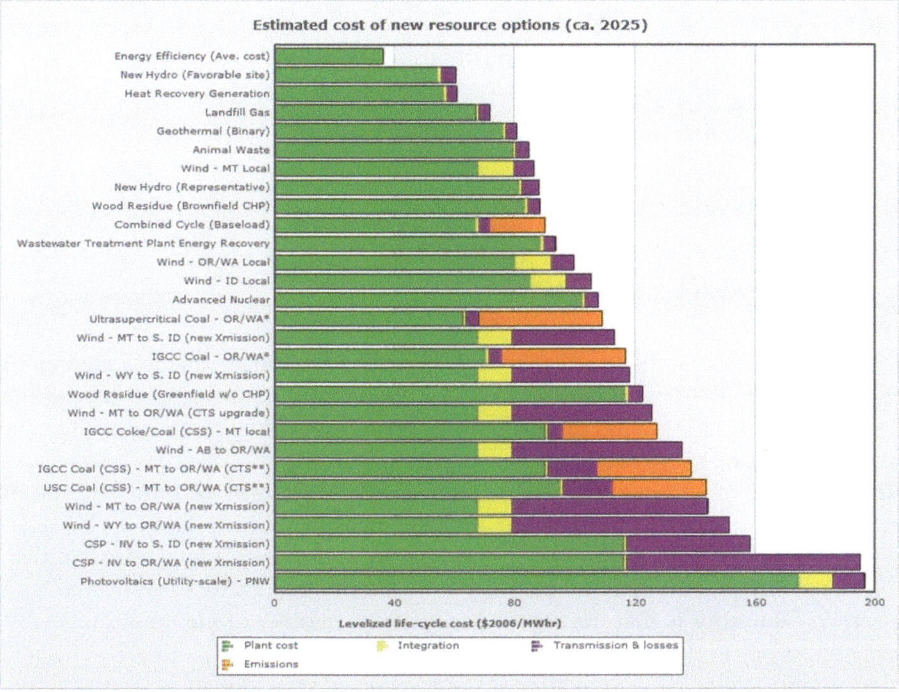

Fig. 10.4 Estimated levelized cost of new resource options

10.5.2 Potential to Mitigate Carbon Policies

A total of 38 % of the nationwide carbon dioxide emissions are related to electricity generation whereas this share is 23 % for the Pacific Northwest. This rate is observed to be lower in the case of Oregon due to big share of electricity generation through hydroelectric system. In order to address carbon emission issues, there have been ongoing discussions on implementing carbon tax or carbon trade policies. Purpose of such initiatives is to internalize cost of environmental impacts caused by emission of carbon dioxide which constitutes to the highest share of greenhouse gas emissions. Since fossil fuels are significant part of the existing electricity generation operations, potential impacts of carbon policies are crucial. As the debate goes on, uncertainties regarding the cost of carbon dioxide remain; however, scenario studies approach the issue by considering carbon mitigation costs from a range of 0 to $100 per ton. It is further expected that carbon costs will climb up from $0 to $47 per ton by 2030. Accordingly, impact of carbon policies is expected to increase wholesale electricity prices from $30 to $74 per megawatt-hour by 2030 ($2,006) [4].

Although there is no national level carbon policy yet, regional policies have started to be formed. Western Climate Initiative has been established in 11 US states and Canadian provinces in an attempt to set greenhouse gas emission goals. Defined goals are to be achieved using a market-oriented cap and trade processes, aiming to force 40 % reduction from 2005 levels of emission by 2030 horizon [4].

10.5.3 Impacts of Increasing and Changing Electricity Demand

Electricity demand in the Pacific Northwest region has changed due to a number of reasons in the last 30 years and is expected to continue to change in the coming decades. For instance, in the following paragraphs you will be provided with major historical demand changes caused by energy price hikes as well as projected changes stemmed from increasing population and changes in age distribution, diffusion of new appliances, and new construction projects.

Historically, Pacific Northwest region has been more electricity intensive than the rest of the US states due to relatively low electricity rates. Reasons behind high electricity intensity is that the region has hosted a number of electricity intensive industries such as aluminum smelting, lumber, food processing industries, etc., and needs of residential and commercial sector have mostly been supplied by electricity. As a result, historical electricity consumption per capita in the region has always been above US average [4]. Due to recent events, gap between national and regional electricity consumption per capita is shrinking, although the region still provides some of the lowest electricity rates in the nation. Reason behind the changing trend is related to two major price hikes that occurred in 1979s and 2001. Both of the events were observed to shift end-use energy choices toward natural gas from electricity, especially in space and water heating end uses.

Projections show that if the current trend goes on, electricity load is going to grow by about 7,000 average megawatts from 2009 to 2030. This growth accounts for 335 average megawatts per year which accounts for 1.4 % yearly increase. Much of this demand increase is expected to occur in residential and commercial sectors as there is no expected increase in industrial output throughout the region. Electricity consumption growth in residential sector is believed to be caused by the fast diffusion of air conditioning and consumer electronics products. Especially, air conditioning units are diffusing in the region at a very high pace; since these systems reach their peak consumption in summers, they constitute significant implications on utility operations [4].

There are a number of drivers that cause increases and changes in demand load in the Pacific Northwest region. For instance, these are population increase and changes in age distribution, new residential and commercial construction projects, and diffusion of new appliances.

10.5.3.1 Population Increase and Change in Age Distribution

Over the next 20 years, the Pacific Northwest is expected to develop and expand. Regional population is expected to increase from 12.7 million in 2007 to 16.7 million by 2030. While the regional population is expected to increase by 28 %, older population (65 and older) is expected to increase by 50 %. Such significant shift in population distribution is expected to impact electricity consumption behavior and have implications on utility operations. For instance, such impacts are expected to occur as construction of elder care facilities and smaller sized homes as well as increased leisure activities [4].

10.5.3.2 New Residential and Commercial Construction Projects

Annual growth rate for residential units in the Pacific Northwest is expected to be around 1.4 % annually between 2010 and 2030. Based on the building stock research, 5.7 million homes exist in the region as of 2008 and this number is expected to reach 7.6 million by 2030 with the addition of 83,000 new homes every year [4].

Annual growth rate of commercial footage in the Pacific Northwest is expected to be around 1.2 % annually between 2010 and 2030. Building stock research in 2007 has revealed that existing commercial square footage is about 2.9 million square feet and expected to grow to 3.9 million square feet by 2030. This forecast implies yearly addition of 40 million square feet of space that needs lighting, air conditioning, and similar electricity consuming services. Most of the growth is expected to occur due to aging population and potential need for elder care facilities [4].

10.5.3.3 Diffusion of New Appliances

Technological improvements directly impact people's life styles and electricity demand. As the internet, personal electronic devices, and air conditioning units become widespread, residential electricity demands increase. For instance, during the last 12 years, diffusion rate of electronics appliances such as smart phones, large screen televisions, personal computers, etc., has been 6 % per year. This growth is expected to go on at the rate of 5 % per year in the coming years. Moreover; although Pacific Northwest has quite mild summers and did not have major summer peak issues, it has been reported that 80 % of all new residential places have air conditioning units. Taking new home constructions into consideration, this situation has important implications on both load growth and shape [4].

Northwest hydropower system has traditionally been a winter peaking system due to decreased water run in the winter time. As mentioned before, widespread diffusion of air conditioning units in the residential and commercial sectors implies potential changes in summer load shape. For instance, summer peak demand is expected to grow from 29,000 MW in 2010 to 40,000 MW by 2030, corresponding to 1.7 % annual growth whereas; winter peak demand is expected to grow from about 34,000 MW in 2010 to 43,000 MW by 2030, corresponding to 1 % annual growth. Projections show that although the hydrosystem will still remain a winter peaking system, gap between winter and summer peaks will shrink overtime. Summer peaks will start creating more stress over the system considering the shrinking operating flexibility in summer due to fish and wildlife regulations. As a result, sixth power plan suggests that next generation of resource planning needs to be more focused on required meeting of peak load capacity and operational flexibility [4].

10.5.4 Reduced Operating Capacity, Flexibility, and Reliability

Power generation systems need to be synchronized with annual, seasonal, hourly, and subhourly scale demand in order to function reliably. If Pacific Northwest's generation history is analyzed it is observed that different generation resources have been favored throughout the history due to changing favorable resource options. As a result, existing generation system in the region has become quite diverse in resource options as of now whereas hydropower was the only main resource in 1960s.

Early capacity extensions were focused on coal power plants which were the least cost alternative at the time; whereas in late 1990s and 2000s, natural gas was the most favorable resource option. Recently, due to renewable energy portfolio standards and incentives, wind energy has become favorable and been aggressively included in the energy portfolio. Please refer to Fig. 10.5 for changing energy portfolio of Pacific Northwest in the last 50 years.

Electricity rates in the Pacific Northwest remain remarkably low compared to the rest of the nation. For instance, it was reported that, as of 2007 Idaho ranked as the lowest, Washington as the seventh, Oregon as the 15th, and Montana as the 22nd lowest states based on average retail electricity prices. Although electricity prices in Pacific Northwest are below US average, California's average electricity rate was significantly higher. Important factor behind this is the fact that peak loads in California are about 70 % higher than average annual electricity use whereas it is 25 % in the Pacific Northwest. Accordingly, California has to bear high costs of fossil-based peak load generation units which are significantly underutilized due to very short peak demand load hours. Contrary to California, peak load hours in the Pacific Northwest are supplied by hydropower which is an inexpensive and non-carbon-emitting alternative. As a result, hydropower stands out as one of the keys in keeping region's rates low due to its unique feature.

As the electricity portfolio of the region expanded, capacity planning became more complex due to special conditions involved with each resource. For instance, in order to eliminate misleading assessment of electricity supplies, Pacific Northwest Power and Conservation Council adopted an adequacy standard in 2008. For instance, aforementioned standard is used as an early warning system for detecting when a specific power supply can no longer meet the annual or peak load

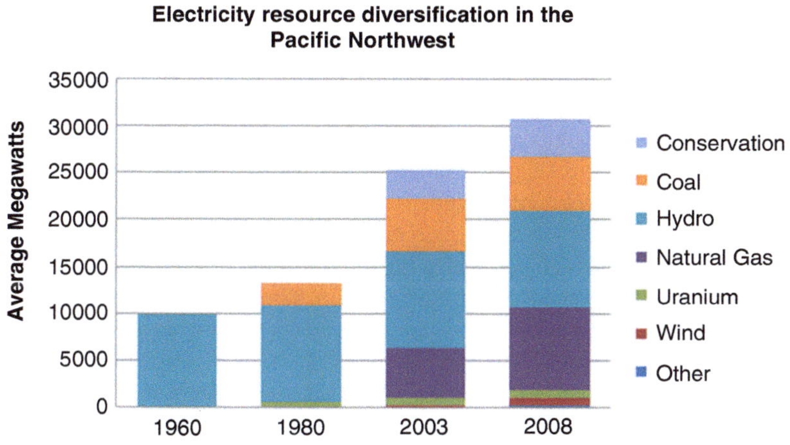

Fig. 10.5 Electricity resource diversification in the Pacific Northwest [75]

requirements reliably. Accordingly, importance of two metrics, dependable and installed capacity, is emphasized in the standard. Dependable capacity refers to capacity that can be used from an energy source during peak demand hours whereas installed capacity refers to generation potential of a given resource at full capacity. For example, dependable capacity of wind resource is only 5 % of the actual installed capacity since it cannot be used as a reliable resource due to intermittency of wind.

Hydroelectric power system has historically been used as a balancing authority to meet peak loads and provide flexibility as long as there was available water stored. Recently, existing system is coming close to its ancillary service limitations due to several reasons such as growing and changing seasonal electricity needs (caused by large-scale diffusion of air conditioning units), shrinking share of hydroelectricity in meeting total demand, growing gap between average load and peak load, reducing generation capacity and flexibility due to compliance with fish and wildlife regulations, and growing share of variable electricity generation caused by integration of wind generation. It should be kept in mind that if the regional demand characteristics goes beyond hydropower's peak capacity limitations, more and more fuel-based peaking resources might need to be added to the system. As observed from the case of California, this may result in significantly higher electricity rates.

Considering the aforementioned reasons, there is an increasing need for creating additional capabilities that can increase capacity and flexibility of power generation system. Accordingly, energy efficiency, energy storage technologies, and demand response are some of the potential areas that solutions can be created within. In the following sections, you will be briefly informed about two of the major factors that are likely to cause capacity, flexibility, and reliability issues in the medium term.

10.5.4.1 Impacts of Renewable Energy Portfolio Standards

Although there is an ongoing debate on causes of climate change, it is evident that greenhouse gas emissions caused by combustion of fossil fuels is a significant contributor. Concerns on climate change have resulted in variety of policies throughout the world. One of the most significant one of all in the USA is renewable energy portfolio standards which are designed to impact resource choices of utilities. Although timing and goal levels for each state vary, it is clear that movement toward integration of environmentally more friendly resources will have significant impacts on utility operations. In the Pacific Northwest region, renewable energy portfolio standards in Montana, Oregon, and Washington will require significant part of the utilities' energy portfolio to be supplied by renewable energy resources. For instance, 15 % and 25 % of electricity load needs to be met with renewable energy alternatives in Montana and Oregon respectably by 2025. Similarly, by 2020 15 % of Washington's electricity load must be coming from renewable alternatives as well.

Analyses show that there is readily accessible wind power potential of 5,300 average megawatts in the Pacific Northwest region [4]. Currently, there is more than 3,000 MW of wind power connected to the BPA's grid, and this amount is expected to double by 2013 [74]. Although these statistics provide favorable information for compliance with renewable energy portfolio standards, they also bring out new integration problems. Energy efficiency programs may reduce the need for costly solutions by reducing demand load where system reliability is in danger.

10.5.4.2 Impacts of Fish and Wildlife Programs

Pacific Northwest Electric Power Planning and Conservation Act of 1980 recognized that region's hydropower dams had negative impacts on migratory fish and wildlife. Accordingly, the Columbia River Basin Fish and Wildlife Program is by law incorporated into the power planning operations in the Pacific Northwest aiming to assure minimizing impacts on fish and wildlife. Fish and wildlife program has sizeable impact on the Columbia River power system operations. For instance, since 1980s, hydroelectric generation has been reduced about 1,200 average megawatts compared to the system operations without any fish and wildlife constraints. BPA has managed to address the impact by making secondary power purchases, promoting conservation programs, adding new generation resources, and developing resource adequacy standards. All the fish and wildlife program costs across the Columbia River Basin are absorbed by BPA ratepayers and it is estimated to account for $750–$900 million per year.

In the sixth power plan, it is stated that there will be evident challenges in meeting both new power and fish and wildlife program requirements. Considering the other challenges such as flexibility reserves for wind integration and other renewable resources, potential changes of water supply to hydrosystem due to climate changes, and conflicts between fish and wildlife operations and climate change policies, it might be impossible for existing hydrosystem to meet all needs of power, fish, navigation, irrigation, recreation, and flood control.

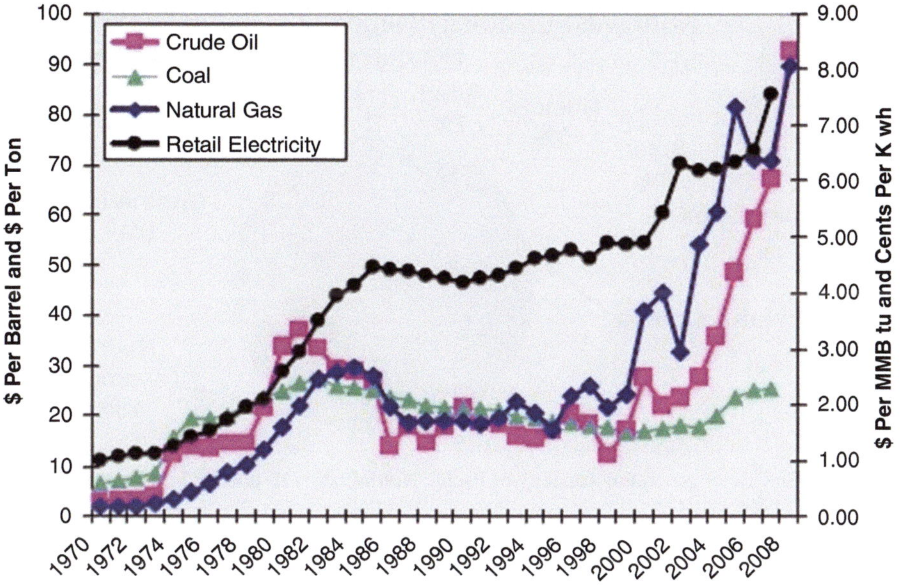

Fig. 10.6 Energy price trends between 1970 and 2008, Figure adapted from NWPCC [4]

10.5.5 Uncertainties in Fuel Prices and Growing Cost of Electricity

Throughout the history energy prices have been subject to cycles due to macroeconomic events. For instance, one of the biggest events has been OPEC policies that caused dramatic oil price increases in 1973. A more recent price hike for natural gas prices has been experienced between late 1990s and early 2000s when price of natural gas has risen from $2 to $6 per million Btu. Since then, energy prices have become more volatile and its detrimental impacts are felt at greater levels due to increasing reliance on coal and natural gas in the region. Energy price trends are presented in the Fig. 10.6 for further information.

Cost of electricity generated from fossil fuels is expected to be significantly higher than the price levels in 1990s. Although energy prices have dropped as of 2008 due to economic downturn and increased use of nonconventional natural gas production, cost of extraction has increased slightly. With the recovering world economies in mind, this situation is expected to contribute to fuel price increases in the medium term. Considering the fact that natural gas and coal sourced power plants account for 23 % and 17 % of the total electricity generation respectably, potential disturbances or price hikes in supply may have significant implications in the Pacific Northwest region [75] (Fig. 10.7).

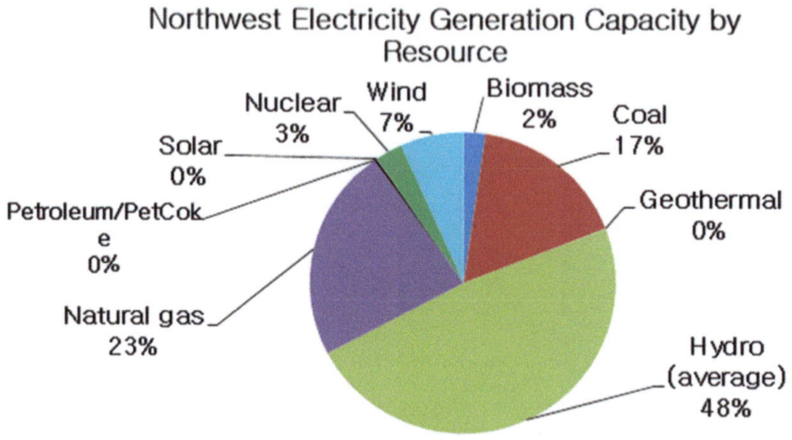

Fig. 10.7 Electricity generating capacity of Pacific Northwest by resources, Figure adapted from NWPCC [75]

Detrimental consequences of rapidly increasing electricity prices in the region were first experienced 30 years ago. For instance,between 1960 and 1980, electricity demand growth had been around by 5 % each year. In order to meet the highly growing demand, large-scale coal and nuclear power plants were put in operation; however, cost of electricity generation from the newer capacity extensions was significantly higher than existing hydroelectric power system. As the cost of capacity extensions was reflected on the electricity rates, aluminum smelting plants in the region, which accounted for 20 % of the demand, lost their competitiveness around the global market. At the same time, other users also reacted to the price increases by changing their consumption behavior. As a result, expected rate of demand growth for the following decades, 1980–2000, dropped from 5 to 1 % per year, causing nuclear facilities to be abandoned at very high costs which are still incurred to electricity rates. In 2000 and 2001, the region experienced a second round of price increase which was caused by underinvestment in generation resources. Moreover, failure of power market design in California and poor water year experienced by the hydrosystem strengthened the negative impacts. Price increase caused closure of majority of the aluminum smelting plants and cutback of other energy intensive industries in the region. As a result, regional demand load dropped 16 % between 1999 and 2001, regressing back to 1980s demand load [4].

As the Pacific Northwest energy portfolio becomes more dependent on the fossil fuel-based resources, electricity rates will be affected more by the national and global energy price volatilities. Especially, development of active trading markets for energy commodities has strengthened this relationship to a greater extent. For instance, oil prices have become a global commodity and price levels are reluctant to global events such as diplomatic frictions, wars, sanctions, supply disturbances, etc., whereas coal tends to be a regional commodity due to relatively more difficult logistics involved. Due to prohibiting legislations, possibility of oil and new coal-based power plant investments remains very low in the region, thus natural gas attracts special focus in energy planning. As of now, natural gas stands as a North American commodity; however, statuesque is expected to change as use of natural gas reaches higher levels. For instance, expected energy prices tend to be between 3.5 and 10 dollars per million Btu for natural gas, \$55 and \$120 per barrel for oil, and \$0.52 and \$1.05 per million Btu by 2030 (\$2,006) [4].

10.5.6 Regional Development and Nonenergy Benefits

It has been reported that in the last 30 years of Pacific Northwest history, two major electricity price hikes have been experienced to damage energy intensive local industry. First price hike has occurred between 1979 and 1981 due to overinvestment on nuclear facilities that were not utilized to its full potential. Price increases have been felt notably by the local electricity intensive industries that were pushed to their limits in competing with producers in the world market. The second electricity price hike has occurred between 2000 and 2001 due to underinvestment in electricity generation and caused permanent closure of many of the aluminum plants. For instance, number of operating aluminum plants in the region has dropped from ten to three which are actually partially functioning. In addition to aluminum plants, there have been permanent closures in other electricity intensive industries in the last 10 years. Moreover, it has further been emphasized that some of the energy intensive industries are provided with low rate electricity in order to keep local industry competitive and keep local jobs remaining.

Value of energy conservation does not solely come from the savings associated with the power and transmission system, but also the residents of the region as well. For instance, sixth power plan has considered avoiding cost of detergents, water, and waste water treatment savings as benefits in cost–benefit calculations of conservation resources. This perspective is supported by the statement that not all of the costs and benefits are paid or received by the region's power system, but it is the consumers where ultimately the costs and benefits end up.

10.6 Regional Energy Efficiency Collaboration in the Pacific Northwest

Energy conservation potential in the Pacific Northwest region is assessed within the power plans that are conducted periodically. These plans are prepared taking whole region into perspective and set the planning goals for the next 20 years. Based on these goals, roles and responsibilities are split up among the regional energy efficiency actors in an attempt to coordinate the workload. For instance, Bonneville Power Administration, NEEA, RTF, and ETO, and other public utilities take the outcome of the power plans and align their operations accordingly.

In order to better understand conservation potential in the Pacific Northwest region and roles and responsibilities of the regional organizations, it would be beneficial to mention about conservation resource types. Accordingly, there are three primary conservation sources that can be utilized to meet the specified targets set in the sixth power plan. For instance, these conservation resources are nonprogrammatic savings, market transformation, and programmatic savings.

10.6.1 Nonprogrammatic Savings

Nonprogrammatic conservation approaches are maintained by the actors outside of BPA and its participating utility programs. For instance, nonprogrammatic savings may come from market-induced adoption, and codes and standards. Savings from market-induced adoption may come from tax credits, social and environmental pressures, influence of utility activities, and end user economic considerations. Savings from codes and standards occur due to changing their energy codes that require more building efficiency performance levels. BPA accounts for energy efficiency savings through nonprogrammatic approach, however does not provide reimbursements.

10.6.2 Market Transformation

As fulfillment of regional collaboration; NEEA, funded by BPA, is responsible for market transformation activities in the region. Market transformation programs use variety of tools such as marketing; training and outreach; upstream activities to influence manufacturers, contractors, dealers, retailers, consumers, and similar entities.

10.6.3 Programmatic Savings

Programmatic savings are achieved through the programs supported by BPA and its customer utilities; and accordingly, funding for the programs might either come from BPA and participating utilities. Programmatic savings can be delivered through various ways based on energy savings measurement and verification processes associated with energy efficiency measures. For instance, these are deemed savings, calculated measures, custom projects, and third-party programs. In the next section, you will be presented each of delivery mechanisms in greater detail.

10.7 Energy Efficiency Delivery Mechanisms Pursued by Bonneville Power Administration

BPA energy efficiency department delivers energy efficiency services to six distinct sectors. Each sector has a lead that is responsible for developing and implementing a strategy in order to meet the energy saving targets set by the sixth power plan. Each sector collaborates with marketing, planning, and engineering departments to deliver these programs. Furthermore, they also collaborate with customer utilities, third-party contractors, NEEA, and other stakeholders as necessary. Energy efficiency reimbursements can be delivered through five distinct ways which are deemed measure reimbursement, deemed calculating savings, custom projects, third-party programs, and upstream incentives (market transformation). Appropriateness of each delivery mechanism for each measure depends on the generalizability of the savings. For instance, generalizability of delivery mechanisms, deemed measure reimbursement, deemed calculated savings, and custom projects decreases, respectively. RTF is the responsible organization for verification of savings and incremental costs associated with energy efficiency measures. It should be noted that each energy efficiency measure has a useful lifetime which is also referred as measure life. If a measure's lifetime ends it needs to be redeemed by RTF process again.

10.7.1 Deemed Measure Reimbursement

Deemed measure reimbursements are the most generalizable reimbursement type without requiring any specific information about the application. BPA offers these reimbursements to its customer utilities which can determine their own incentive levels by mixing and matching multiple measures. This way they can create programs that best meet their interests and end users.

10.7.2 Deemed Calculated Savings

Energy savings estimations for some measures need to be determined by using site-specific information. In such cases, BPA develops calculators and provides them to its customer utilities for their use. For instance, savings estimations for commercial lighting projects are determined through a lighting calculator. Reimbursement levels through this delivery method mainly depend on a number of measures installed rather than kWh savings.

10.7.3 Custom Projects

In some cases it is not possible to provide deemed or calculated reimbursements due to excess variability associated with energy savings in each site. Such projects are considered custom project offerings. Custom project proposals can come from both BPA's customer public utilities and end users. BPA selects among the submitted proposals and has the authority to approve or disapprove the results. Each project is kept subject to measurement and verification process for savings determination, thus custom projects are resource intensive.

10.7.4 Third-Party Programs

BPA also works with third-party contractors to deliver energy efficiency programs. Third-party contractors maintain all activities required for program delivery such as marketing, technical support, installation, and savings analyses.

10.7.5 Upstream Incentives (Market Transformation)

Funded by BPA, market transformation activities are mainly managed by NEEA in the Pacific Northwest region. BPA works with NEEA, public utilities, and other stakeholders to decide on prioritization of market transformation projects. Upstream incentives may focus on manufacturers, dealers, retailers, and end users.

10.8 Energy Conservation Potential in the Pacific Northwest

Sixth power plan has identified electricity conservation potential for each of the end-use sectors which are residential, commercial, industrial, agriculture, and distribution efficiency. In addition to those, consumer electronics have been given special attention due to its increasing energy savings potential. Identified energy savings resources have scattered over wide range of energy conservation measures which are available in a variety of applications. To name a few, new and existing residential and commercial buildings, commercial and residential appliances, street lighting, sewage treatments, industrial and irrigational processes are some of them.

As mentioned before, by 2030, there are around 7,000 average megawatts of energy efficiency potential in the region. Although not all of the conservation potential is cost effective for now, around 5,900 average megawatts of conservation can be captured at the cost of $200 or less per average megawatt. To breakdown the potential across sectors, residential sector has 2,600 average megawatts of potential where most of the savings are expected to come through energy efficiency improvements

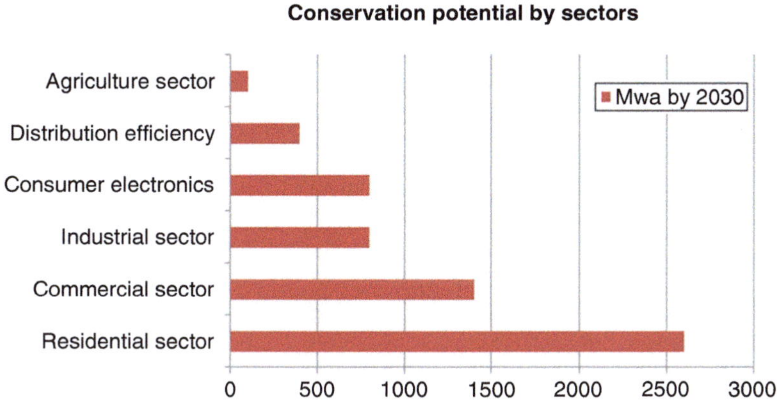

Fig. 10.8 Conservation potential in the Pacific Northwest by sectors

on water heating and heating, ventilation, and air conditioning units. Commercial sector has around 1,400 average megawatts of potential where majority of the savings are from improvements on lighting systems such as LEDs, improved lighting fixtures, and controls. Around 800 average megawatts worth of potential is expected to be exploited from electronics appliances such as televisions, set top boxes, desktop computers, and monitors. Agriculture sector has almost 100 average megawatts of energy conservation potential through irrigation system efficiency improvements, water management, and dairy milk processes. Energy efficiency savings potential in the industrial sector is projected to be around 800 average megawatts which can be achieved through equipment and system optimization measures. Lastly, utility distribution systems also promise significant energy conservation potential which is around 400 average megawatts that can be captured through better distribution management practices [4] (Fig. 10.8).

In Fig. 10.9, you can see a more precise breakdown of energy efficiency resources that have been revealed in the sixth Power Plan.

In the following sections, you will be presented about the programmatic energy efficiency programs for each sector which are residential, commercial, industrial, agriculture, federal, distribution system efficiency sectors, and consumer electronics.

10.8.1 Conservation Potential in the Residential Sector

A total of 2,600 average megawatts worth of conservation potential is projected to be exploited through energy efficiency programs in the residential sector. Majority of the conservation potential is to come from three major areas which are water heating, HVAC, and lighting. There is also significant amount of energy efficiency savings potential available from more efficient building shells and residential appliances [4] (Fig. 10.10).

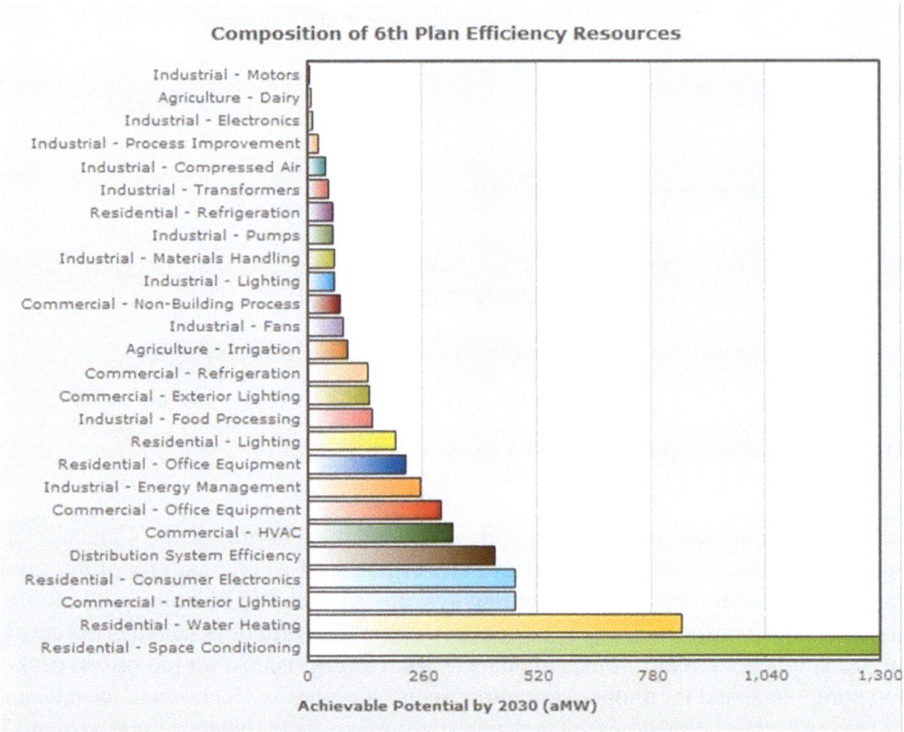

Fig. 10.9 Conservation resources in the Pacific Northwest in the sixth Power Plan, Figure adapted from NWPCC [4]

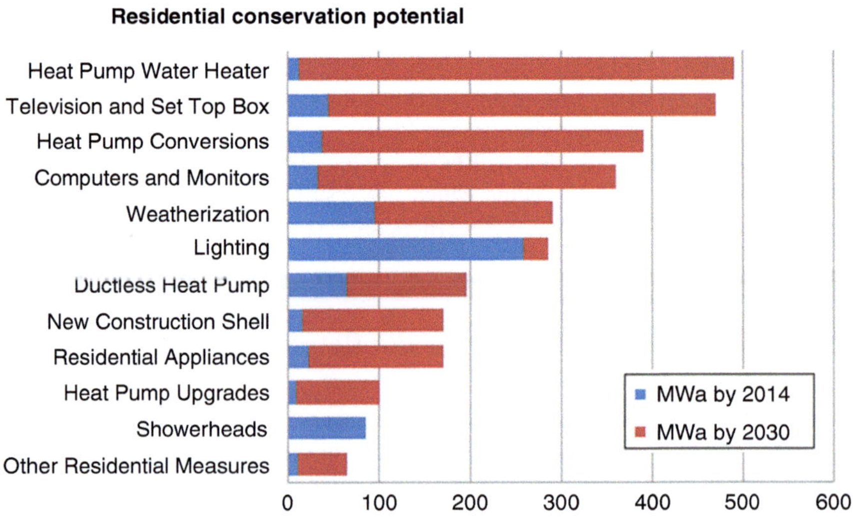

Fig. 10.10 Residential sector conservation potential in the Pacific Northwest, Figure adapted from NWPCC [4]

10.8.2 Conservation Potential in the Commercial Sector

Developments in lighting technology have enabled significant increases in energy efficiency potential. These developments are improvements on fluorescent lights, fixture efficiency, lighting controls, and improved lighting design practices. Moreover, availability of LEDs and ceramic metal halide lighting also has increased savings potential. As a result, streetlights, parking lots, and outdoor area lighting areas have become new areas of improvement that can provide between 25 and 50 % energy savings. In the sixth power plan, energy efficiency potential through the lighting efficiency improvements has become the top ranked priority in commercial sector. Apart from lighting improvements, there is significant area of improvement in HVAC applications; however, it is reported that available conservation potential in this area has been dropping since the previous power plan. It is projected that 1,200 average megawatts of energy efficiency savings is available at a cost of $40 per megawatt hour. The remaining energy efficiency savings, however, are hard to capture due to complex nature of decision making in the commercial sector. Moreover, some of the technologies require upfront design requirements, constant maintenance or control in order to ensure energy savings occur as planned. Such requirements are mainly significant in the area of HVAC systems and process-related improvements [4] (Fig. 10.11).

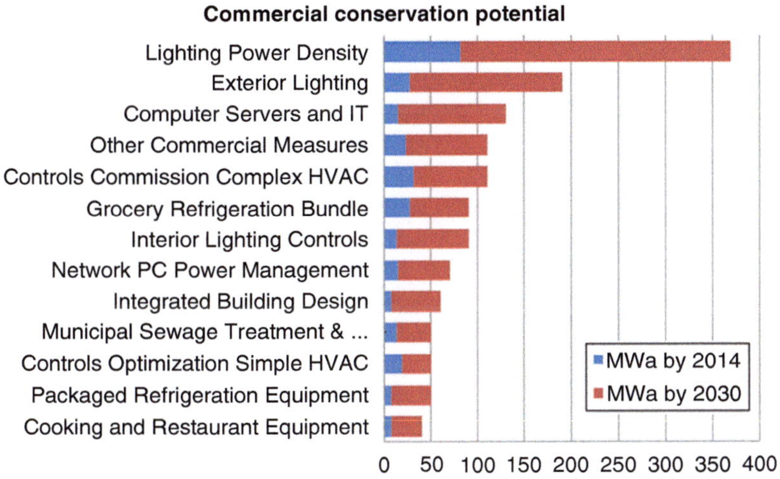

Fig. 10.11 Commercial sector conservation potential in the Pacific Northwest, Figure adapted from NWPCC [4]

10.8.3 Conservation Potential in the Industrial Sector

Total conservation potential in the industrial energy sector is around 800 average megawatts at a cost of almost $50 per megawatt hour whereas 400 average megawatts of this potential can be captured at a cost of $20 per megawatt hour. Energy efficiency improvement areas are quite diverse such as improved business management practices, system sizing and optimization, demand reduction, and use of more efficient equipment. Around a quarter of the projected potential is expected to be captured through improvements within specific subsectors such as refiner plate improvements in mechanical pulping and refrigeration improvements in frozen food processes. Pulp and paper and food processing industries promise the biggest energy efficiency savings which are 300 and 400 average megawatts of savings respectably. The rest of the savings are expected to be captured across all industry subsectors such as pump, fan, compressed air, lighting, and material handling systems [4].

It is worth noting that most of the industrial energy efficiency improvements are subject to human factors, require upfront design requirements, and need continuous control in order to make sure savings occur as planned. Furthermore, industrial decision makers are more risk averse than the rest of the sectors due to criticality of process continuity. These are some of the important considerations in further program design approaches (Fig. 10.12).

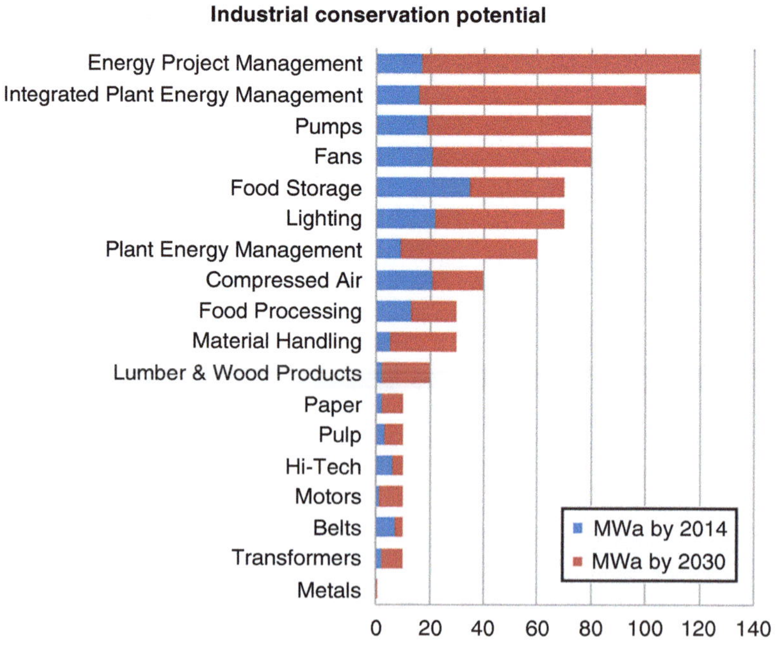

Fig. 10.12 Industrial sector conservation potential in the Pacific Northwest, Figure adapted from NWPCC [4]

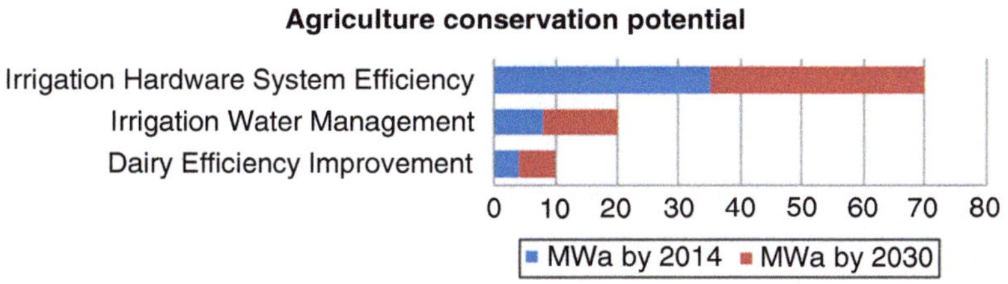

Fig. 10.13 Agriculture sector conservation potential in the Pacific Northwest, Figure adapted from NWPCC [4]

10.8.4 Conservation Potential in the Agriculture Sector

In the sixth power plan, it is reported that there are three major improvement areas in the agricultural sector. These are improvements on irrigation management practices, irrigation hardware, and dairy processes. Hardware improvements promise 75 average megawatts of conservation potential at a cost of less than $100 per megawatt hour. Potential energy efficiency programs are suggested focusing on efficiency improvements such as pump efficiency, leak reduction, promotion of lower pressure irrigation applications, and better sprinkler management practices. Better irrigation management practices are expected to contribute around 15 average megawatts energy efficiency savings. This amount is stated to be limited due to existence of state specific water laws. Lastly, it has been reported that there is significant dairy milk production growth within the last decade. As a result small amount of energy efficiency savings remain in dairy processes. Potential energy efficiency programs are expected to provide ten average megawatts of energy savings through efficient lighting, variable speed drives on milking machine vacuum pumps, and flat plate heat exchangers for precooling milk [4] (Fig. 10.13).

10.8.5 Conservation Potential in the Utility Distribution Systems

Energy efficiency savings can also be realized from optimization of utility distribution systems. For instance, this set of measures is called conservation voltage reduction and attempts to lower acceptable voltage levels to 114–120 V range without harming customer. Typical improvements reduce primary and secondary line losses, optimize reactive power management on substation feeders and transformers, as well as balance feeder voltage and current. NEEA's pilot project has demonstrated that there is also 2–5 % reduction in peak load demand which has positive implications for system reliability matters. Overall, voltage reduction system improvements promise 10–40 % savings for utilities whereas the rest of the savings benefit consumers. Potential savings are projected to be about 400 average megawatts by 2030 at a cost of $30 per megawatt hour [4].

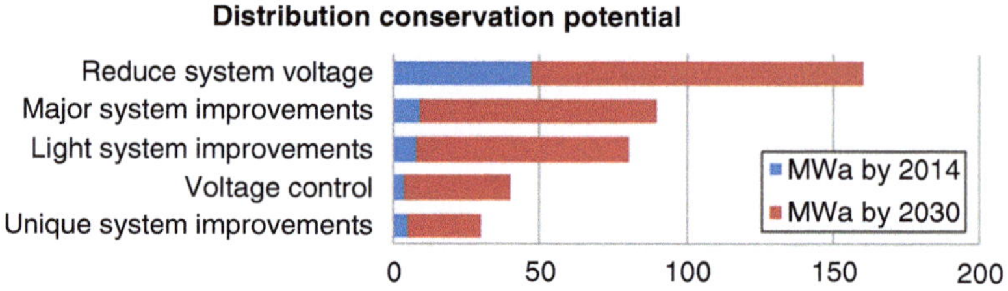

Fig. 10.14 Distribution efficiency sector conservation potential in the Pacific Northwest, Figure adapted from NWPCC [4]

Important considerations with regard to this measure are that there are regulatory disincentives, need for technical assistance, lack of verification methods and organizational challenges within the utilities in the region. A potential energy efficiency program may go beyond BPA's service and require a regional participation over the issue (Fig. 10.14).

10.8.6 Conservation Potential in Consumer Electronics

In the last decade, there is huge market growth in consumer electronics in the Pacific Northwest region. Since energy savings through consumer electronics scatter across all sectors with a variety of products, sixth power plan has placed a special focus on consumer electronics as a conservation resource. Energy efficiency improvements in consumer electronics are expected to provide 800 average megawatts of energy savings by 2030 at a cost of less than $60 per megawatt hour. Energy savings in the consumer electronics is projected to be achieved through labeling programs such as Energy Star. For instance, these programs cover commercial and residential desktop computer monitors, commercial and residential desktop computers, set top boxes, and televisions. Efficiency of consumer electronics can be realized through embedding power management systems into products or using technologies that use less electricity. For example, monitors using LEDs are reported to be 40 % more efficient than same sized alternatives using LCDs or plasma screens.

It has been reported that average size of the televisions has been increasing and the biggest energy savings potential can be captured through labeling programs on televisions. It is expected that approximately 90 % of the televisions are going to exceed 32 in. in size and there is around 400 average megawatts of energy savings potential. Similarly, second highest potential is expected to be exploited through diffusing more efficient residential desktop computers which promise around 200 average megawatts of savings [4].

10.9 Energy Conservation Plan for the Pacific Northwest for 2010–2014 Period

In the sixth power plan, energy conservation has been identified as the one of the least risky and cheapest resource in meeting the regions increasing demand. Sixth power plan has identified total of 1,200 MWa of savings targets in total for 2010–2014 planning period. A total of 42 % of this target corresponds to public's power share which is around 504 MWa and almost one and half times more than previous achievements between 2005 and 2009. It is recognized there are multiple ways of achieving these targets which fall under three major categories: programmatic conservation, market transformation, and nonprogrammatic conservation. Targets set for the new planning period are very challenging due to increased savings expectations in each sector. For instance, total savings targets for 2010–2014 have been determined as 670 MWa for residential sector, 254 MWa for commercial sector, 212 MWa for industrial sector, 46 MWa for agriculture sector, 25 MWa for federal sector, and 67 MWa for distribution efficiency [76].

BPA accounts for the savings gained through nonprogrammatic measures such as codes and standards and tax credits driven by the federal or state governments, however does not offer incentives. Apart from nonprogrammatic measures, BPA, NEEA, and other public utilities in the region are the major drivers behind programmatic energy efficiency efforts. NEEA, funded by BPA, carries out the majority of the market transformation programs, and it is expected that nonprogrammatic and market transformation-related energy efficiency activities will cover 26 % of the conservation target (133 MWa) for 2010–2014 period. Programmatic activities account for the rest of the energy conservation target which is approximately 371 MWa, and it will require BPA and its customers to increase their historical programmatic savings achievement by 68 % [76].

In the following section, you are going to be presented existing programmatic energy efficiency efforts pursued by Bonneville Power Administration as well as new program and emerging technology considerations.

10.10 Energy Efficiency Programs Pursued by Bonneville Power Administration

Bonneville Power Administration pursues a number of energy efficiency programs in variety of sectors which are residential, commercial, industrial, agriculture, distribution efficiency, and federal. In the following sections, existing and new program considerations will be presented with respect to each sector.

Table 10.4 Existing and new energy efficiency program efforts in residential sector

		Nature of program efforts	
Focus area	Opportunity	Existing efforts	New efforts
Residential lighting and shower heads	ENERGY STAR CFL fixtures	x	
	LEDs		x
	Showerheads	x	x
Appliances & water heaters	Appliance (deemed)	x	
	Electric resistance water heater	x	
	Heat pump water heater		x
	Fridge/freezer decommissioning	x	
Consumer electronics	TVs 30 % above ENERGY STAR	x	
	Computers and other		x
HVAC	Air source and geo heat pump	x	x
	Ductless heat pump	x	x
	PTCS installation	x	
	Weatherization	x	x
	ENERGY STAR New Homes	x	
	Low-income weatherization	x	
	Manufactured homes	x	x
	Multifamily program		x

10.10.1 Residential Energy Efficiency Programs

BPA's energy efficiency savings target for the residential sector is 132 MWa for the next 5 years. This target corresponds to 110 % increase from 2005 to 2009 achievement level. Overall strategy for this sector has been identified as improving existing programs in the areas of residential lighting, appliances and electric water heaters, consumer electronics, weatherization, HVAC, Energy Star new homes, and low income programs. New activities have been identified as development of new residential measures and programs for manufactured homes. For the long term, BPA will focus on new measures in the areas of lighting and showerhead for multifamily housing, consumer electronics, heat pump water heaters, and DHPs [77, 78]. Please refer to Table 10.4 for breakdown of existing and new energy efficiency program efforts in residential sector.

10.10.1.1 Existing Residential Energy Efficiency Programs

Existing residential energy efficiency programs pursued by BPA cover areas of lighting, showerheads, weatherization, heating ventilation and air conditioning, refrigeration and water cooling, and manufactured homes. Please refer to Appendix 2 for all existing residential sector program components and measures with corresponding reimbursement levels [77].

10.10.5.1.1 Residential Lighting Program

Over the recent years, residential lighting programs have focused on promoting both standard and specialty screw in CFLs. With the new standard in place which will increase the baseline performance levels, potential savings from CFLs will slightly reduce. Despite the detrimental effects of the standard, CFLs will still be cost effective. In the next 5-year plan for 2010 through 2014, BPA is planning to increase distribution of all types of CFLs among low income and rural markets. BPA will also continue maintaining its efforts with ENERGY STAR qualified fixtures and LEDs in its lighting program offerings. Program continues its operations by focusing on retail promotion, direct mail, and direct installation efforts [78].

10.10.5.1.2 Simple Steps, Smart Savings Program

BPA carried out a regional CFL promotion program named "Change-a-light" between 2006 and 2010 and extended its operations under a new program named "Simple Steps, Smart Savings." Simple steps, smart savings program is promoting the market penetration of CFLs, lighting fixtures, LEDs, and showerheads by collaborating with other public utilities. Program focuses on retail promotions, builder channel development, training and infrastructure support, and direct mail and installation models [79].

10.10.5.1.3 Weatherization Program

Residential weatherization measures have considerable energy efficiency savings potential throughout the region. BPA has been focusing on weatherization programs over multiple decades now and continues the initiative by working with other public utilities and third-party supporters. Focus of this program is to improve energy efficiency standards of residential houses in the areas of insulation, windows, and air sealing. Program employs incentive approach for window replacement and custom project approach for air sealing and weatherization projects. In the next operational period, BPA is planning to change its existing custom project approach to a more streamlined approach in order to reduce procedural complexity of savings calculations and increase customer participation [78].

10.10.5.1.4 Heating, Ventilation, and Air Conditioning Programs

BPA has two major programs in HVAC focus area in order to improve efficiency of heating and cooling systems in the residential sector. These programs are ductless heat pump systems, and duct sealing and ducted heat pumps. For instance, residential HVAC measures include Performance Tested Comfort Systems (PTCS), high efficiency air source heat pump upgrades and conversions, PTCS ground source heat pumps, duct sealing with PTCS protocols and, DHPs in single family homes with zonal electric heat.

PTCS, sponsored by BPA and public utilities, is a regional program aiming to make sure residential ducted systems perform at the optimum level. Program focuses on training technicians on sealing leaky ductwork, installation specifications, performance testing, and adjusting for improved operating performance and life.

In the coming years, opportunities from air source heat pumps and DHPs will be exploited further by identifying new applications for program offerings [80].

10.10.5.1.5 Refrigeration and Water Heating Appliances

Water heating constitutes more than 15 % of the electric energy use in the residential sector. BPA offers incentives for ENERGY STAR qualified clothes washers, refrigerators, and freezers, as well as refrigerator decommissioning, resistance storage water heaters, and heat pump water heaters. BPA is also improving its connections with major retailers and increase diffusion of high efficiency water heater technologies. Moreover, BPA will extended its existing decommissioning program by helping solid waste removal companies build capacity for waste removal services in rural areas. Program will be extended to include clothes washers as well [81].

10.10.5.1.6 Manufactured Homes Program

BPA has been heavily involved in promoting energy efficient homes through number of programs such as Manufactured Home Acquisition Program, Super Good Cents, and Northwest Energy Efficient Manufactured Homes. Due to economic downturn in the recent years, diffusion of energy efficient manufactured homes has slowed down. Currently, BPA is working toward improving energy efficiency of lighting fixtures and heating systems in manufactured housing as well as developing measurement and verification protocols. Over the recent years, BPA worked with NEEA to expand the scope of ENERGY STAR Homes program by incorporating Built Green and LEED homes which meet required efficiency standards. Similar to this effort, BPA will also look for opportunities in an electric home model named as Montana House (MVP Home).

10.10.5.1.7 Tribal and State Low Income Public Purpose Program

Community Action Program implements Federal Weatherization Assistance Program and Low Income Energy Assistance Program. BPA's Public Purpose Program funds qualifying low income households in its service territory through Community Action Program. In addition to that, local tribes in the region are trained on low income weatherization in order to equip them knowledge skills to start their own weatherization programs. Some of the funding is also used to support energy efficient measures such as refrigerators, clothes washers, and CFLs.

10.10.1.2 Residential Energy Efficiency Program Considerations

New residential program considerations in the next 5 years are identified as extending the existing programs to also serve multifamily end use and develop reimbursements for electronics measures. Furthermore, long-term program considerations are focused on exploiting energy savings from heat pump water heaters and DHPs [77].

10.10.5.1.1 Multifamily Program Considerations

Sixth power plan has indicated that multifamily sector has 5.4 MWa of weatherization energy savings potential in the next 5 years. Although savings are significant, it is very challenging to exploit these opportunities. In the next 5-year period, BPA has developed plans that will gradually allow for more savings realizations. For instance, initially residential CFL lighting and showerhead measures are going to be supported through direct installation programs. Additionally, savings potentials of new multifamily measures will be assessed for program consideration. These measures are identified as weatherization, space heating, water heating, lighting, and plug load measures [76].

10.10.5.1.2 Consumer Electronics Program

As of 2008, BPA, ETO, and NEEA have started working toward transforming consumer electronics market. NEEA has started an initiative to provide upstream market incentives to manufacturers for development of more efficient televisions. Initially, the program is focused on televisions which are 30 % or above ENERGY STAR performance levels [82].

For the next 5 years, BPA is going to maintain its existing initiative with NEEA on energy efficient televisions. Under the same initiative, BPA and NEEA will develop more measures such as personal home computers, computer monitors, set top boxes, game consoles, and DVD players [76].

10.10.5.1.3 Heat Pump Water Heaters

Energy efficient electric storage water heaters and heat pump water heaters are interests of water heating program at BPA. It is stated that heat pump water heaters are almost 50 % more efficient than conventional electric storage water heaters; however, its effectiveness in the cold climate regions has not been fully understood. For instance, collaborating with Electric Power Research Institute (EPRI), water heating program focuses on coordinating lab testing and technology demonstration efforts which started in 2009. Furthermore, supported by NEEA, program efforts also focus on development of manufacturer specifications that will ensure the next generation of products will meet or exceed ENERGY STAR standards [76].

10.10.5.1.4 Ductless Heat Pumps

Almost 50 % of the residential energy use belongs to heating and cooling systems which constitute significant importance. For instance, DHPs have been identified as superior to wall heaters, base boards, and other forms of zonal electric resistance heating units. Moreover, since DHP applications do not require expensive and invasive ductwork, and work quietly, they have potential to diffuse in most of the residential end use in the region. Many utilities in the region have started providing incentives for these systems in their own program initiatives.

NEEA and BPA have started a pilot study for single family homes with zonal heat and forced air furnaces in 2009. Pilot study is going to help collect data for developing deemed savings offer for DHPs. Currently, BPA's emerging technologies group is looking for more applications where DHP can create savings. Areas of research have been identified as manufactured houses with electric forced furnaces and small commercial buildings [76].

10.10.2 Commercial Energy Efficiency Programs

BPA's energy efficiency savings target for the commercial sector is 100 MWa for the next 5 years. Strategies identified on improving existing programs are in the areas of commercial lighting, Trade Ally Network, and Energy Smart Grocer program. BPA is going to continue investing its lighting program in order to cope with federal standard changes which are to increase baseline performance levels and shrink energy savings potential. Energy Smart Grocer program is stated to reach a saturation point; BPA will put more effort toward HVAC and new construction measures in order to capture remaining potentials in this area.

Strategies on new activities have been identified as new and improved savings calculators, PC power management, commercial kitchens, and commercial electronics. Long-term research initiatives have been identified as analysis of a similar industrial program for a more streamlined commercial program design; and emerging technologies such as demand-controlled ventilation, and smart monitoring and diagnostic systems for rooftop HVAC systems [77, 83]. Please refer to Table 10.5 for breakdown of existing and new energy efficiency program efforts in commercial sector.

10.10.2.1 Existing Commercial Energy Efficiency Programs

Existing commercial energy efficiency programs cover areas of lighting, showerheads, plug load, kitchen and food services, heating, ventilation and air conditioning, new construction, and weatherization. Please refer to Appendix 3 for all existing commercial sector program components and measures with corresponding reimbursement levels [77].

Table 10.5 Existing and new energy efficiency program efforts in commercial sector

	Nature of program efforts	
Opportunity	Existing efforts	New efforts
Commercial lighting and TAN	x	
Grocery refrigeration	x	
Custom projects	x	
Small commercial	x	
Network PC power management		x
Hospitality and commercial kitchens		x
Commercial electronics		x
Commissioning		x
New construction		x
New fast-track measures		x
Emerging technologies research		x

10.10.5.1.1 Commercial Lighting Program

BPA has been maintaining commercial lighting program for several years now, and it has become the sector's largest initiative. With the participation of customer utilities, 6 MWa electricity savings have been realized in 2008. BPA runs the program with Northwest Trade Ally Network (NTAN) and helps utilities design and implement programs through providing trainings. Program also provides end users with necessary information and tools to detect potential ways of reducing energy consumption in a cost-effective way. Due to changing lighting standards, emergence of new technologies and growing interests in large-scale retrofit projects, BPA has decided to extend its program offerings. In the next 5 years, BPA is planning to improve its Commercial Lighting Program and Trade Ally Network by simplifying its commercial lighting calculator tool for increased participation. Lighting program will also start promoting a limited selection of LED lighting units. Moreover, exterior lighting opportunities are considered as an area of emerging technologies [84].

10.10.5.1.2 Grocery Refrigeration Program

Supported by a third-party contractor named Portland Energy Conservation Inc., BPA maintains its refrigeration conservation efforts under a program named Energy Smart Grocer. This program has been in place since 2007 and has delivered 16 MWa of savings. Energy Smart Grocer program offers incentives for more than 50 measures to grocers and other entities with commercial refrigeration loads. Program also supports interior and exterior lighting, new construction, building commissioning, and HVAC-related energy efficiency improvements as long they are affiliated with a refrigeration project [85, 86].

10.10.5.1.3 Hospitality and Commercial Kitchen Program

Commercial kitchens in restaurants, hospitals, and similar facilities have been reported to be the most energy intensive areas per square foot and account for almost 60 % of the total energy use. Potential savings from more efficient appliances and practices is expected to be around 10–30 %. BPA has included energy efficient commercial kitchen measures into its portfolio as of 2010. These measures cover appliances such as refrigerators, freezers, ice makers, steamers, hot food holding cabinets, prerinse spray wash valves, convection and combination ovens, electric fryers, and commercial dishwashers. BPA provides reimbursement for these measures as long as they satisfy the energy efficiency performance levels determined by CEE tier 2 standards. For instance, list of these appliances can be found in ENERGY STAR Commercial Kitchen package [87, 88].

10.10.5.1.4 Heating, Ventilation, and Air Conditioning Program

BPA offers reimbursements for HVAC measures through its custom project incentive process. Custom project process requires technical staff to determine savings by conducting measurement and verification procedures at actual place. Thus, amount of reimbursements is determined based on the savings performance. HVAC program has started providing incentives for rooftop AC units since 2010 and DHPs in small commercial spaces as of 2011. Despite the recent progress, measure development is still considered in progress. For instance, in the next operation period, BPA is planning to simplify its custom project procedures starting with its HVAC measures. Simplification will be toward developing standardized evaluation, measurement, and verification process in order to increase participation. In the longer time period, BPA is planning to develop new measures such as demand-controlled ventilation, premium ventilation for rooftop packaged units, rooftop unit servicing and upgrades, and smart monitoring and diagnostic systems for rooftop HVAC systems [89].

10.10.5.1.5 Commercial Custom Projects

Many of the HVAC systems projects, building shell projects, and some small commercial applications are given reimbursements under custom project proposals due to complexity and variability associated with each project. As a result, participation in custom projects is rather slow. In order to obtain faster realization of savings, BPA is going to create alternate ways of providing reimbursements. For instance, these alternatives will be more standardized approaches as long as they are cost effective.

10.10.5.1.6 Commercial Building Shell Program

Existing BPA program offerings for commercial building shell area are for retrofit insulation projects and small commercial window retrofits. This measure is aimed at commercial buildings with less than 5,000 sq. ft. with specified window characters. BPA aims to improve the existing program by adding more robust measure offerings which will support customer utilities to capture more of building shell opportunities [90].

10.10.5.1.7 New Construction Program

Due to economic downturn, there is no significant new construction expectation in the region. New construction program aims to capture design-related energy efficiency opportunities in commercial buildings by providing incentives and custom project offerings. Since custom projects are time and resource intensive, BPA staff is working toward streamlining the existing approach for reaching wider participation. Moreover, BPA will also look into LEED accreditation process and align its programmatic approaches accordingly.

10.10.2.2 Commercial Energy Efficiency Program Considerations

Energy efficiency program considerations in the commercial sector are electronics equipment and commercial commissioning projects. Currently, E3T program is evaluating a number of applications for commercial sector. These applications are control commissioning, demand-controlled ventilation, premium ventilation packages for rooftop packaged units, rooftop unit servicing/upgrades, smart monitoring and diagnostic systems for rooftop HVAC systems, VRF systems, energy management, integrated building design, and daylighting controls [77].

10.10.5.1.1 Commercial Electronics Program

Commercial electronics program aims to exploit energy efficiency opportunities from idle office equipment such as scanners, computers, monitors, printers, and desk lighting. It is estimated that smart power strips have potential to reduce energy consumption in an office at around 100 kWh per year. Program provides reimbursement for smart power strips which monitor and actively manage plug load usage from electrical equipment, turning off the power supply when devices are idle. Currently, BPA is collecting data for deeming this measure in order to enable wider diffusion throughout the region. In the next 5 years, BPA is planning to start a new initiative to capture energy savings from computers which are idle. Network PC Power Management initiative will focus on direct installation of a power supply control software package. This program will be delivered from the same marketing channel as smart power strips. Similarly, BPA will also promote new initiatives toward promoting more efficient office equipment such as computers, monitors, servers, and data centers. In order to deliver the best savings, BPA is carrying on emerging technologies effort to exploit opportunities from computer servers for data servers [91].

10.10.5.1.2 Commissioning Program

Sixth power plan has indicated that building commissioning has significant energy savings potential; however, commissioning projects are difficult to implement due to excess staff time requirement. In the next 5-year operation period, BPA is planning to incorporate building commissioning for commercial sector by learning lessons from its industrial commissioning program. In the long run, BPA will work with NEEA on improving codes and standards for existing commercial new constructions in order to increase savings potential.

10.10.3 Industrial Energy Efficiency Programs

BPA's energy efficiency savings target for the industrial sector is 74 MWa for the next 5 years. Energy Smart Industrial program is an initiative established in 2009 and designed to replace a previous custom project initiative by providing one-on-one relationship with utility customers and end users. This approach is experienced to be more streamlined than the earlier program design. New activities will be focused on expanding available measures such as energy management pilot studies, trade ally for small industrial measures, and technical service providers. Long-term industrial energy efficiency research initiatives will be focused on energy management pilots [77, 92]. Please refer to Table 10.6 for breakdown of existing and new energy efficiency program efforts in industrial sector.

10.10.3.1 Existing Commercial Energy Efficiency Programs

Existing commercial energy efficiency programs cover areas of lighting, showerheads, plug load, kitchen and food services, heating, ventilation and air conditioning, new construction, and weatherization. Please refer to Appendix 3 for all existing commercial sector program components and measures with corresponding reimbursement levels [77].

10.10.5.1.1 Commercial Lighting Program

BPA has been maintaining commercial lighting program for several years now, and it has become the sector's largest initiative. With the participation of customer utilities, 6 MWa electricity savings have been realized in 2008. BPA runs the program

Table 10.6 Existing and new energy efficiency program efforts in industrial sector

	Nature of program efforts	
Opportunity	Existing efforts	New efforts
Commercial lighting and tan	x	
Grocery refrigeration	x	
Custom projects	x	
Small commercial	x	
Network PC power management		x
Hospitality and commercial kitchens		x
Commercial electronics		x
Commissioning		x
New construction		x
New fast-track measures		x
Emerging technologies research		x

with Northwest Trade Ally Network (NTAN) and helps utilities design and implement programs through providing trainings. Program also provides end users with necessary information and tools to detect potential ways of reducing energy consumption in a cost-effective way. Due to changing lighting standards, emergence of new technologies and growing interests in large-scale retrofit projects, BPA has decided to extend its program offerings. In the next 5 years, BPA is planning to improve its Commercial Lighting Program and Trade Ally Network by simplifying its commercial lighting calculator tool for increased participation. Lighting program will also start promoting a limited selection of LED lighting units. Moreover, exterior lighting opportunities are considered as an area of emerging technologies [84].

10.10.5.1.2 Grocery Refrigeration Program

Supported by a third-party contractor named Portland Energy Conservation Inc., BPA maintains its refrigeration conservation efforts under a program named Energy Smart Grocer. This program has been in place since 2007 and has delivered 16 MWa of savings. Energy Smart Grocer program offers incentives for more than 50 measures to grocers and other entities with commercial refrigeration loads. Program also supports interior and exterior lighting, new construction, building commissioning, and HVAC-related energy efficiency improvements as long they are affiliated with a refrigeration project [85, 86].

10.10.5.1.3 Hospitality and Commercial Kitchen Program

Commercial kitchens in restaurants, hospitals, and similar facilities have been reported to be the most energy intensive areas per square foot and account for almost 60 % of the total energy use. Potential savings from more efficient appliances and practices are expected to be around 10–30 %. BPA has included energy efficient commercial kitchen measures into its portfolio as of 2010. These measures cover appliances such as refrigerators, freezers, ice makers, steamers, hot food holding cabinets, prerinse spray wash valves, convection and combination ovens, electric fryers, and commercial dishwashers. BPA provides reimbursement for these measures as long as they satisfy the energy efficiency performance levels determined by CEE tier 2 standards. For instance, list of these appliances can be found in ENERGY STAR Commercial Kitchen package [87, 88].

10.10.5.1.4 Heating, Ventilation, and Air Conditioning Program

BPA offers reimbursements for HVAC measures through its custom project incentive process. Custom project process requires technical staff to determine savings by conducting measurement and verification procedures at actual place. Thus, amount of reimbursements is determined based on the savings performance. HVAC program has started providing incentives for rooftop AC units since 2010 and DHPs in small commercial spaces as of 2011. Despite the recent progress, measure development is still considered in progress. For instance, in the next operation period, BPA is planning to simplify its custom project procedures starting with its HVAC measures. Simplification will be toward developing standardized evaluation, measurement, and verification process in order to increase participation. In the longer time period, BPA is planning to develop new measures such as demand-controlled ventilation, premium ventilation for rooftop packaged units, rooftop unit servicing and upgrades, and smart monitoring and diagnostic systems for rooftop HVAC systems [89].

10.10.5.1.5 Commercial Custom Projects

Many of the HVAC systems projects, building shell projects, and some small commercial applications are given reimbursements under custom project proposals due to complexity and variability associated with each project. As a result, participation in custom projects is rather slow. In order to obtain faster realization of savings BPA is going to create alternate ways of providing reimbursements. For instance, these alternatives will be more standardized approaches as long as they are cost effective.

10.10.5.1.6 Commercial Building Shell Program

Existing BPA program offerings for commercial building shell area are for retrofit insulation projects and small commercial window retrofits. This measure is aimed at commercial buildings with less than 5,000 sq. ft. with specified window characters. BPA aims to improve the existing program by adding more robust measure offerings which will support customer utilities to capture more of building shell opportunities [90].

10.10.5.1.7 New Construction Program

Due to economic downturn, there is no significant new construction expectation in the region. New construction program aims to capture design-related energy efficiency opportunities in commercial buildings by providing incentives and custom project offerings. Since custom projects are time and resource intensive, BPA staff is working toward streamlining the existing approach for reaching wider participation. Moreover, BPA will also look into LEED accreditation process and align its programmatic approaches accordingly.

10.10.3.2 Commercial Energy Efficiency Program Considerations

Energy efficiency program considerations in the commercial sector are electronics equipment and commercial commissioning projects. Currently, E3T program is evaluating a number of applications for commercial sector. These applications are control commissioning, demand-controlled ventilation, premium ventilation packages for rooftop packaged units, rooftop unit servicing/upgrades, smart monitoring and diagnostic systems for rooftop HVAC systems, VRF systems, energy management, integrated building design, and daylighting controls [77].

10.10.5.1.1 Commercial Electronics Program

Commercial electronics program aims to exploit energy efficiency opportunities from idle office equipment such as scanners, computers, monitors, printers, and desk lighting. It is estimated that smart power strips have potential to reduce energy consumption in an office at around 100 kWh per year. Program provides reimbursement for smart power strips which monitor and actively manage plug load usage from electrical equipment, turning off the power supply when devices are idle. Currently, BPA is collecting data for deeming this measure in order to enable wider diffusion throughout the region. In the next 5 years, BPA is planning to start a new initiative to capture energy savings from computers which are idle. Network PC Power Management initiative will focus on direct installation of a power supply control software package. This program will be delivered from the same marketing channel as smart power strips. Similarly, BPA will also promote new initiatives toward promoting more efficient office equipment such as computers, monitors, servers, and data centers. In order to deliver the best savings, BPA is carrying on emerging technologies effort to exploit opportunities from computer servers for data servers [91].

10.10.5.1.2 Commissioning Program

Sixth power plan has indicated that building commissioning has significant energy savings potential; however, commissioning projects are difficult to implement due to excess staff time requirement. In the next 5-year operation period, BPA is planning to incorporate building commissioning for commercial sector by learning lessons from its industrial commissioning program. In the long run, BPA will work with NEEA on improving codes and standards for existing commercial new constructions in order to increase savings potential.

10.10.4 Industrial Energy Efficiency Programs

BPA's energy efficiency savings target for the industrial sector is 74 MWa for the next 5 years. Energy Smart Industrial program is an initiative established in 2009 and designed to replace a previous custom project initiative by providing one-on-one relationship with utility customers and end users. This approach is experienced to be more streamlined than the earlier program design. New activities will be focused on expanding available measures such as energy management pilot studies, trade ally for small industrial measures, and technical service providers. Long-term industrial energy efficiency research initiatives will be focused on energy management pilots [77, 92]. Please refer to Table 10.7 for breakdown of existing and new energy efficiency program efforts in industrial sector.

10.10.4.1 Existing Industrial Energy Efficiency Programs

BPA hosts an industrial energy efficiency program which is referred as "Energy Smart Industrial Program" and maintains industrial energy efficiency projects. Industrial energy efficiency projects are mostly custom project based, thus Energy Smart Industrial program provides a number of services. Please refer to Appendix 4 for all existing industrial sector program components and measures with corresponding reimbursement levels [77].

10.10.5.1.1 Energy Smart Industrial Program

Energy Smart Industrial program has recently been launched in 2009 and delivered 12 and 15 MWa in 2010 and 2011, respectively. According to sixth power plan, this amount needs to be doubled in the coming years. Maintained by a third-party program delivery parent, Cascade Engineering, Energy Smart Industrial program offers a wide range of support for BPA utility customers and their industrial end users. Program has been designed in a way that it will have a one-on-one contact with end users and BPA engineering service. This structure has been so far experienced to be

Table 10.7 Existing and new energy efficiency program efforts in industrial sector

Opportunity	Nature of program efforts	
	Existing efforts	New efforts
Energy smart industrial	x	
Energy smart industrial partners		x
Energy management pilot		x
Trade ally delivered small industrial measures		x
Northwest trade ally network (C&I lighting)		x
BPA-funded technical service providers		x
Target segment strategies		x
Emerging technologies research		x

more streamlined in a way that it functions more effectively. Energy Smart Industrial Program has five components which are named as Energy Smart Industrial Partner (ESIP), Energy Management, small industrial measures, enhanced lighting, and technical service provider.

- Energy smart industrial partner
  ESIP is an industrial energy efficiency expert who acts as a contact person between utility and industrial customer. Partner helps customers in identifying energy efficiency goals and meeting them. In addition to those, ESIP also facilitates development and implementation of industrial projects. Cascade engineering fulfills this program component.
- Energy management pilot
  Energy management is a pilot component of Energy Smart Industrial program which attempts to improve energy management practices of the customers. Energy management pilot program provides cofunding for energy efficiency projects in order to make them more competitive against other facility specific projects, provides ways of tracking energy consumption and optimized solutions, as well as training on how to implement energy management practices.
- Trade ally delivered small industrial measures
  Majority of the energy efficiency improvements in the industrial sector are custom projects which require constant measurement and verification for assuring savings. Small industrial measures component attempts to identify measures that may provide small savings, however can be streamlined to a wider population. BPA has been offering incentives for compressed air measures which contain variable frequency drive air compressors, load and unload air compressors, zero loss drains, cycling refrigeration dryers, increased air storage, and efficient filter.
- Northwest trade ally network (nonresidential lighting)
  Industrial lighting projects are supported through NTAN. Program assigns industrial energy efficiency specialists to help customers with related lighting projects. As described earlier, specialists work one on one with the customers in order to effectively manage the projects.

– Technical service provider

  BPA also assists its customers with project scoping, measurement and verification analyses, specific assessments, completion reports, and similar tasks.

10.10.4.2 Industrial Energy Efficiency Program Considerations

BPA is working in collaboration with Cascade engineering that has developed a strategy for exploiting energy efficiency savings from wastewater treatment facilities. As a result, IT data center efforts in this sector have been pushed to a later stage [77].

10.10.5 Agriculture Energy Efficiency Programs

BPA's energy efficiency savings target for the agriculture sector is 20 MWa for the next 5 years. Agriculture end users reside across the rural areas widespread within BPA's service territory. In order to capture these savings, BPA will focus on marketing and regional coordination services as well as improving the existing programs and activities such as variable frequency drivers (VFDs) and irrigation and hardware system measures. In the long term, BPA is going to evaluate available technologies in the area of pump impeller and adjustment, and sprinkler heads for landscaping. Moreover, new measures such as impeller optimization and computerized sprinkler heads are going to be developed as well. BPA will also deliver a program named Turf Irrigation program which will exploit energy and water savings from commercial landscaping, golf courses, and properties owned by government and local municipalities [77, 93]. Please refer to Table 10.8 for breakdown of existing and new energy efficiency program efforts in agriculture sector.

10.10.5.1 Existing Agriculture Energy Efficiency Programs

Existing agriculture energy efficiency programs pursued by BPA provides services in the areas of scientific irrigation, wineries, and diaries as well as high efficiency motors and irrigation hardware. Please refer to Appendix 5 for all existing agriculture sector program components and measures with corresponding reimbursement levels [77].

10.10.5.1.1 Scientific Irrigation Scheduling Program

BPA supports use of scientific methods in farming through its programs called Scientific Irrigation Scheduling (SIS) and Scientific Irrigation Scheduling Light (SISL). These programs help BPA save water and energy along with some side

Table 10.8 Existing and new energy efficiency program efforts in agriculture sector

	Nature of program efforts	
Opportunity	Existing efforts	New efforts
Irrigation hardware	x	
Irrigation systems	x	
Scientific irrigation scheduling light	x	
Federal irrigation districts		x
Custom projects		x
BPA-funded TSP		x
Pump testing		x
Scientific irrigation scheduling	x	
Turf pilot		x

benefits such as labor, fertilizer, increased crop yield, and quality to participating customers. Both SIS and SISL are applicable to irrigation systems where there is excess pumping capacity than required crop needs. Under the program, participants are asked to provide data about the amount of water applied, moisture level in the soil, etc., that can potentially help regions' water management efforts. BPA is also developing an online tool which will integrate all data gathered from farmers and utilities for better water management planning. In the next 5 years, BPA is going to extend its scientific irrigation program by providing incentives to smaller and federal irrigation districts. Existing programs will be further extended through identification of end users types and specific offerings for them [93].

10.10.5.1.2 Irrigation Hardware and Systems Program

BPA's existing program offers incentives for irrigation system upgrades such as irrigation piping, sprinklers, regulators, nozzles, and green motors.

10.10.5.1.3 Wineries and Diaries

Wineries constitute significant energy conservation potential in the agriculture sector. For instance, there are almost 1,000 wineries and vineyards in the Pacific Northwest, and it has been the fastest growing of all segments in agriculture sector. BPA provides incentives on energy savings enhancements for wineries. These incentives are provided in the areas of lighting, HVAC, pipe insulation, compressed air, VFDs, and refrigeration.

BPA has been supporting energy conservation activities in diaries through more efficient barn area lighting, flat plate chiller improvements, and variable frequency drive applications in milking processes [93].

10.10.5.1.4 Premium Efficiency Motors Program

Premium efficiency program focuses on replacing older rewound agriculture motors with rewind motor alternatives.

10.10.5.1.5 Pump Testing Program

Pump testing program focuses on performance evaluation of pumping motors. In case of an upgrade need detected and implemented, BPA supports the qualifying projects by reimbursing the cost of pump testing.

10.10.5.2 Energy Efficiency Program Considerations

Agricultural energy efficiency programs in consideration are dedicated to capture energy efficiency savings in irrigation districts as well as small-sized irrigation applications, variable frequency drive, and low-energy precision applications [77].

10.10.5.1.1 Irrigation Districts

BPA's customer utilities are servicing considerable amount of irrigation districts which have significant energy savings potential. Since the nature of irrigation district projects requires large and long-term funding, savings realizations have been rather low. BPA has recently added couple of irrigation district projects in its portfolio and will focus on identification and elimination of program diffusion barriers for further program considerations.

10.10.5.1.2 Variable Frequency Drives for Hardware

VFDs can control operation performance of a motor or pump and provide energy efficiency savings. Energy efficiency improvements under this area are mainly regarded as custom projects; however, BPA plans to provide more streamlined approach for faster diffusion. Currently, BPA has been gathering data in order to create VFD deemed measures.

10.10.5.1.3 Low-Energy Precision Applications

Low-energy precision applications aim to support market for agriculture energy efficiency opportunities in the region. Particularly, the initiative will focus on both human and animal food producers that use irrigation services. First step of the program is to assess viability of such applications in the region for further program considerations.

10.10.5.1.4 Turf Pilot

In order to address capacity issues at peak irrigation season between May and September, BPA is running a pilot study to find out proper residential and commercial irrigation applications. For instance, these applications include residential and commercial landscaping, gold courses, and government and municipality properties. Currently, BPA is collecting data in an attempt to assess savings potential and cost effectiveness.

10.10.6 Distribution System Efficiency Programs

Electric utilities are having big issues with respect to maintaining system reliability and performance due to aging infrastructure. In order to address the issue, BPA has recently started a new program called Energy Smart Utility Efficiency. Through the program BPA is planning to achieve 20 MWa of energy savings in the next 5 years. Aforementioned program aims to improve distribution system efficiency in a variety of ways such as high efficiency transformer replacements, load balancing, line reconductoring, and voltage optimization [77]. Please refer to Table 10.9 for breakdown of existing and new energy efficiency program efforts in distribution efficiency sector.

10.10.6.1 Existing Distribution System Efficiency Programs

Currently, there are a number of distribution efficiency related programs pursued by BPA. These programs mostly provide technical trainings, educations, and pilot studies [77].

Table 10.9 Existing and new energy efficiency program efforts in distribution efficiency sector

	Nature of program efforts	
Opportunity	Existing efforts	New efforts
Simplified voltage optimization M&V protocols	x	
Stability thresholds	x	
Calculator improvements/enhancements	x	
Business case model	x	
Education and training workshops	x	x
Utility efficiency technical work group		x
Expanded technical services pool		x
TrakSmart		x
Program implementation		x
Utility efficiency technical work group		x

10.10.5.1.1 System Improvement Program

Distribution infrastructure becomes less efficient as it becomes older. BPA delivers a program to help utilities' renew some of the distribution hardware. For instance, program offers power transformer replacements, service conductor replacement, higher distribution primary voltage, transformer load management, balancing loads and phases, adding parallel feeders, operation improvements, deenergizing seasonally unloaded transformers, and service of distribution transformers [77].

10.10.5.1.2 Voltage Optimization Program

In simple terms, voltage optimization program attempts to keep voltage between 114 and 120 V which does not hurt customers, but saves utilities energy due to reduced line loss. Voltage optimization methods have been in use for almost 30 years. Due to perceived cost, negative customer impacts, and complex design and operation issues, utilities have been hesitant about exploiting efficiency savings in this sector. In order to address the issue, BPA has been working on simplifying measurement and verification process for enabling rapid diffusion.

In order to make voltage optimization program successful, BPA is going to develop a set of system reliability procedures for reducing low voltage problems. BPA will also deliver a calculator for easier analysis of distribution system savings. Moreover, BPA is also planning to provide education services for the utilities' staff in order to address knowledge-related barriers [77].

10.10.5.1.3 Education and Training Workshops

BPA has started offering education and training workshops to distribution engineers in its service territory as of 2009. The workshops aims to provide information about voltage optimization design principles, effective approaches, and impacts on equipment. Further education and training workshops will focus on modeling and mapping for voltage optimization analyses [77].

10.10.5.1.4 Distribution System Studies and Program Partner

As of 2010, BPA has started working with 15 of its public utility customers on assessing distribution system energy efficiency potential. Based on the assessment results, custom project proposals have been received and five of the utilities have been selected for implementation. BPA has also brought in a third-party contractor to manage the projects' scopes, performance, and required documentation necessities [77].

10.10.6.2 Distribution System Efficiency Program Considerations

BPA is planning to extend its existing programs to smaller utilities as well; in order to do that number of technical service providers will be increased. Through the workshops and training programs, awareness about the efficiency potential in distribution systems will be outreached [77].

10.10.7 Federal Energy Efficiency Programs

BPA provides federal facilities with energy efficiency-related services in the areas of energy auditing, project scoping, engineering, project financing and management, and implementation and verification. End use in this sector is usually military bases or other large federal facilities. BPA's energy efficiency savings target for the federal sector is 25 MWa for the next 5 years. Plans for the next period have been identified as whether to focus on large custom projects or small measures that can be provided to a larger group of end users [77].

BPA has been carrying out energy efficiency efforts with federal facilities under custom projects. Since custom projects are site specific and procedurally costly and complex, BPA is planning to focus its effort on developing small measures that can be streamlined easily over wider range of participants.

10.10.7.1 Existing Federal Energy Efficiency Programs

BPA will continue working with a number of federal agencies in meeting their strategic energy management plans, accessing financing opportunities, and managing implementation projects. For instance, BPA is about to complete a multimillion dollar projects with Puget Sound Naval Bases and Joint Base Lewis McChord. BPA will continue working toward exploiting energy efficiency savings by targeting key federal subsectors which are utility served federal, direct served federal, and reserve power federal [77].

10.10.7.2 Federal Energy Efficiency Program Considerations

Majority of the federal energy efficiency projects are custom projects which require significant amount of resources to implement. Currently, BPA is screening a number of energy efficiency measures that can provide energy efficiency savings without requiring large amount of resources. Accordingly, BPA is working with a number of agencies to uncover additional federal energy saving opportunities [77].

Appendix 1: List of Emerging Technologies Under Review at BPA

This list has been acquired from Bonneville Power Administration Energy Efficiency Group website. Link: http://www.bpa.gov/energy/n/emerging_technology/pdf/E3TNW_TechList_2012_01_30.pdf

No	Emerging technology	Focus area
1	Food cooking	Commercial kitchen
2	Computer power supply ET	Consumer electronics
3	High power density in data centers	Consumer electronics
4	Intelligent power strip	Consumer electronics
5	Advanced network power management	Consumer electronics
6	High-efficiency set-top boxes	Consumer electronics
7	Appliance power monitoring device	Consumer electronics
8	OLED screens	Consumer electronics
9	Efficient office equipment	Consumer electronics
10	Flash memory for hard drives computers	Consumer electronics
11	Copier of the future	Consumer electronics
12	ENERGY STAR standby in electronics	Consumer electronics
13	Integrated processor chips	Consumer electronics
14	Efficient power supplies	Consumer electronics
15	Heat pump dryer	Consumer electronics
16	Low swirl combustion	Consumer electronics
17	Smart grid	Consumer electronics
18	Best-in-class efficient equipment	Consumer electronics
19	Youlet glowing wall outlet	Consumer electronics
20	Smart charger controller	Consumer electronics
21	Indoor plants to improve indoor air quality	Energy management
22	Ceramment (TM) ceramic cement-based concrete replacement	Energy management
23	ASHRAE Advanced energy design guides (AEDG)	Energy management
24	Building energy performance analytics software and services	Energy management
25	Nonintrusive load monitoring	Energy management
26	Meter each building on university campuses	Energy management
27	Wireless pneumatic thermostat	Energy management
28	Advanced rooftop unit controls with remote access and energy monitoring	Energy management
29	Air flow management for data centers	Energy management
30	Effective customer engagement programs for home energy management	Energy management
31	Innovative behavior change techniques	Energy management
32	On-bill financing of energy efficiency projects	Energy management
33	Continuous energy improvement (CEI)	Energy management
34	Flower pod	Energy management

(continued)

(continued)

No	Emerging technology	Focus area
35	Utility-to-building connection	Energy management
36	Hotel room automation	Energy management
37	Home displays	Energy management
38	Server virtualization	Energy management
39	Continuous energy management	Energy management
40	Advanced metering	Energy management
41	Wireless building automation	Energy management
42	Dashboard systems and continuous monitoring-based commissioning	Energy management
43	Smart circuit controllers	Energy management
44	Residential occupancy based HVAC energy management	Energy management
45	Meter each space in multitenant leased property	Energy management
46	Home energy automation with interface to the smart grid	Energy management
47	Creative financing for emerging technology projects	Energy management
48	EnergyCAP energy management software	Energy management
49	Master control for residential electrical appliances	Energy management
50	Home energy wireless controls	Energy management
51	Wastewater treatment best practices trainings	Energy management
52	National sharing of custom project lessons learned	Energy management
53	Energy saving competitions among businesses	Energy management
54	Promote ASHRAE 100	Energy management
55	Piggyback on other programs for program outreach	Energy management
56	Infrared drive-by building envelope assessments	Energy management
57	Energy tracking and cost accounting software	Energy management
58	Internet-based refrigeration system monitoring system	Energy management
59	Training on energy efficient product selection	Energy management
60	Increased use of key performance indicators in HVAC system optimization	Energy management
61	ASD trainings	Energy management
62	Integration of technologies	Energy management
63	Techniques associated with technologies for success	Energy management
64	Wireless, web-enabled monitoring	Energy management
65	Integrated and turnkey energy management and control solutions	Energy management
66	Data center energy management tools	Energy management
67	Nonintrusive load monitoring	Energy management
68	Zero-based benchmarking tools	Energy management
69	Web-enabled thermostats	Energy management
70	ISO 50,001	Energy management
71	Home energy use displays	Energy management
72	Handheld audit devices	Energy management
73	Programs supporting energy savings indirectly	Energy management
74	Energy use benchmarking tool using utility meter data	Energy management

(continued)

(continued)

No	Emerging technology	Focus area
75	Regression analysis of utility bills to identify savings opportunities	Energy management
76	High-performance window value purchase program	Energy management
77	Prorated tenant energy billing	Energy management
78	Incentivizing use of energy star's portfolio manager	Energy management
79	Whole-building energy monitoring	Energy management
80	Integrate industrial energy management systems into existing equipment tracking systems	Energy management
81	Increased feedback from vendors to building managers	Energy management
82	Data translators for energy management system inputs	Energy management
83	List of low/no-cost measures	Energy management
84	Logic flow diagrams for sequence of operations	Energy management
85	Enterprise information systems	Energy management
86	Incorporate energy management into continuous improvement programs	Energy management
87	Guidelines for building energy management	Energy management
88	Shared savings with commercial leasing	Energy management
89	Innovative behavior change techniques	Energy management
90	Noninvasive energy meter	Energy management
91	Passive house	Energy management
92	Web-enabled thermostat for small commercial application	Energy management
93	Smart residential thermostat	Energy management
94	Low-cost energy management and control system for small to medium commercial buildings	Energy management
95	Vacancy sensors	Energy management
96	Biobased window foam	Envelope
97	Cool roofs	Envelope
98	Smart garage	Envelope
99	Thin buildings	Envelope
100	Low cost, stable switchable mirrors: lithium ion mirrors with improved stability	Envelope
101	DOE high performance windows volume purchase program	Envelope
102	Window films	Envelope
103	Aerogel insulation—piping, ducts, and buildings	Envelope
104	Smart windows	Envelope
105	Super windows	Envelope
106	Insulated vinyl siding	Envelope
107	Reverse cycle chiller application in multifamily dwelling	HVAC
108	Advanced design rooftop HVAC unit	HVAC
109	Pressure independent flow control valves	HVAC
110	Retrofit roof top HVAC w/ heat recovery economizer	HVAC
111	Ductless mini-split heat pump for small commercial applications	HVAC

(continued)

(continued)

No	Emerging technology	Focus area
112	Super energy saver heat pump	HVAC
113	Desiccant dehumidifiers for grocery and cold storage	HVAC
114	Active and multiservice chilled beams	HVAC
115	Sewage heat recovery	HVAC
116	Variable-speed pool pumps	HVAC
117	Advanced refrigeration control	HVAC
118	Variable capacity compressor in packaged rooftop units	HVAC
119	Hybrid DX packaged RTU, staged compressors	HVAC
120	Domestic water heating from ac condensing unit	HVAC
121	Prevention of overventilating (during noneconomizer times)	HVAC
122	Economizer modification via VFD on blower	HVAC
123	PAX Principle flow technology	HVAC
124	Ultraviolet germicidal irradiation	HVAC
125	Passive options for heating and ventilation	HVAC
126	Ductless heat pump with OSA ventilation	HVAC
127	Underfloor air distribution	HVAC
128	Low pressure duct design	HVAC
129	Displacement ventilation: underfloor, etc.	HVAC
130	CO_2 sensor control (instead of heat exchangers)	HVAC
131	Heat recovery ventilator for commercial application	HVAC
132	Direct–indirect evaporative cooler	HVAC
133	Freezer defrost controller	HVAC
134	Field vacuum precooling (for produce storage)	HVAC
135	District system well water with WSHP	HVAC
136	HRV need low pressure considerations in sizing	HVAC
137	Improving thermal distribution systems for small commercial buildings	HVAC
138	AHU designed for low pressure	HVAC
139	Variable refrigerant flow (VRF) w/o internal/external heat recovery	HVAC
140	Commercial heat pump water heaters	HVAC
141	Potential worse performance at peak load, optimized controls	HVAC
142	Vapor injection for scroll compressors in air conditioning (subcool refrigerant to get below ambient)	HVAC
143	Premium ventilation package for rooftop units	HVAC
144	Hotel/motel room PTACs with key card or occupancy sensors	HVAC
145	Thermal displacement ventilation	HVAC
146	Coolerado cooler 5-t rooftop unit (RTU)	HVAC
147	Natural-gas-fired heat pump, toyota ecovaire	HVAC
148	Integral-collector-storage (ICS) solar water heater	HVAC
149	Reuse of heat from server	HVAC
150	Use lighting occupancy sensors for offices to control HVAC	HVAC

(continued)

(continued)

No	Emerging technology	Focus area
151	Commercial ECM motors on series VAV for speed control on terminal units	HVAC
152	Set residential and/or commercial HVAC equipment tightness requirements per ASHRAE Std 193 MOT	HVAC
153	UVC treatment of HVAC coils	HVAC
154	Hybrid direct expansion with evaporative cooling	HVAC
155	Air flow management in high density data centers	HVAC
156	eCube temperature sensor	HVAC
157	Direct server cabinet cooling	HVAC
158	Solar preheater for ventilation air	HVAC
159	Micro-CHP	HVAC
160	Radiant heating and cooling with dedicated outside air (OSA) ventilation system	HVAC
161	Operation and maintenance rating system for commercial buildings	HVAC
162	Smart defrost kits	HVAC
163	Self-commissioning using building diagnostic software	HVAC
164	Retrocommissioning major equipment such as chillers	HVAC
165	Residential water cooled condenser	HVAC
166	Verified air conditioner refrigerant charge and air flow	HVAC
167	Motor by nova torque 1/3 to 10 HP, super efficiency	HVAC
168	Integrated night ventilation cooling	HVAC
169	Premium HVAC equipment	HVAC
170	Centrifugal compressors with magnetic bearings and speed control	HVAC
171	Spray cooling for CPUs	HVAC
172	Occupancy sensor HVAC controls	HVAC
173	Ultrasonic humidification	HVAC
174	Fleet ventilation	HVAC
175	Air conditioning for climates with high or low sensible heat ratios	HVAC
176	RTU economizer embedded diagnostics with anomaly detection	HVAC
177	Variable refrigerant flow (VRF) heat pumps with internal/external heat recovery	HVAC
178	Airblade hand dryer	HVAC
179	Single-phase motor speed control	HVAC
180	Antifog film	HVAC
181	Frigitek evaporator fan controller	HVAC
182	Insulated heat tape for freeze protection for water pipes	HVAC
183	Aqua chill/cool 3–5 t water cooled DX	HVAC
184	Water treatment strategies for evaporative cooling systems	HVAC
185	Ducts in the conditioned space	HVAC

(continued)

(continued)

No	Emerging technology	Focus area
186	Demand-controlled ventilation for commercial kitchens	HVAC
187	Function heat pump: hot water, radiant floors, and cooling	HVAC
188	Drain-water heat recovery	HVAC
189	In-rack data center cooling	HVAC
190	Desiccant cooling	HVAC
191	Embedded diagnostics	HVAC
192	Demand control on item for morning warm-up	HVAC
193	Demand-controlled ventilation	HVAC
194	Heat pump water heater-residential integrated units	HVAC
195	FANWALL technology	HVAC
196	Aerosol-based duct sealing for residential and commercial buildings	HVAC
197	Heat pump maximizer	HVAC
198	Combined heat and power	HVAC
199	Low-lift cooling technology option set	HVAC
200	Efficient laboratory hoods	HVAC
201	HRV for common spaces in multifamily	HVAC
202	Evaporation-cooled ceiling tiles	HVAC
203	Air to water and water to water heat pumps, turbo loop	HVAC
204	Low energy comfort using solar, night cooling, etc.	HVAC
205	Thermal energy storage (to reduce peak load)	HVAC
206	Premium efficiency motor for furnace PSC vs. ECM motors change-outs	HVAC
207	Duct design with ECM motors as a package	HVAC
208	Filter efficiency requirements: set static pressure limit for flow (lower pressure ventilation)	HVAC
209	Integrated design strategies: duct sizing, coils, number of zones, filter sizing; a package of design measures; keep it simple	HVAC
210	Heat home with dehumidifier	HVAC
211	Condensing dryers, hp water heater, water heater extracts heat from dryer for DHW	HVAC
212	Ductless heat pumps, low temperature, for residential application	HVAC
213	Variable speed compressors	HVAC
214	Integrated bldg. systems	HVAC
215	Absorption chiller with solar hw and digital controls	HVAC
216	Low temp supply air to reduce volumetric flow rate	HVAC
217	RTU simplified energy meter and embedded diagnostics with anomaly detection	HVAC
218	ECM for exhaust fans and terminal units	HVAC
219	Cromer under ASHRAE review	HVAC
220	Microchannel evaporator	HVAC

(continued)

(continued)

No	Emerging technology	Focus area
221	Phase-change drywall	HVAC
222	Solid state economizer controller and sensor	HVAC
223	Diagnostic software for HVAC systems	HVAC
224	Dedicated outside air system (DOAS) for VAV HVAC systems	HVAC
225	Single zone VAV	HVAC
226	Tighter HVAC boxes	HVAC
227	High efficiency chiller	HVAC
228	High-performance indirect evaporative cooler	HVAC
229	Improved duct sealing standards	HVAC
230	Partial OSA economizer for loop-type systems	HVAC
231	Ground heat exchanger	HVAC
232	Low pressure drop piping system	HVAC
233	Low temperature air for airside economizer	HVAC
234	Packaged CHP	HVAC
235	Cooling tower slab cooling	HVAC
236	Desiccant-enhanced evaporative air conditioner (DEVap)	HVAC
237	Differential enthalpy control of economizers, not dry bulb or simple enthalpy	HVAC
238	Energy enhancing controller for constant volume rooftop units	HVAC
239	Microheater microchannel combustor	HVAC
240	Circulating pumps with variable speed permanent magnet motors	HVAC
241	Ductless mini-split heat pumps with variable speed compressors for residential application	HVAC
242	Ductless mini-split heat pump for multifamily housing	HVAC
243	Ductless mini-split heat pump for manufactured homes	HVAC
244	Low temperature VRF with simultaneous heating and cooling for large commercial applications	HVAC
245	Low temperature VRF without simultaneous heating and cooling for large commercial	HVAC
246	DualCool evaporative cooler to precool ventilation air	HVAC
247	Ground source variable refrigerant flow systems	HVAC
248	Quiet climate 2: efficient heat pump for portable classrooms	HVAC
249	Variable-speed compressor single-zone ducted heat pump	HVAC
250	Modular HVAC equipment	HVAC
251	Nanotechnology membrane-based dehumidifier	HVAC
252	Air-side economizers for data centers	HVAC
253	Smart thermostat	HVAC
254	Variable volume dust collection system	Industrial/agriculture
255	Energy efficient motors	Industrial/agriculture
256	Heat pump water heaters using CO_2 for commercial and industrial applications	Industrial/agriculture

(continued)

(continued)

No	Emerging technology	Focus area
257	Heat-of-compression desiccant dryers (replacing existing desiccant compressed air dryers that use electric resistance heat)	Industrial/agriculture
258	Ultrananocrystalline diamond (UNCD®) mechanical seals	Industrial/agriculture
259	Wireless sensors	Industrial/agriculture
260	Electrodialysis for wine industry	Industrial/agriculture
261	Replacing desiccant compressed air dryers with refrigerated dryers	Industrial/agriculture
262	Compressors with variable speed drives	Industrial/Agriculture
263	New variable speed drives	Industrial/agriculture
264	Industrial pumps and fans	Industrial/agriculture
265	System optimization	Industrial/agriculture
266	Positive displacement pump and control for injection molding	Industrial/agriculture
267	Automatic reset of dryer dewpoint setpoint	Industrial/agriculture
268	Thermosorber for food processing	Industrial/agriculture
269	Conversion of industrial equipment from compressed air to direct drive	Industrial/agriculture
270	Clean-on-demand compressed air controls for baghouses	Industrial/agriculture
271	Variable speed die casting machine	Industrial/agriculture
272	Near net-shape casting	Industrial/agriculture
273	Interlock of pulsation unit with dairy milking vacuum system	Industrial/agriculture
274	Advanced industrial controls	Industrial/agriculture
275	Emissivity sensors	Industrial/agriculture
276	Moisture sensors	Industrial/agriculture
277	Power electronics	Industrial/agriculture
278	Smart controllers—sensorless	Industrial/agriculture
279	Temperature sensors	Industrial/agriculture
280	Low-energy precision application (LEPA) irrigation	Industrial/agriculture
281	On-demand ventilation controls for dust and fume collection systems	Industrial/agriculture
282	Add pumps to reduce block heater loads	Industrial/agriculture
283	Office of the future technologies	Integrated design
284	New technology education and awareness	Lighting
285	LED high-bay lighting fixture with built-in daylight dimming	Lighting
286	LED task lighting	Lighting
287	Mesopic lighting for low light visibility	Lighting
288	High-bay fluorescent lighting	Lighting
289	Modular CFL downlights	Lighting
290	High color temperature lighting	Lighting
291	HID electronic ballast	Lighting
292	Lighting occupancy sensors for linear fluorescent, hid, and solid state lighting	Lighting
293	Lighting fixtures with advanced local lighting controls	Lighting

(continued)

(continued)

No	Emerging technology	Focus area
294	Metal halide retail lighting ceramic	Lighting
295	Reduced wattage, high color temperature T5 lamps	Lighting
296	Level lighting with occupancy sensors in parking lots and garages	Lighting
297	SSL display lighting	Lighting
298	Scotopic lighting	Lighting
299	Daylighting solar canopy	Lighting
300	LED street lighting	Lighting
301	Solid state outdoor area lighting	Lighting
302	Solid state plasma lighting	Lighting
303	Fluorescent dimming controls	Lighting
304	CFL downlights	Lighting
305	LED refrigerated and freezer case lighting	Lighting
306	Induction lighting	Lighting
307	Advanced occupancy controls for lighting and HVAC	Lighting
308	OLED lighting	Lighting
309	LED downlights for residential retrofit applications	Lighting
310	Continuous dimming with daylight harvesting	Lighting
311	Electron-stimulated luminescence lighting	Lighting
312	Level stairwell lighting control	Lighting
313	Directional solid state lighting for interior spaces	Lighting
314	LED linear signage lighting	Lighting
315	Circadian lighting color tuning	Lighting
316	LEDs for commercial signs (other than linear)	Lighting
317	Integral LED nightlight/occupancy sensor control for hotel bathroom lighting	Lighting
318	Wireless lighting controls	Lighting
319	Mirrored light pipes	Lighting
320	5/T-8 HO wall packs for 12/7 or 24/7 outdoor applications	Lighting
321	Advanced lens fluorescent fixtures	Lighting
322	Integrated classroom lighting system	Lighting
323	LED elevator lighting	Lighting
324	Lighting control commissioning	Lighting
325	Nondaylight sensors for outdoor lighting	Lighting
326	Dimmable fluorescent lighting	Lighting
327	Personal lighting system	Lighting
328	Photoluminescent exit signs	Lighting
329	Fiber optic daylighting	Lighting
330	Simplified daylight controls	Lighting
331	Street sign lighting	Lighting
332	LED for covered parking lots	Lighting

(continued)

(continued)

No	Emerging technology	Focus area
333	LED under-cabinet lighting	Lighting
334	LED luminaires, residential	Lighting
335	Improved CFLs	Lighting
336	Hotel room lighting	Lighting
337	Intelligent hotel bathroom lights	Lighting
338	LED directional replacement lamps	Lighting
339	LED A-lamp replacement	Lighting
340	LED testing	Lighting
341	Low wattage metal halide	Lighting
342	Increased task lighting, reduced ambient lighting	Lighting
343	Bi-level office lighting with occupancy sensors, auto-on 50 %	Lighting
344	Cold-cathode lighting dimmable	Lighting
345	Lighting controls for outdoor lighting	Lighting
346	LED T8 lamp replacement	Lighting
347	Cold cathode compact fluorescent lamps for general illumination	Lighting
348	Cold cathode fluorescent sign lighting	Lighting
349	Street lighting using t5 fluorescent lamps	Lighting
350	Microplasma lighting	Lighting
351	Sulfur plasma lighting	Lighting
352	LED high bay with wireless control	Lighting
353	Highly efficient incandescent light bulbs	Lighting
354	Hybrid halogen-CFL light bulb	Lighting
355	Skylights and windows with aerogel filling	Lighting
356	LED retrofit kits for outdoor lighting applications	Lighting
357	LED outdoor wall packs	Lighting
358	LED bollards	Lighting
359	LED outdoor wall wash luminaires	Lighting
360	LED parking garage lighting	Lighting
361	LED track lighting	Lighting
362	LED display case lighting	Lighting
363	LED linear panels (troffers)	Lighting
364	LED interior high bay and low bay lighting	Lighting
365	LED canopy lighting	Lighting
366	LED indoor wall wash luminaires	Lighting
367	LED downlights (recessed cans) for commercial applications	Lighting
368	High temperature water and space heating for peak demand management	Water heating
369	Heat pump water heaters for commercial kitchens	Water heating
370	Pump controls	Water heating
371	EcoCute carbon-dioxide-based heat pump water heater	Water heating

Appendix 2: BPA Residential Sector Energy Efficiency Program Reimbursements

Program component or measure	Reimbursement level
Lighting	
Specialty CFLs	$2.25–$5.50
Standard twister CFLs	$1.00–$4.00
CFL fixtures	$10.00
Showerheads	$12.00–$20.00/unit
Appliances (new)	
ENERGY STAR clothes washers	$20.00–$70.00/washer
ENERGY STAR freezers	$15.00/freezer
ENERGY STAR refrigerators	$15.00/refrigerator
Refrigerator and freezer decommissioning	$100.00/unit
Electric water heating	
Electric storage water heaters	$25.24–$105.77
Gravity-film heat exchangers	$159.77–$228.24
Pipe insulation	$4.31–$12.92
HVAC measures	
Ductless heat pumps	$1,500.00/unit
Ducted systems with PTCS	
PTCS air-source heat pump upgrade	$500.00–$1,00,000/unit
PTCS air-source heat pump conversions	$1,400.00–$1,900.00/unit
PTCS heat pump commissioning and controls	$300.00/unit
PTCS geothermal (ground-source) heat pump systems (new)	$2,400.00–$3,000.00/unit
PTCS duct sealing	$400.00–$500.00/unit
Line voltage electronic thermostats	$115.00–$160.00/unit
New construction	
New ENERGY STAR manufactured homes	$850.00–$1,450.00/unit
New ENERGY STAR/built green site-built homes	$200.00–$1,800.00/unit
Montana house (v 2.0)	$200.00–$1,500/unit
New multifamily construction	$80.00–$140.00/unit
Weatherization (standard income)	
Insulation	No info available
Prime window replacement	$6.00/square foot
Air sealing	No info available
Low-income weatherization and duct sealing	<$20.00/ square foot for prime window replacement
Residential custom projects	Subject to custom project reimbursements

Appendix 3: BPA Commercial Sector Energy Efficiency Program Reimbursements

Program component or measure	Reimbursement level
Commercial custom projects—existing buildings	Subject to custom project reimbursements
Commercial lighting	Subject to multisector reimbursements
LED traffic signals	$50.00–$115.00
Commercial HVAC	
Unitary air-conditioning	No info available
Commercial ductless heat pump	$750.00–$1,000.00/unit
Web-enabled programmable thermostats in modular classrooms	$100.00–$275.00/unit
Commercial shell measures	
Insulation in existing small office or retail	$0.13 per kWh
Small commercial retrofit windows	$3.00–$6.00/square foot
Commercial refrigeration	
BPA Energy Smart Grocer Program	No info available
Deemed refrigeration retrofit measures	No info available
Commercial kitchen and food service equipment	
Commercial food service reach-in refrigerators and freezers	$50.00–$350.00/refrigerator or freezer
Commercial ice makers	$100.00–$300.00/ice maker
Commercial kitchen steamers	$100.00–$200.00/steamer
Commercial kitchen hot food holding cabinets	$200.00–$400.00/cabinet
Combination ovens	$1,750.00/oven
Convection ovens	$200.00/oven
Electric fryers	$125.00/installation
Dishwashers	$100.00–$750.00/dishwasher
Prerinse spray wash valves	$100.00/installation
Refrigerator/freezer recycling	$100.00/unit
Additional deemed offerings	
Network computer power management	$10.00/workstation
ENERGY STAR commercial clothes washer	$25.00–$200.00/washer
Electric storage water heaters	$25.00/unit
Smart power strips	$20.00/strip
Commercial showerheads	$30.60/showerhead
Commercial new construction	
Energy smart design—office and trade-offs	$0.25–$0.50/square foot
New commercial construction/major renovation	Subject to custom project reimbursements

Appendix 4: BPA Industrial Sector Energy Efficiency Program Reimbursements

Program component or measure	Reimbursement level
Energy management pilot	
Energy project manager	Subject to custom project reimbursements
Track and tune projects	Subject to custom project reimbursements
High performance energy management	Subject to custom project reimbursements
Trade ally delivered small industrial measures	Subject to custom project reimbursements
Northwest trade ally network (nonresidential lighting)	Subject to custom project reimbursements
BPA funded technical service providers (TSP)	Not applicable
Variable frequency drives in spud and onion storage facilities	$200.00/hp

Appendix 5: BPA Agriculture Sector Energy Efficiency Program Reimbursements

Program component or measure	Reimbursement levels
Freeze-resistant stock water tanks/fountains	$140.00–$225.00/tank or fountain
Irrigation-related measures	
Irrigation system upgrades	$0.75–$175.00/sprinkler equipment
Irrigation system motors replacement	$75.00–$7,500.00/unit
Scientific irrigation scheduling	$5.20/acre
Irrigation pump testing and system analysis	$50.00–$300.00/test or analysis
Variable frequency drives	
Small milking machines in dairies and other approved applications	$2,200.00/verified installation
Agricultural turbine pump applications	$80.00/hp
Transformer deenergization	$0.15/kWh or 70 % of project cost
New agricultural construction	Subject to custom project reimbursements
Other agricultural measures	Subject to custom project reimbursements

References

1. BPA. LED replacement lamps. Bonneville Power Administration Website (2012), http://www.bpa.gov/energy/n/emerging_technology/LED_Lamps.cfm. Accessed 15 Mar 2012
2. AHSD. Fluorescent lamp. The American Heritage Science Dictionary Website (n.d.), http://dictionary.reference.com/browse/fluorescent lamp. Accessed 15 Mar 2012

3. Merriam-Webster. Light emitting diodes. Merriam Webster Dictionary Website (2012), http://www.merriam-webster.com/dictionary/led. Accessed 15 Mar 2012
4. NWPCC. Northwest Power and Conservation Council Sixth Northwest Conservation and Electric Power Plan. Northwest Power and Conservation Council Website (2010), http://www.nwcouncil.org/energy/powerplan/6/default.htm. Accessed 20 Feb 2012
5. Energy Star. How it works-heat pump water heaters. Energy Star Website (2012), http://www.energystar.gov/index.cfm?c=heat_pump.pr_how_it_works. Accessed 15 Mar 2012
6. Energy Star. High efficiency water heaters. Energy Star Website (2006), http://www.energystar.gov/ia/new_homes/features/WaterHtrs_062906.pdf. Accessed 17 Mar 2012
7. DOE. Conventional storage water heaters. U.S. Department of Energy Website (2011), http://www.energysavers.gov/your_home/water_heating/index.cfm/mytopic=12980. Accessed 15 Mar 2012
8. ORNL. Heat recovery from waste water using gravity-film heat exchanger. Oak Ridge National Laboratory Website (2001), http://gfxtechnology.com/Femp.pdf. Accessed 17 Mar 2012
9. DOE. Insulate hot water pipes for energy savings. U.S. Department of Energy Website (2011), http://www.energysavers.gov/your_home/water_heating/index.cfm/mytopic=13060. Accessed 17 Mar 2012
10. M. Shealy, Insulating your hot water pipes can save money and protect the environment. Malcolm Shealy Website (2011), http://www.leaningpinesoftware.com/hot_water_pipes.shtml. Accessed 17 Mar 2012
11. BPA. Ductless, mini-split heat pumps. Bonneville Power Administration Website. (2012), http://www.bpa.gov/energy/n/emerging_technology/DHP.cfm. Accessed 18 Mar 2012
12. BPA. Two-stage Heat Pumps for Cold Climates. Bonneville Power Administration Website (n.d.), http://www.bpa.gov/energy/n/emerging_technology/CCHP.cfm. Accessed 18 Mar 2012
13. ODOE. Ductless heat pumps—a new option for home heating. Oregon Department of Energy Website (2007), http://www.oregon.gov/ENERGY/CONS/RES/tax/docs/Ductless_Heat_Pumps_Fact_Sheet.pdf?ga=t. Accessed 18 Mar 2012
14. NAHB. Ductless (Mini-split) heat pumps. National Association of Home Builders Research Center Website (2001), http://www.toolbase.org/TechInventory/techDetails.aspx?ContentDetailID=746. Accessed 18 Mar 2012
15. DOE. Ductless, mini-split heat pumps. U.S. Department of Energy Website (2011), http://www.energysavers.gov/your_home/space_heating_cooling/index.cfm/mytopic=12630. Accessed 18 Mar 2012
16. DOE. Air source heat pumps. U.S. Department of Energy Website (2011), http://www.energysavers.gov/your_home/space_heating_cooling/index.cfm/mytopic=12620. Accessed 18 Mar 2012
17. NREL. Air source heat pumps. National Renewable Energy Laboratory Website (2001), http://www.nrel.gov/docs/fy01osti/28037.pdf. Accessed 18 Mar 2012
18. DOE. Geothermal heat pumps. U.S. Department of Energy Website (2011), http://www.energysavers.gov/your_home/space_heating_cooling/index.cfm/mytopic=12640. Accessed 19 Mar 2012
19. DOE. Benefits of geothermal heat pumps. U.S. Department of Energy Website (2011), http://www.energysavers.gov/your_home/space_heating_cooling/index.cfm/mytopic=12660. Accessed 19 Mar 2012
20. E3T. Web-Enabled Thermostat for Small Commercial Application. Bonneville Power Administration Emerging Technologies Website (2011), http://www.e3tnw.org/ItemDetail.aspx?id=247. Accessed 19 Mar 2012
21. NAHB. Aerosol duct sealing. National Association of Home Builders Research Center Website (2001), http://www.toolbase.org/Technology-Inventory/HVAC/aerosol-duct-sealing. Accessed 19 Mar 2012
22. DOE. Tips: air ducts. U.S. Department of Energy Website (2012), http://www.energysavers.gov/tips/ducts.cfm. Accessed 19 Mar 2012
23. LBNL. Aerosol duct sealing. Lawrence Berkeley National Laboratory Website (2012), http://eetd.lbl.gov/l2m2/aerosol.html. Accessed 19 Mar 2012

24. Energy Star. Benefits of duct sealing. Energy Star Website (n.d.), http://www.energystar.gov/index.cfm?c=home_improvement.hm_improvement_ducts_benefits. Accessed 19 Mar 2012
25. CEE. High efficiency specifications for commercial refrigerators and freezers. Consortium for Energy Efficiency Website (2010), http://www.cee1.org/com/com-kit/files/Refrigeration Specification.pdf. Accessed 20 Mar 2012
26. CEE. High efficiency specifications for commercial icemakers. Consortium for Energy Efficiency Website (2010), http://www.cee1.org/com/com-kit/files/IceSpecification01Jul2011.pdf. Accessed 20 Mar 2012
27. CEE. High efficiency specifications for commercial steamers. Consortium for Energy Efficiency Website (2010), http://www.cee1.org/com/com-kit/files/SteamerSpecification.pdf. Accessed 20 Mar 2012
28. CEE. High efficiency specifications for hot food holding cabinets. Consortium for Energy Efficiency Website (2007), http://www.cee1.org/com/com-kit/files/HotFoodHolding CabinetSpecification.pdf. Accessed 20 Mar 2012
29. CEE. High efficiency specifications for commercial convection ovens. Consortium for Energy Efficiency Website (2011), http://www.cee1.org/com/com-kit/files/ConvectionOvens Specification.pdf. Accessed 20 Mar 2012
30. CEE. High efficiency specifications for commercial fryers. Consortium for Energy Efficiency Website (2012), http://www.cee1.org/com/com-kit/files/FryerSpecification.pdf. Accessed 20 Mar 2012
31. CEE. High efficiency specifications for commercial dishwashers. Consortium for Energy Efficiency Website (2008), http://www.cee1.org/com/com-kit/files/DishwasherSpecification.pdf. Accessed 20 Mar 2012
32. CEE. Program guidance on pre-rinse spray valves. Consortium for Energy Efficiency Website (n.d.), http://www.cee1.org/com/com-kit/prv-guides.pdf. Accessed 20 Mar 2012
33. CEE. Horizontal-axis technology enables washers to save energy and water. Consortium for Energy Efficiency Website (2011), http://www.cee1.org/com/cwsh/howduzit.php3. Accessed 20 Mar 2012
34. DOE. How insulation works. U.S. Department of Energy Website (2011), http://www.energysavers.gov/your_home/insulation_airsealing/index.cfm/mytopic=11330. Accessed 20 Mar 2012
35. DOE. Blanket (batt and roll) insulation. U.S. Department of Energy Website (2011), http://www.energysavers.gov/your_home/insulation_airsealing/index.cfm/mytopic=11520. Accessed 20 Mar 2012
36. DOE. Foam board insulation. U.S. Department of Energy Website (2011), http://www.energysavers.gov/your_home/insulation_airsealing/index.cfm/mytopic=11620. Accessed 20 Mar 2012
37. DOE. Insulating concrete forms. U.S. Department of Energy Website (2011), http://www.energysavers.gov/your_home/insulation_airsealing/index.cfm/mytopic=11640. Accessed 20 Mar 2012
38. DOE. Loose fill insulation. U.S. Department of Energy Website (2011), http://www.energysavers.gov/your_home/insulation_airsealing/index.cfm/mytopic=11650. Accessed 20 Mar 2012
39. DOE. Reflective insulation. U.S. Department of Energy Website (2011), http://www.energysavers.gov/your_home/insulation_airsealing/index.cfm/mytopic=11670. Accessed 20 Mar 2012
40. DOE. Rigid fiber board insulation. U.S. Department of Energy Website (2011), http://www.energysavers.gov/your_home/insulation_airsealing/index.cfm/mytopic=11690. Accessed 20 Mar 2012
41. DOE. Sprayed foam and formed in place insulation. U.S. Department of Energy Website (2011), http://www.energysavers.gov/your_home/insulation_airsealing/index.cfm/mytopic=11510. Accessed 20 Mar 2012
42. DOE. Structural insulated panels. U.S. Department of Energy Website (2011), http://www.energysavers.gov/your_home/insulation_airsealing/index.cfm/mytopic=11740. Accessed 20 Mar 2012

43. DOE. Caulking. U.S. Department of Energy Website (2011), http://www.energysavers.gov/your_home/insulation_airsealing/index.cfm/mytopic=11270. Accessed 20 Mar 2012
44. DOE. Weatherstripping. U.S. Department of Energy Website (2011), http://www.energysavers.gov/your_home/insulation_airsealing/index.cfm/mytopic=11280. Accessed 20 Mar 2012
45. LBNL. Selecting windows for energy efficiency. Lawrence Berkeley National Laboratory Website (1997), http://windows.lbl.gov/pub/selectingwindows/window.pdf. Accessed 20 Mar 2012
46. EWC. Window technologies: glazing types. Efficiency Windows Collaborative Website (2012), http://www.efficientwindows.org/gtypes.cfm. Accessed 20 Mar 2012
47. EWC. Window technologies: low-e coatings. Efficiency Windows Collaborative Website (2012), http://www.efficientwindows.org/lowe.cfm. Accessed 20 Mar 2012
48. EWC. Window technologies: argon and krypton gas fills. Efficiency Windows Collaborative Website (2012), http://www.efficientwindows.org/gasfills.cfm. Accessed 20 Mar 2012
49. EWC. Window technologies: low conductance spacers. Efficiency Windows Collaborative Website (2012), http://www.efficientwindows.org/spacers.cfm. Accessed 20 Mar 2012
50. Energy Star. Energy efficient homes features. Energy Star Website (2012), http://www.energystar.gov/index.cfm?c=bldrs_lenders_raters.nh_features. Accessed 20 Mar 2012
51. NREL. Elements of an energy efficient house. National Renewable Energy Laboratory Website (2000), http://www.nrel.gov/docs/fy00osti/27835.pdf. Accessed 20 Mar 2012
52. NRC. Variable frequency drives. Natural Resources of Canada Website (2009), http://oee.nrcan.gc.ca/industrial/equipment/variable-frequency-drives/1812. Accessed 20 Mar 2012
53. CEC. Variable frequency drive. The California Energy Commission Website (n.d.), http://www.energy.ca.gov/process/pubs/vfds.pdf. Accessed 20 Mar 2012
54. S. Rossiter, Benefits of variable speed drives. Energy Management Corporation Website (2011), http://www.emcsolutions.com/article_vfd_benefits.html. Accessed 20 Mar 2012
55. CREIA. New water conservation for shower heads. California Real Estate Inspection Association Website (2012), http://www.creia.org/August-2010-Inspector-eNews/new-water-conservation-for-shower-heads.html. Accessed 15 Mar 2012
56. BPA. Smart power strips. Bonneville Power Administration Website. (2012), http://www.bpa.gov/energy/n/strips.cfm. Accessed 15 Mar 2012
57. NEEA. The NEEA 2010–2014 business plan. Northwest Energy Efficiency Alliance Website (2012), http://neea.org/participate/reviewanddrafts.aspx. Accessed 16 Mar 2012
58. BPA. Northwest energy efficiency technology roadmap portfolio. Bonneville Power Administration Website (2012), http://www.bpa.gov/energy/n/emerging_technology/pdf/NW_EE_Tech_Roadmap_Portfolio_March_2012.pdf. Accessed 29 Mar 2012
59. S.K. Melink, Kitchen hoods using demand ventilation. ASHRAE J., 24 (2003)
60. E3T. Demand controlled ventilation for commercial kitchens. Bonneville Power Administration Emerging Technologies Website (2011), http://www.e3tnw.org/ItemDetail.aspx?id=155. Accessed 22 Mar 2012
61. E3T. Demand controlled ventilation. Bonneville Power Administration Emerging Technologies Website (2011), http://www.e3tnw.org/ItemDetail.aspx?id=166. Accessed 22 Mar 2012
62. E3T. Variable capacity compressor in packaged rooftop units. Bonneville Power Administration Emerging Technologies Website (2011), http://www.e3tnw.org/ItemDetail.aspx?id=6. Accessed 22 Mar 2012
63. E3T. Air side economizers for data centers. Bonneville Power Administration Emerging Technologies Website (2011), http://www.e3tnw.org/ItemDetail.aspx?id=119. Accessed 22 Mar 2012
64. E3T. Variable refrigerant flow (VRF) heat pumps with internal/external heat recovery. Bonneville Power Administration Emerging Technologies Website (2011), http://www.e3tnw.org/ItemDetail.aspx?id=126. Accessed 22 Mar 2012
65. E3T. Advanced rooftop unit controls with remote access and energy monitoring. Bonneville Power Administration Emerging Technologies Website (2011), http://www.e3tnw.org/ItemDetail.aspx?id=338. Accessed 22 Mar 2012

66. E3T. Direct-indirect evaporative cooler. Bonneville Power Administration Emerging Technologies Website (2011), http://www.e3tnw.org/ItemDetail.aspx?id=23. Accessed 22 Mar 2012
67. E3T. Bi-level stairwell lighting control. Bonneville Power Administration Emerging Technologies Website (2011), http://www.e3tnw.org/ItemDetail.aspx?id=108. Accessed 22 March 2012
68. E3T. Integrated classroom lighting system. Bonneville Power Administration Emerging Technologies Website (2011), http://www.e3tnw.org/ItemDetail.aspx?id=133. Accessed 22 Mar 2012
69. E3T. Wireless lighting controls. Bonneville Power Administration Emerging Technologies Website (2011), http://www.e3tnw.org/ItemDetail.aspx?id=117. Accessed 22 Mar 2012
70. E3T. Low-cost energy management and control system for small to medium commercial buildings. Bonneville Power Administration Emerging Technologies Website (2011), http://www.e3tnw.org/ItemDetail.aspx?id=347. Accessed 22 Mar 2012
71. E3T. Building energy performance analytics software and services. Bonneville Power Administration Emerging Technologies Website (2011), http://www.e3tnw.org/ItemDetail.aspx?id=353. Accessed 22 Mar 2012
72. E3T. Non-intrusive load monitoring. Bonneville Power Administration Emerging Technologies Website (2011), http://www.e3tnw.org/ItemDetail.aspx?id=294. Accessed 22 Mar 2012
73. NWPCC. Estimated cost of new resource options (ca. 2025). Northwest Power and Conservation Council Website (2012), http://www.nwcouncil.org/energy/powerplan/6/newresourcecosts.htm. Accessed 10 Feb 2012
74. BPA. Wind power. Bonneville Power Administration Website (n.d.), http://www.bpa.gov/corporate/WindPower/index.cfm. Accessed 30 Mar 2012
75. NWPCC. Power Generation in the Northwest. Northwest Power and Conservation Council Website (2012), http://www.nwcouncil.org/energy/powersupply/projects/Projects.xlsm. Accessed 11 Mar 2012
76. BPA. BPA 2010–2014 action plan for energy efficiency. Bonneville Power Administration Website (2012), http://www.bpa.gov/energy/n/pdf/BPA_Action_Plan_FINAL_20120301.pdf. Accessed 28 Feb 2012
77. BPA. BPA energy efficiency implementation manual. Bonneville Power Administration Website (2011), http://www.bpa.gov/energy/n/archives/pdf/October_2011_Implementation_Manual_%28updated_12-19-11%29.pdf. Accessed 18 Feb 2012
78. BPA. Residential sector. Bonneville Power Administration Website. (n.d.), http://www.bpa.gov/energy/n/residential.cfm. Accessed 11 Mar 2012
79. BPA. Residential products promotion. Bonneville Power Administration Website (n.d.), http://www.bpa.gov/energy/n/residential_pro.cfm. Accessed 10 Mar 2012
80. BPA. Residential heating, ventilation and air conditioning. Bonneville Power Administration Website (n.d.), http://www.bpa.gov/resHVAC/. Accessed 10 Mar 2012
81. BPA. Heat pump water heaters. Bonneville Power Administration Website (n.d.), http://www.bpa.gov/Energy/N/HPWH.cfm. Accessed 10 Mar 2012
82. BPA. Consumer electronics. Bonneville Power Administration Website (n.d.), http://www.bpa.gov/Energy/N/consumerelectronics.cfm. Accessed 10 Mar 2012
83. BPA. Commercial sector. Bonneville Power Administration Website (n.d.), http://www.bpa.gov/energy/n/commercial.cfm. Accessed 10 Mar 2012
84. NTAN. Expand your business through public utility lighting efficiency incentive programs. Northwest Trade Ally Network Website (n.d.), http://northwest-lighting.org/about.html. Accessed 10 Mar 2012
85. BPA. BPA energy smart grocery program. Bonneville Power Administration Website (n.d.), http://www.bpa.gov/energy/n/esgprogram.cfm. Accessed 10 Mar 2012
86. Energy Smart. BPA and energy smart grocer bring energy savings to life. Energy Star Website (2011), http://www.energysmartonline.org/utilities/bpa_index.html. Accessed 10 Mar 2012
87. BPA. Commercial kitchen and food service equipment. Bonneville Power Administration Website (n.d.), http://www.bpa.gov/energy/n/commercial_kit.cfm. Accessed 10 Mar 2012

88. Energy Star. Commercial kitchen package for businesses and operators. Energy Star Website (n.d.), http://www.energystar.gov/index.cfm?fuseaction=find_a_product.showProductGroup&pgw_code=CKP. Accessed 10 Mar 2012
89. BPA. HVAC program details. Bonneville Power Administration Website (n.d.), http://www.bpa.gov/energy/n/HVAC.cfm. Accessed 10 Mar 2012
90. BPA. Small commercial window retrofits. Bonneville Power Administration Website (n.d.), http://www.bpa.gov/energy/n/small_commercial.cfm. Accessed 10 Mar 2012
91. BPA. Smart power strips. Bonneville Power Administration Website (n.d.), http://www.bpa.gov/energy/n/strips.cfm. Accessed 10 Mar 2012
92. BPA. Energy smart industrial. Bonneville Power Administration Website (n.d.), http://www.bpa.gov/energy/n/industrial/index.cfm. Accessed 10 Mar 2012
93. BPA. Agriculture sector. Bonneville Power Administration Website (n.d.), http://www.bpa.gov/energy/n/agriculture.cfm. Accessed 10 Mar 2012

Chapter 11
Evaluating Energy Efficiency

Ibrahim Iskin and Tugrul U. Daim

Abstract Nature of resource planning has changed dramatically since 1970s due to increased diversity in resource options such as renewable alternatives, demand-side management, space conditioning, cogeneration of heat and electricity in industrial applications, and deregulation of the energy market. Along with that new objectives have been added to the utilities' decision-making processes beyond cost minimization (Hobbs, Eur. J. Oper. Res., 83:1–20, 1995). Moreover; technological development, instability in fuel markets, and government regulations are taking place faster than ever before and as a result complexity and uncertainty involved in decision-making practices have become increasingly significant. This chapter provides an approach based on hierarchical decision modeling to consider new factors in evaluating energy efficiency technologies.

As mentioned, energy efficiency programs have been considered as a resource and part of integrated resource planning since 1970s; however, its value role as a resource has been approached from a limited point of view. Current approaches, employing cost–benefit ratio method, have been observed to take only the quantifiable variables such as avoided peak demand, increased reliability, avoided investment costs, program implementation costs, etc., into consideration and miss taking some of the other variables that cannot be easily quantified, but have important impacts on utility operations. Moreover, due to nature of cost–benefit ratio-based methods decision makers are not provided with enough information to enable decision analysis at the variable level, but rather given single data point. This gap has also been supported by

I. Iskin (✉)
Portland State University, Portland, OR, USA
e-mail: ibrahimiskin@gmail.com

T.U. Daim
Department of Engineering and Technology Management, Portland State University, 1900 SW 4th Avenue 50-02, Portland, OR 97201, USA
e-mail: tugrul@etm.pdx.edu

T.U. Daim et al. (eds.), *Policies and Programs for Sustainable Energy Innovations*, Innovation, Technology, and Knowledge Management,
DOI 10.1007/978-3-319-16033-7_11

Gellings and Smith [8] who claim that energy efficiency programs need to be assessed considering their implications on utility operational objectives such as operational flexibility, use of critical fuels, environmental damage, job creation, public and employee health, etc.

Although there is significant amount of cost-effective energy efficiency opportunities, exploitation of these has been rather slow due to various barriers. One of the significant and mainly studied barriers is associated with user heterogeneity. User heterogeneity impacts the relevancy of adoption barriers and drivers associated with an energy-efficient technology in different end uses. This point has also been supported by the studies: Iskin [11] in the case of wider energy efficiency literature and Iskin et al. [12] in the case of an energy efficiency program development for variable capacity heat pumps in Pacific Northwest US. Thus, user heterogeneity prevents energy efficiency program planning from making accurate judgments about program success. As a result, energy efficiency programs' use as a resource in integrated resource planning cannot be exploited to its full potential. Moreover, since user heterogeneity is not considered as a market failure, but a normal market phenomenon, potential policy solutions through market interventions cannot be justified. As a result, energy efficiency program planning needs to address potential impacts of user heterogeneity in decision-making practices.

11.1 Research Approach and Methodology

Objective of proposed research is to develop an energy efficiency program assessment model for energy efficiency program planning in electric utilities. Proposed model will expand the existing assessment models by incorporating utility operational goals and objectives rather than just financial metrics. Incorporation of operational goals and objectives is expected to enable trade-off analysis at the utility goal level and provide a more comprehensive decision analysis. Decision alternatives used in the model will be developed by incorporating technology- and -related variables together in order to decrease uncertainties associated with user heterogeneity. Such an approach is expected to create decision alternatives homogeneous enough to enable more accurate assessment. Overall, proposed improvements are expected to contribute to existing level of knowledge by enabling a more accurate energy efficiency technology evaluation and planning approach that can provide better understanding of the potential implications of the program decisions at the utility goal and objective level.

First contribution of the proposed research effort is to enable assessment of energy efficient technologies with respect to their impacts on utility objectives and goals. Having looked at the current literature, it has been observed that there is no research study taking utility objectives and goals into account in DSM decision-making practices although quite a few research studies have identified these points as potential improvement areas. Accordingly, Gellings and Smith [8] have suggested that evaluation of DSM technologies in utility planning operations needs to be conducted by considering their implications on operational objectives.

The reason behind aforementioned suggestion is based on the fact that given the variables associated with an energy efficiency program and service area of a utility, impact of the aforementioned given program on utility operations would vary to a great extent. By utilizing utility objectives and goals as decision variables; proposed assessment model is proposed to enable trade-off analysis and contribute to decision-making practices in the field.

In the literature, user heterogeneity has often been considered as a significant barrier to diffusion of energy efficiency technologies. Accordingly, Sanstad and Howarth [27] have suggested that in order to increase accuracy of decision-making practices an approach that can link technology parameters with diffusion parameters. Accordingly, second contribution of the proposed model is to enable better assessment of energy efficiency programs through more accurate understanding of interactions between energy efficiency technologies and diffusion-related variables. Such an approach taking end use heterogeneity into consideration will make decision alternatives to be more homogeneous. A more precisely developed decision alternatives are proposed to provide better linkage between technology alternatives and their implications from perspective of utility operational objectives and enable decision makers to give better decisions.

Overall, proposed improvements are expected to contribute to existing level of knowledge by enabling a more accurate energy efficiency technology evaluation and planning approach that can provide better understanding of the potential implications of the strategic decisions.

Proposed approach consists of several phases which involve multiple techniques and methodologies. Proposed research approach starts with development of a case study through literature reviews and semistructured interviews with subject matter experts. Second and third phases involve in development of hierarchical decision-making model and model validation with respect to construct and content validity. Fourth phase deals with data collection and validation followed with the fifth phase where results and sensitivity analyses are presented. Lastly, case study is validated with respect to criteria validity with subject matter experts. In Fig. 11.1, major steps of the proposed research approach can be observed.

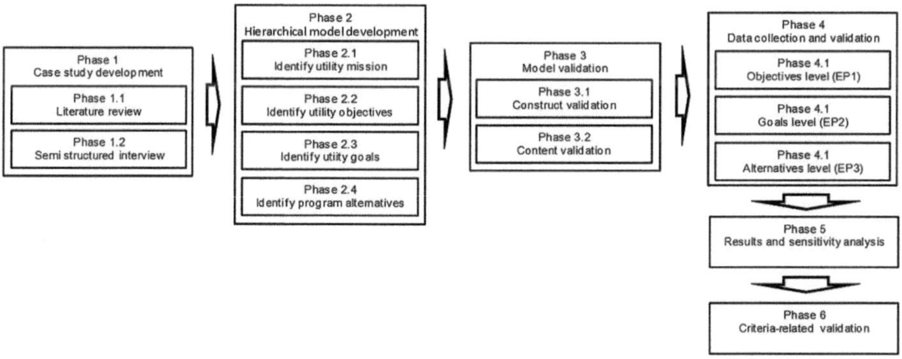

Fig. 11.1 Comprehensive approach for assessment of energy efficiency technologies from a utility perspective

11.2 Case Study Development

Proposed research approach is going to be demonstrated through a case study in the Pacific Northwest USA region. In order to develop the case study, the following steps have been fully or partially completed.

11.2.1 Literature Review

Assessment variables used in the proposed hierarchical decision model have been identified from a number of literature reviews. These literature reviews have been conducted on the following topics: multiperspective energy technology assessment (please refer to Independent study 1) and adoption of energy efficient technologies from utility demand-side management (DSM) perspective (please refer to Independent study 2 and 3). For the purpose of this analysis, you will be presented a brief literature review on multicriteria assessment models for utility DSM and energy generation technologies below.

Literature review has revealed that multicriteria decision-making models have been widely used in the utility DSM decision-making literature due to existence of wide range of objectives associated with utility operations. For instance, specific areas covered in the literature are value-focused objectives in electric utilities [13, 14] integrated resource planning [18], assessment of lighting energy efficiency measures [24], prioritization of energy efficiency barriers [30, 19], assessment of DSM programs [16, 17, 9] and strategies [10, 29].

Keeney and McDaniels [13] have attempted to identify hierarchy of objectives in the electric utilities by employing a value-focused perspective. Developed hierarchy has been proposed to address all major assessment issues that a utility organization would face due to assessment criteria's coverage of high level objectives. Accordingly, you can see the complete list below. This list gives a quick glimpse at wide range of utility objectives; however, more comprehensive studies can also be found in the literature [14].

1. Maximize contribution to economic development
   (a) Minimize cost of electricity use
   (b) Maximize funds transferred to government
   (c) Minimize economic implications of resource losses
2. Act consistently with the public's environmental values
   (a) About local environmental impacts
      - To flora
      - To fauna
      - To wildlife ecosystems

    - To limit recreational use
    - To esthetics

   (b) About global impacts

3. Minimize detrimental health and safety impacts

   (a) To the public

    - Mortality
    - Morbidity

   (b) To employees

    - Mortality
    - Morbidity

4. Promote equitable business arrangements

   (a) Equitable pricing to different customers
   (b) Equitable compensation for concentrated local impacts

5. Maximize quality of service

   (a) To small customers

    - Minimize outages
    - Minimize duration of outages

   (b) To large customers

    - Minimize outages
    - Minimize duration of outages

   (c) Improve new service
   (d) Improve response to telephone inquiries

6. Be recognized as public service oriented

Hobbs and Horn [9] have employed multicriteria decision-making method to develop an energy portfolio at BC Gas. One of the significant contributions of the model has been stated to be its ability to provide significance analysis and provide better communication with the stakeholders. Alternatives assessed in the study are programs that are not necessarily related to energy efficiency programs, thus assessment criteria used in the study have been kept rather general in order to be able to assess wide range of program alternatives.

A more energy efficiency specific decision model has been developed by Ramanathan and Ganesh [24] who have developed an energy efficiency measure selection model by combining analytic hierarchy process with goal programming. AHP has been used to find relative weights of assessment criteria as can be seen in the table below and goal programming has been used to optimize decision-making model by minimizing life cycle cost, use of petroleum products, use of fuelwood products, emission of carbon, sulfur, and nitrogen oxides and maximize system efficiency, employment generation, and use of locally available resources. Based on the results, suggestions regarding weak points of the measured alternatives have been proposed. Authors have indicated that a follow-up model could improve the existing model by incorporating issues faced during the implementation of the energy efficiency programs such as adoption barriers, market penetration-related variables [26, 28]. At this point it would be beneficial to mention about related variables that Reddy [26] has mentioned in his study. Reddy [26] has proposed that four essential factors that would enable better facilitate diffusion of energy efficient technologies. These are using combinations of energy efficiency measures at a strategic level, employing policy-assisted or market-oriented mechanisms, and promoting technological innovation for reaching improved energy efficiency.

A complementary study, to understand how different energy efficiency barriers affect diffusion of energy efficient technologies, has been conducted by Wang et al. [30]. Study has attempted to explore the interactions between 13 energy efficiency barriers identified in Chinese industry by employing interpretive structural modeling (ISM). ISM has been observed to provide information in exploring influential relationship between various barriers based on expert judgments. Accordingly, barriers such as lack of strategic planning, lack of awareness, and limited policy framework have been identified as being driving power to the rest of the barriers without any dependencies. Thus, they are suggested to be treated as strategic issues and root of the barrier problem. Barriers such as lack of funding or finance difficulties, lack of research personnel or trained manpower, and inadequate data and information have been identified as strong drivers with less dependency on other barriers. Moreover, barriers such as lack of experience in technology and management, reluctance to invest due to high risk, lack of appropriate production technologies, objections from different interest groups, lack of public participation, and inappropriate industrial framework have been identified as highly dependent barriers whose solutions lay behind more driving barriers. Aside from ISM, MICMAC analysis has been conducted and identified barriers have been clustered into four categories [25]. These clusters are named as autonomous, dependent, linkage, and independent barriers [30]. This method has been claimed to provide valuable information in strategizing development of policy tools for removing market barriers. Another complementary study has been conducted by Nagesha and Balachandra [19] who have utilized analytic hierarchy process to prioritize some of the previously determined market barriers among small-sized foundry and brick and tile manufacturing industrial firms in India. Model consists of three levels of which are objective as the first level, assessment factors as the second level, and barriers as the third level. For instance, assessment factors are stated as intensity of the barrier, easiness of removal, impact of barrier removal on energy efficiency, impact of barrier removal on economic performance. Barriers that were studied are awareness and information barriers, financial and economic barriers, structural and institutional barriers, policy and regulatory barriers, behavior and personal barriers. Firm owners from 44 organizations, representing 25 and 50 % of the brick and tile and foundry organization population in the case study, have been identified as experts for data acquisition.

Pairwise comparison method has been used to quantify expert judgments on energy efficiency barriers with respect to assessment factors and assessment factors with respect to objective. Interestingly, in order to make judgment process more accurate, experts have been provided with data that describe the important dimension of each barrier [19].

Lee et al. [16, 17] have employed analytical hierarchy process method to manage planning, execution, and assessment of DSM investment programs. Proposed model consists of four main criteria which are proposed to address planning, implementation, effectiveness, and usefulness of a given DSM program. Proposed criteria address some of the potential improvement areas pointed out by Ramanathan and Ganesh [24], such as incorporation of program implementation-related variables; however, diffusion-related variables still remain missing.

A complementary study to Lee et al. [16, 17] has been carried out by Vashishtha and Ramachandran [29] who have attempted to develop a model for selecting DSM implementation strategies, policy tools, by employing analytic hierarchy process. Decision model has been applied to experts from three stakeholder perspectives which are utilities, regulators, and consumers. Similarities and differences in implementation strategy desires have been observed by using Spearman's correlation coefficient and it has been indicated that certain DSM programs were favored by all three stakeholders. A weakness of the model has been indicated by the authors as that the model cannot help utilities transform into energy efficiency service providers due to potential high transactions costs, necessity of establishing new functions, and developing procedures for continuous operations.

In the case of generation technologies, there have been considerable amount of multiperspective assessment studies in the literature. For instance, these studies are generic assessment framework for power plant assessment [4], assessment of hydrogen and natural gas power plants [22, 23], evaluation of CHP systems [21], assessment of natural gas supply options [1, 2], and evaluation of liquid biofuels [20]. Furthermore, a few studies have been identified to assess impact of power plants on living standards [5, 6] and nonradioactive waste emission [3].

11.2.2 Semistructured Interview

In order to uncover assessment criteria that might have been missed from the literature review, semistructured interviews have been conducted with three energy efficiency program design and implementation experts. As part of the examination, results extracted from the interviews have been integrated into further analysis. Please note that final model needs to be modified with the suggestions from the expert interviews. Briefly, these changes involve in separation of existing variable "Reduce/postpone capital investments" into two distinct variables such that "Reduce capital investments" and "Postpone capital investments." Also, separation of an existing variable "Create or retain job opportunities" into two distinct variables such as "Create job opportunities" and "Retain job opportunities." In addition to new variable additions, one of the experts suggested eliminating the variable "Improve life standards (non energy benefits)" claiming that the variable is useful for determining programs' diffusion to the market, but not contribution to the utility objectives. Moreover, it was further suggested that the model could be strengthened by adding diffusion and program development-related variables since those are also significant considerations from a utility standpoint. Lastly, it has been stated that objective "Minimize adverse effects on public" and utility goals under it do not have direct contribution to utility missions in the context of energy efficiency. Moreover, it has been further emphasized that it would be difficult to find sufficient expertise to assess impact of energy efficiency programs with respect to aforementioned variables.

11.3 Hierarchical Decision Model Development

As part of the research approach, assessment variables in the decision model have been kept subject to a content validity test. Accordingly, objectives "Increase operating flexibility and reliability" and "Reduce system costs" have been selected due to their relatively high content validity scores. Furthermore, objective "Promote regional development" has been selected since it has received one of the lowest content validity scores among the proposed utility objectives. By selecting mix objectives with high and low content validity scores, author has aimed to compare and contrast the results, and how content validity scores impact the local contribution values and final results. You will be presented with more information regarding how content validation was conducted in the following sections.

Proposed demonstrative hierarchical decision model is presented in Fig. 11.2. Following sections will briefly provide information about the hierarchical levels and assessment variables used in the model.

- Level 1 Mission: Identify the energy efficiency programs (measures) which have the highest contribution potential to overall goals and objectives of a utility
- Level 2 Objectives: Utility objective k supporting the mission: O_k, $k=1, \ldots, K$
- Level 3 Goals: Utility goal j under objective k: G_{jk} $j=1, \ldots, J$
- Level 4 Alternatives: Energy efficiency program alternative i supporting goal j under objective k: A_{ijk}, $i=1, \ldots, I$

11.3.1 Mission Statement

Proposed hierarchical decision model is intended to identify energy efficiency measures which have the highest contribution potential to overall goals and objectives of a utility. Proposed research model attempts to reach its goal by expanding existing assessment models by employing a multiperspective view allowing trade-off analysis at the utility objective and goal levels.

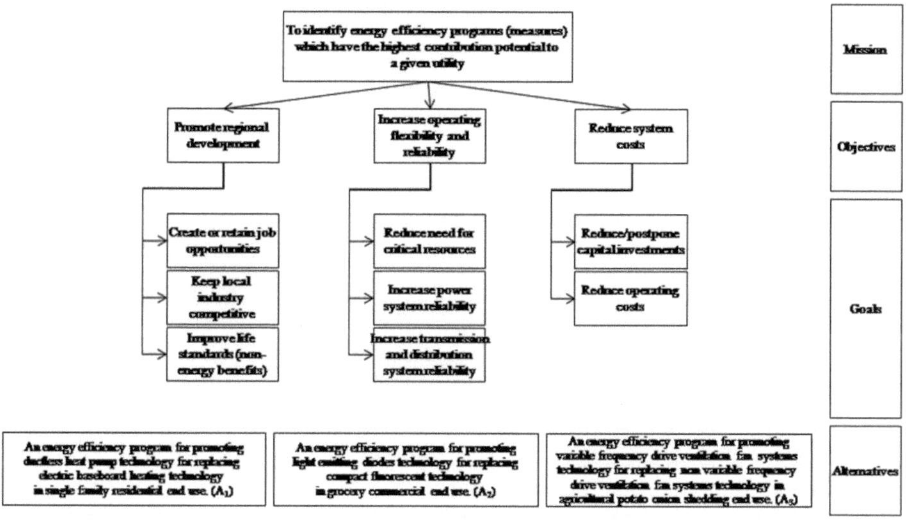

Fig. 11.2 Hierarchical decision model constructed for demonstration purpose

11.3.2 Assessment Variables in the Objectives Level

Variables at this hierarchy level are utilized to understand the relative importance of utility objectives in achieving the mission. These objectives are treated as the general roles and responsibilities of utilities and help understand trade-off at the utility strategy level.

- Promote regional development: New job opportunities emerging from utility operation supply chain, maintenance of existing infrastructure as well as new infrastructure additions, strengthening local industry by providing low rate energy, helping diffusion of better technology alternatives are some of the examples of utilities contributions to regional development. Variable "Promote regional development" is intended to capture importance of improving quality of life in their area of operation.
- Increase operating flexibility and reliability: Due to increasing load demand and changing load shapes, generation variability of renewable energy alternatives, aging generation and distribution systems, volatilities in supply of conventional energy resources, utilities are looking for new opportunities that would enable them to respond rapidly to changing environments faster and accurately without compromising quality of utility operations. Variable "Increase operating flexibility and reliability" is intended to capture the importance of maximizing system reliability in general.
- Reduce system cost: Given the challenges such as increasing population, seasonality in both off peak and peak demands, generation variability in renewable

energy alternatives, and aging generation and transmission systems, utilities need to sustain their operations by making minimum cost decisions. Variable "Reduce system cost" is intended to capture importance of providing the least cost rate within the range of sound business practices.

11.3.3 Assessment Variables in the Goals Level

Variables at this hierarchy level are utilized to understand the relative importance of utility goals with respect to corresponding utility objectives. These variables are treated as goals which help align utility operations with higher level utility objectives. Variables at the goals level can be achieved through capacity extensions as well as more efficient use of existing resources through energy efficiency programs. Accordingly, energy efficiency programs are going to be assessed with respect to their potential contribution to each of these goals.

Promote regional development: Utility objective "Promote regional development" is defined with following utility goals.

- Create or retain job opportunities: Energy efficiency programs directly and indirectly impact magnitude of economic activity through promotion of newer technologies to the market. For instance, various actors throughout the supply chain such as manufacturers, designers, contractors, retailers, etc., have chance to participate in development and delivery of energy efficiency programs and position themselves in growing markets. Variable "Create or retain job opportunities" is intended to capture magnitude of all these direct and indirect job opportunities created or retained through delivery of an energy efficiency program alternative.
- Keep local industry competitive: Energy efficiency programs can enable rapid diffusion of some of the new manufacturing technologies by eliminating implementation and operation-related concerns through demonstration projects. Variable "Keep local industry competitive" is intended to capture degree of both energy savings and competitive advantages provided to local industrial actors through delivery of an energy efficiency program alternative.
- Improve life standards (nonenergy benefits): New technology alternatives provide not only energy savings, but also improve life standards of public through newly added functions embedded in new products such as programs for replacing less efficient heating units with more efficient heating and cooling units in residential end use. Variable "Improve life standards (nonenergy benefits)" is intended to capture degree of improved life standards of public through delivery of an energy efficiency program alternative.

Increasing operating flexibility and reliability: Utility objective "Increasing operating flexibility and reliability" is defined with following utility goals.

- Reduce need for critical resources: Significant percentage of world's energy generation is supplied from fossil fuel resources whose supply chain has been subject to disturbances due to political instabilities. Consequences of such events

have had negative impacts on both developed and developing countries around the globe. Variable "Reduce need for critical resources" is intended to capture degree of energy resource diversity gained by reducing need for some of the critical resources (oil, water, etc.) through delivery of an energy efficiency program.
- Increase power system reliability: Utilities can make use of energy efficiency programs to reduce/alter loads on critical pieces of power generation systems in order to cope with some of the challenges such as increasing population, seasonality in magnitude off peak load and peak loads, generation variability in renewable energy alternatives, and aging power generation systems. Variable "Increase power system reliability" is intended to capture degree of reliability improvement over the power generation systems through delivery of an energy efficiency program.
- Increase transmission and distribution system reliability: Utilities can make use of energy efficiency programs to reduce/alter loads on critical parts of power transmission systems in order to cope with some of the challenges such as increasing population, seasonality in load and peak demands, and aging transmission and distribution systems. Variable "Increase power system reliability" is intended to capture degree of reliability improvement over the power transmission and distribution systems through delivery of an energy efficiency program.

Reduce system cost: Utility objective "Reduce system cost" is defined with the following utility goals.

- Reduce/postpone capital investments: One of the most important factors taken into consideration in utility energy planning operations is the projection of peak load demand which sets the minimum generation capacity for reliable services. Energy efficiency programs can be utilized to reduce magnitude of peak loads in order to reduce or postpone some of the capital investments which require high upfront capital requirements. Variable "Reduce/postpone capital investments" is intended to capture magnitude of new construction investments avoided through delivery of an energy efficiency program alternative.
- Reduce operating costs: Depending on the efficiency rate of generation technology in operation, unit price of energy output varies. For instance, marginal cost of generation increases as the demand load increases and reaches peak load where less efficient, but more responsive generation units become operational. Variable "Reduce operating costs" is intended to capture degree of high cost generation units avoided through delivery of an energy efficiency program alternative.

11.3.4 Energy Efficiency Program Alternatives in the Alternative Level

For demonstration purposes, following energy efficiency program alternatives have been selected from a list posted on Regional Technical Forum's (RTF) website. These programs have already been undertaken and supported by utilities and public organizations operating in the Pacific Northwest U.S.A. region. Motivation behind picking existing energy efficiency programs is to test the proposed assessment model with alternatives that experts have readily available information and get feedback for criteria validation purposes.

- Alternative 1: An energy efficiency program for promoting ductless heat pump technology for replacing electric baseboard heating technology in single family residential end use.
- Alternative 2: An energy efficiency program for promoting light emitting diodes technology for replacing compact fluorescent technology in grocery commercial end use.
- Alternative 3: An energy efficiency program for promoting variable frequency drive ventilation fan systems for replacing nonvariable frequency drive ventilation fan systems in agricultural potato onion shedding end use.

11.4 Model Validation

Constructed model has been validated using two distinct validity types which are construct and content validity.

11.4.1 Content Validity

Content validity refers to the degree to which the content of the items reflects the domain of interest. In the context of hierarchical decision modeling, researcher uses content validity to refer to each decision variable's degree of appropriateness to measure a higher level decision variable it is proposed for.

In this research, group disagreements in content validity are proposed to be resolved by using the Delphi technique. The Delphi technique is selected by the researcher since it allows for data collection through questionnaires eliminating scheduling conflicts among various participants. For demonstration purposes, the Delphi process is limited to only one round; however, results are discussed with an expert selected as Delphi decision maker for further articulation. Selected expert works as a supervisory engineer in energy efficiency group at Bonneville Power Administration.

For each level of the hierarchical model, relevant experts have been identified and assigned as panel respondents. For instance, for the second level of decision hierarchy where utility objectives are validated against the mission, researcher has made sure the respondents have a higher level of perspective on utility-specific objectives. Accordingly, these experts are currently employed as vice president, supervisory engineer, and program implementation manager in energy efficiency groups, as well as academic scholars that have relevant experience in energy technology assessment. Expertise from wide range of utility operations would be needed for validation of third level decision hierarchy, where utility goals are validated against corresponding utility objectives. For the purpose of this analysis, experts who are currently employed as project managers, program managers, and planning specialists in energy efficiency groups.

A survey has been constructed to gather data for content validity analysis. Necessary definitions and instructions have been attached to the survey and sent to experts via e-mails. As described earlier, questions were developed in a way that each decision element in the hierarchy is validated against a higher level corresponding variable. Experts were asked to state their judgment on each variable using a Likert scale of 1–3. (1 indicating that proposed decision variables were "Essential," 2 indicating that proposed decision variables were "Not essential but useful," and 3 indicating that proposed decision variables were "Not necessary"). Please see Appendix 1 for the content validation survey.

Please note that variables in the demonstrative model are relevant to expertise of the aforementioned expert group; however, as mentioned earlier complete research model will require experts from other fields. For instance, in order to validate goal variables under utility objective "Minimize adverse effects on public," experts from public affairs. Similarly, experts from environment, fish, and wildlife service might be a good fit for validating utility goals under utility objective "Minimize environmental impacts." For validation of the complete research model, experts that researcher does not have connections to will be identified through the expert suggestions. Expert identification and selection will go on until there is sufficiently enough expertise covering all the decision variables in the research model.

11.4.1.1 Content Validity Ratio

In order to detect disagreement among experts, content validity ratio, developed by Lawshe [15], will be used. Content validity ratio will be calculated for each decision variable based on the formula below. This formula yields values between +1 and −1; positive values indicate that at least half of the experts have indicated that the item in question as essential.

$$\text{CVR} = \frac{n_e - N/2}{N/2}$$

n_e: number of experts in the panel indicating "essential"
N: total number of panelists

Calculated CVR values are going to be compared to a threshold value that is a function of the number of participants [15]. Please see below for the threshold values for statistical significance at $P < .05$ with respect to size of panel (Table 11.1).

Using the formula stated earlier, content validity ratios for each of the decision variables are calculated.

Table 11.1 Threshold values for CVR ratio testing

Number of panelists	Min value
5	0.99
6	0.99
7	0.99
8	0.75
9	0.78
10	0.62
11	0.59
12	0.56
13	0.54
14	0.51
15	0.49

Survey responses and CVR values for each decision element are presented in Table 11.2.

Values obtained from the CVR formula are compared against the threshold value of 0.75 which corresponds to panelist number of 8. Accordingly, as observed from Table 11.1, only utility goals "Increase power system reliability," "Increase transmission and distribution system reliability," and utility objective "Reduce/postpone capital investments" have CVR ratio greater than or equal to the threshold value. In addition to those, utility objective "Reduce system cost" and utility goal "Reduce operating costs" have received CVR value of 0.5 which is relatively close to the threshold value.

It is observed that majority of the decision variables have received very low scores using the CVR ratio method. Author acknowledges this as a potential improvement area which needs to be resolved. Reasons behind the low content validity ratio with all the other variables may emerge from multiple stems. Based on the content validity results, it is observed that experts have consistently emphasized the importance of finance-related variables which are widely used by the current practices. For instance, since it is the first time this type of multicriteria assessment model is proposed for assessment of energy efficiency measures, experts might have been biased toward conventional assessment criteria. Second, as the evaluation was only conducted for only a single round, potential feedback between the experts is missing. Two or more rounds of evaluation might create better convergence among the panel participants and increase CVR values. Third reason might be because of the nature of the method employed. For instance, CVR ratio has mainly been used in survey research studies where the number of respondents is relatively greater than the number of respondents required for the Delphi technique. As realized from the table where the threshold values with respect to number of panelists are presented, threshold values decrease as the number of respondents increase. Thus, researcher believes inclusion of more experts might provide better CVR ratio results. This perspective is also in parallel to the earlier notification that the original research approach will not be limited to experts within energy efficiency group, but also with experts from other utility departments. This will both increase the number of participants and balance potential bias.

Table 11.2 Content validity results—first round

	Expert 1	Expert 2	Expert 3	Expert 4	Expert 5	Expert 6	Expert 7	Expert 8	Mean	Standard Dev.	CVR
Promote regional development	1	1	1	2	3	2	2	2	1.75	0.71	−0.25
Create or retain job opportunities	1	1	1	2	2	3	2	1	1.625	0.74	0
Keep local industry competitive	1	2	1	2	2	2	2	2	1.75	0.46	−0.5
Improve life standards (nonenergy benefits)	1	2	2	2	2	2	1	3	1.875	0.64	−0.5
Minimize environmental impacts	1	1	1	1	2	2	2	2	1.5	0.53	0
Reduce GHG emissions	1	1	1	1	2	2	1	2	1.375	0.52	0.25
Reduce emission of soil, air, and water contaminants	1	1	1	1	2	2	2	2	1.5	0.53	0
Avoid flora and fauna habitat loss	1	2	1	1	2	2	3	2	1.75	0.71	−0.25
Increase operating flexibility and reliability	1	1	2	2	2	1	1	1	1.375	0.52	0.25
Reduce need for critical resources	1	1	1	2	2	1	2	1	1.375	0.52	0.25
Increase power system reliability	1	1	1	1	1	1	1	1	1	0.00	1
Increase transmission and distribution system reliability	1	1	1	1	1	1	1	1	1	0.00	1
Reduce system cost	1	1	2	2	1	1	1	1	1.25	0.46	0.5
Reduce/postpone capital investments	1	1	1	2	1	1	1	1	1.125	0.35	0.75
Reduce operating costs	1	1	2	2	1	1	1	1	1.25	0.46	0.5
Minimize adverse effects on public	1	1	2	1	2	2	2	2	1.625	0.52	−0.25
Avoid noise and odor	1	2	2	2	2	2	3	2	2	0.53	−0.75
Avoid visual impacts	1	2	2	2	2	3	2	2	2	0.53	−0.75
Avoid property damage and impact on lifestyles	1	1	2	1	1	2	2	2	1.5	0.53	0

Please note that since this process was limited to only one round of evaluation, researcher decided to conduct further analyses which will not necessarily be conducted the same way in the original research approach. For instance, using chi-square test, each variable has been tested whether there is any significant expert disagreement in the expert panel or not. Second, using one sample *T*-tests, aggregated expert responses have been characterized. For instance, each decision variable was tested whether they came from normal distributions with mean values of 1, 2, or 3. If aggregated response for a particular variable is proven to come from a normal distribution with mean value of 1, 2, or 3, then respectively the variable is considered as "Essential," "Useful but not essential," and "Not necessary." Although there is no expert agreement on the aggregated results, utility objective "Minimize environmental impacts" and utility goals "Reduce emission of soil, air and water contaminants" and "Avoid property damage and impacts on lifestyles" have passed the *T*-tests for mean values of both 1 and 2. Accordingly, they were considered to have values between "Essential" and "Useful but not essential."

Results are presented in Table 11.3. Please refer to Appendix 2 with details of the analyses.

As observed from Table 11.3, there are about eight decision elements that the experts have reached an agreement on. Considering these, one of the key findings of

Table 11.3 Further analysis of content validity and expert disagreements

	Chi-square agreement (Yes/No)	One sample *T* test
Promote regional development	No	Useful but not essential
Create or retain job opportunities	No	Useful but not essential
Keep local industry competitive	Yes	Useful but not essential
Improve life standards (nonenergy benefits)	No	Useful but not essential
Minimize environmental impacts	No	Essential-useful
Reduce GHG emissions	No	Essential
Reduce emission of soil, air, and water contaminants	No	Essential-Useful
Avoid flora and fauna habitat loss	No	Useful but not essential
Increase operating flexibility and reliability	No	Essential
Reduce need for critical resources	No	Essential
Increase power system reliability	Yes	Essential
Increase transmission and distribution system reliability	Yes	Essential
Reduce system cost	Yes	Essential
Reduce/postpone capital investments	Yes	Useful but not essential
Reduce operating costs	Yes	Useful but not essential
Minimize adverse effects on public	No	Useful but not essential
Avoid noise and odor	Yes	Useful but not essential
Avoid visual impacts	Yes	Useful but not essential
Avoid property damage and impact on lifestyles	No	Essential-Useful

the analysis is that there is no variable that experts have judged as "Not necessary." Discussions with the identified expert and findings of the content validity analysis also reveal that utility experts consider variables as "Essential" if only they are critical in complying with the regulations. For instance, these variables are reported to be related to system reliability, electricity rates, and environmental regulations. Another important finding is that most of the variables associated with "Promote regional development" and "Minimize adverse effects on public" are considered as "Useful, but not essential." Interview with the selected expert provides insight to this situation stating that these variables are taken into consideration in decision-making practices if only the aforementioned critical decision variables comply with regulations.

11.4.2 Construct Validity

In the context of this research, construct validity refers to the extent to which proposed hierarchical decision model complies with its theories. For instance, issues of independency among decision variables on the same level, unidirectional relationships between decision levels are some of the important ones.

Construct validity of the proposed research model was conducted using nominal group technique with the PhD students in Engineering and Technology Management Department at Portland State University. Expert group was selected from those students who have interest in energy-related decision-making research. It was made sure that the panel had enough level of expertise in the area of hierarchical decision modeling. The number of experts participated in the activity was three.

Proposed research model was presented on a visible screen and each panel participant was asked to provide their feedback on the construct validity of the model. Researcher made sure there was special focus on that the proposed hierarchical decision-making model complied with its theories. For instance, issues of independency among decision variables on the same level, unidirectional relationships between decision levels were some of the most important aspects to be elaborated. While a panel expert proposed his/her feedback, panel facilitator took notes on a flip chart which all panel experts were able to see. Please note that nominal group technique requires feedback generation to be carried out in a sequential order without any intervention by the rest of the panelists. Accordingly, it was made sure each participant got to state their feedback without being intervened. After feedback generation phase was completed, each of the proposed feedback was presented by its owner and this time the rest of the panelists were given chance to state their own comments as well. This phase was kept open to face-to-face discussions in order to stimulate critical thinking. After discussions around the proposed feedbacks were completed, panelists were asked to vote whether each particular feedback needed to be incorporated into the model or not.

It has been confirmed with the experts that proposed research approach comply with the theories of hierarchical decision modeling. Accordingly, no change has been proposed to be incorporated into the model.

11.5 Data Collection and Validation

In this section you are going to be informed about the expert panel design and survey instrument development.

11.5.1 Expert Panel Design and Expert Selection

For demonstration purposes, each level of hierarchy in the proposed model has been considered as one expert panel; however, researcher acknowledges that it is often very hard for an expert to comprehend importance of variables outside their field. For instance, this issue becomes significantly clear on goals and alternatives levels where there are diverse range of utility operations and technologies. It has been observed that experts in energy efficiency tend to neglect the fact that utility operations are not entirely providing reliable service at reasonable costs, but also protect environment, comply with public interests, and promote regional development. In order to eliminate this issue, researcher has realized that it would be more accurate to evaluate goals under each objective in different panels where panelists are invited from more relevant departments such as environment fish and wild life, corporate strategy, planning and asset management, system operations, agency compliance, and governance. Since this is outside the scope of the demonstration discussion will be kept rather short.

11.5.1.1 Objectives Level: Expert Panel 1

As mentioned earlier, demonstrative model has only three variables on the objectives level. These objectives are "Promote regional development," "Increase operating flexibility and reliability," and "Reduce system cost." While selecting the experts for panel 1 it was made sure that the experts had general understanding of wide range of utility operations. In order to achieve that, both practitioner and academic perspectives were taken into consideration. Researcher made sure selected experts from academic backgrounds had research interests in energy efficiency and published in related journals. Furthermore, it was also made sure that experts had significant amount of experience in energy efficiency field. Since the experts were selected only from energy efficiency groups it was also made sure that they have had different skill sets such as engineering, management, planning, and finance. For instance, expert panel 1 consisted of three experts from academia and four experts from energy efficiency groups.

11.5.1.2 Goals Level: Expert Panel 2

Although evaluation of utility goals requires more specialized experts than utility objectives, experts selected for expert panel 1 were also selected for expert panel 2 due to time limitations. Impacts of this issue were somewhat realized to a significant degree on the evaluation of utility goals under objective "Promote regional development."

11.5.1.3 Alternatives Level: Expert Panel 3

Evaluation of energy efficiency programs with respect to utility goals requires significantly different skill sets than the expert panels 1 and 2. For instance, experts in this panel were required to have understanding of utility and measure loads, and the interaction between them. For instance, expert panel 3 consisted of four experts, three of which were industry experts who have engineering backgrounds, and one field engineer from an academic institution. All of the industry experts work in energy efficiency group at Bonneville Power Administration as program and project managers, whereas the fourth expert works at Washington State University Energy Extension Program. It should be noted that experts in this group had good understanding about the issues that fall under objectives "Increase operating flexibility and reliability" and "Reduce system cost"; however, they did not have as much understanding with objective "Promote regional development."

11.5.2 Survey Construction and Validation

For each of the expert panels, a pairwise comparison survey has been constructed and tested with students from ETM program. Special focus has been placed to make sure language of the surveys was clear and they did not take more than 15 min to complete. Definitions of the terms used have been included in each survey along with the instructions on pairwise comparison method. Furthermore, a sample question and an answer have been included in order to strengthen the instructions (please refer to Appendix 3–5 for surveys used in the study). Survey has been sent to experts and received back via e-mail; however when required, researcher called the experts and provided necessary instructions about the survey and research model in particular.

11.6 Case Study Results and Sensitivity Analysis

This section is going to present judgment quantification, expert inconsistency, and group disagreement results for each hierarchy level. Please note that if you need further information about data and analysis output please refer to Appendix 6. Appendix 6 includes outputs of judgment quantification, group disagreement, and hierarchical clustering analyses performed. You may see more than one result in cases where second round of expert judgment was required due to data validation purposes.

11.6.1 Objectives Level: Expert Panel 1

Expert panel 1 consisted of seven experts who were asked to evaluate selected utility objectives. Evaluation was based on the relative importance of utility objectives in achieving overall utility mission.

11.6.1.1 Expert Panel 1 Results

According to experts "Reduce system cost" (0.48) and "Promote regional development" (0.42) have the highest relative contributions to overall utility mission whereas "Promote regional development" (0.10) is weighted relatively lower. The summary of the results is presented in Table 11.4.

11.6.1.2 Analysis of Expert Judgment Inconsistencies

Experts reflected their judgment without showing significant degree of inconsistency. Accordingly, all expert judgment was accepted. The summary of consistencies with respect to each expert is presented in Table 11.5.

Table 11.4 Summary of local contributions of objectives to the mission

Objectives (O_k)	Local contributions of objectives to mission (C_k^O)
Promote regional development	0.10
Increase operating flexibility and reliability	0.42
Reduce system cost	0.48

Table 11.5 Inconsistencies and relative contributions of objectives to the mission

Mission				
Expert	Promote regional development	Increase operating flexibility and reliability	Increase operating flexibility and reliability	Inconsistency
Expert 1	0.16	0.38	0.46	0.014
Expert 2	0.16	0.49	0.35	0.001
Expert 3	0.14	0.50	0.36	0.004
Expert 4	0.09	0.24	0.67	0.007
Expert 5	0.09	0.27	0.64	0.015
Expert 6	0.00	0.67	0.32	0.036
Expert 7	0.05	0.41	0.54	0.005
Mean	0.10	0.42	0.48	

11.6.1.3 Analysis of Group Disagreements

It was observed that there was no significant group disagreement among the panel members. Accordingly, all expert judgment was aggregated without requiring second round of judgment quantification. Summary of group disagreement test is presented in Table 11.6.

11.6.2 Goals Level: Expert Panel 2

Expert panel 2 consisted of seven experts who were asked to evaluate utility goals under selected utility objectives. Accordingly, evaluation was based on the relative importance of utility goals in contributing to corresponding utility objectives.

11.6.2.1 Expert Panel 2 Results

It was observed that contributions of all utility goals "Keep local industry competitive" (0.36), "Improve life standards" (0.32), and "Create or retain job opportunities" (0.31) under "Promote regional development" are quite close. Utility goals "Increase power system reliability" (0.37) and "Increase transmission and distribution system reliability" (0.36) under objective "Increase operating flexibility and reliability" have quite similar relative contributions whereas "Reduce need for critical resources" (0.27) has been perceived to have less contribution. Utility goal "Reduce operating costs" (0.60) is scored higher than "Reduce or postpone capital investments" (0.40) in contributing to objective "Reduce system cost." The summary of the results is presented in Table 11.7.

Table 11.6 Final intraclass correlation coefficient and F-value of utility objectives

	r_{ic}			
Utility objectives under	$0<r_{ic}<1$	F-value	F-critical at 0.05 level	F-test result
Mission	0.717	12.830	3.8853	Reject H_0

Table 11.7 Summary of local contributions of goals to corresponding objectives

Objectives (O_k)	Goals (G_{jk})	Local contributions of goals on objectives C_{jk}^{G-O}
Promote regional development	Create or retain job opportunities	0.31
	Keep local industry competitive	0.36
	Improve life standards (nonenergy benefits)	0.32
Increase operating flexibility and reliability	Reduce need for critical resources	0.27
	Increase power system reliability	0.37
	Increase transmission and distribution system reliability	0.36
Reduce system cost	Reduce operating costs	0.60
	Reduce postpone capital investments	0.40

11.6.2.2 Analysis of Expert Judgment Inconsistencies

Expert inconsistencies in evaluation of utility goals were analyzed with respect to each utility objective. You can see each of the analyses below.

Expert Inconsistencies in Evaluation of Utility Goals Under Objective "Promote Regional Development"

Experts reflected their judgment without showing significant degree of inconsistency. Accordingly, all expert judgment was accepted. The summary of consistencies with respect to each expert is presented in Table 11.8.

Expert Inconsistencies in Evaluation of Utility Goals Under Objective "Increase Operating Flexibility and Reliability"

Experts reflected their judgment without showing significant degree of inconsistency. Accordingly, all expert judgment was accepted. The summary of consistencies with respect to each expert is presented in Table 11.9.

Table 11.8 Inconsistencies and relative contributions of goals under objective "Promote regional development"

Expert	Create or retain job opportunities	Keep local industry competitive	Improve life standards (nonenergy benefits)	Inconsistency
Expert 1	0.31	0.38	0.31	0.012
Expert 2	0.27	0.42	0.31	0.002
Expert 3	0.33	0.33	0.33	0.000
Expert 4	0.38	0.31	0.31	0.000
Expert 5	0.25	0.38	0.38	0.000
Expert 6	0.39	0.39	0.21	0.000
Expert 7	0.27	0.33	0.40	0.000
Mean	0.31	0.36	0.32	

Table 11.9 Inconsistencies and relative contributions of goals under objective "Increase operating flexibility and reliability"

Expert	Reduce need for critical resources	Increase power system reliability	Increase transmission and distribution system reliability	Inconsistency
Expert 1	0.27	0.35	0.38	0.001
Expert 2	0.33	0.33	0.33	0.000
Expert 3	0.25	0.38	0.38	0.000
Expert 4	0.25	0.38	0.38	0.000
Expert 5	0.25	0.45	0.30	0.012
Expert 6	0.33	0.33	0.33	0.000
Expert 7	0.21	0.39	0.39	0.000
Mean	0.27	0.37	0.36	

Table 11.10 Inconsistencies and relative contributions of goals under objective "Reduce system cost"

Expert	Reduce operating costs	Reduce postpone capital investments	Inconsistency
Expert 1	0.5	0.5	N/A
Expert 2	0.6	0.4	N/A
Expert 3	0.7	0.3	N/A
Expert 4	0.5	0.5	N/A
Expert 5	0.7	0.3	N/A
Expert 6	0.5	0.5	N/A
Expert 7	0.7	0.3	N/A
Mean	0.60	0.40	

Table 11.11 Intraclass correlation and F-test result of promote regional development

Utility goals under objective	r_{ic} $0<r_{ic}<1$	F-value	F-critical at 0.05 level	F-test result
Promote regional development	−0.252	0.061	3.8853	H_0 cannot be rejected

Expert Inconsistencies in Evaluation of Utility Goals Under Objective "Reduce System Cost"

Since there is only two utility goals associated with objective "Reduce system cost" inconsistency analysis could not be performed. Accordingly, only expert preferences are presented in Table 11.10.

11.6.2.3 Analysis of Group Disagreements

There were three group disagreements observed in expert panel 2. These group disagreements occurred while assessment of utility goals with respect to objectives "Promote regional development," "Increase operating flexibility and reliability," and "Reduce system cost." Details about the process undertaken to cope with the aforesaid group disagreements are presented below.

Group Disagreements in Evaluation of Utility Goals Under Objective "Promote Regional Development"

Results from the first round of expert evaluations revealed that there was a significant degree of group disagreement on relative contributions of utility goals under objective "Promote regional development." Prior results of intraclass correlation coefficient (ICC) and F-tests are presented in Table 11.11.

As observed from the results of hierarchical clustering analysis (please refer to "Analysis and Data Validation of Utility Goals with Respect to "Promote Regional

Table 11.12 ICC and *F*-test of increasing operating flexibility and reliability

Utility goals under objective	r_{ic} $0<r_{ic}<1$	*F*-value	*F*-critical at 0.05 level	*F*-test result
Increase operating flexibility and reliability	−0.139	1.757	3.8853	H_0 cannot be rejected

Development"" in Appendix 6), judgment reflected by Expert 4, Expert 6, and Expert 7 stand out as outliers. Accordingly, these experts were contacted and requested to reconsider their responses. Expert 4 stated that his judgment toward putting the greatest value on "Create or retain job opportunities" was due to ongoing economic downturn in the recent years. In order to be more objective, he has decided to change his judgment and consider a more general view on utility missions. Similar to Expert 4, Expert 6 stated that he was confident about his judgment and added that relative value of variable "Improve life standards" could be increased a little, but it cannot be any greater than the other two variables. Expert 6 agreed to make small adjustments. Different from Expert 4, Expert 7 considered "Improve life standards" as the highest priority goals in promoting regional development. After discussing with Expert 7 it was understood that relative weight of variable "Create or retain job opportunities" was too small. Expert 7 decided to rearrange his judgment allocations based on the feedback. Accordingly, new judgment was incorporated; calculation of relative weights, inconsistency, and group disagreement analyses was conducted.

Group Disagreements in Evaluation of Utility Goals Under Objective "Increase Operating Flexibility and Reliability"

Results from the first round of expert evaluations revealed that there was a significant degree of group disagreement on relative contributions of utility goals under objective "Increase operating flexibility and reliability." Prior results of ICC and *F*-tests are presented in Table 11.12.

As observed from the results of hierarchical clustering analysis (please refer to "Analysis and data validation of utility goals with respect to "Increase operating flexibility and reliability"" in Appendix 6), there appears to be two groups that reflect slightly different opinions. Group 1 consists of Expert 1, Expert 2, and Expert 6, and Group 2 consists of Expert 3, Expert 4, Expert 5, and Expert 7. Since Group 2 has a larger size and more experienced people, experts from Group 1 were asked to reevaluate their judgment. Accordingly, Expert 6 stated that all three variables were very important in running utility operations reliably and wanted to keep his previous judgment. Expert 1 did not make significantly different changes, but decreased the relative importance of "Reduce need for critical resources" slightly. A similar behavior was observed from Expert 2, but the magnitude of change was observed to be greater. It was observed that all the experts put more value on "Increase power system reliability" and "Increase transmission and distribution system reliability."

In group 2, since Expert 4 had significantly different opinion than the rest of the experts he was requested to reevaluate his judgment as well. Through the discussions with Expert 4 it was observed that meaning of "Reduce need for critical resources" was not fully understood. The issue was communicated to the expert through a case study at BPA which emerged due to conflicting interests of environment, fish and wildlife, and power services departments. Accordingly, Expert 4 decided to increase relative weight of variable "Reduce need for critical resources." Accordingly, new judgment was incorporated; calculation of relative weights, inconsistency, and group disagreement analyses was conducted.

Group Disagreements in Evaluation of Utility Goals Under Objective "Reduce System Cost"

Results from the first round of expert evaluations revealed that there was a significant degree of group disagreement on relative contributions of utility goals under objective "Reduce system cost." Prior results of ICC and *F*-tests are presented in Table 11.13.

As observed from the results of hierarchical clustering analysis (please refer to "Analysis and data validation of utility goals with respect to "Reduce system cost"" in Appendix 6), Expert 7 reflects significantly different opinion than the rest of the experts. For instance, from the discussion with Expert 7 it has been observed that there had been a mistake while responding to the survey. Accordingly, new judgment has been incorporated; calculation of relative weights, inconsistency, and group disagreement analyses is conducted.

Final Group Disagreement Analyses Results

Final group disagreement results were calculated based on the modifications made in the second round of expert evaluations. Results are presented in Table 11.14.

Table 11.13 ICC and *F*-test of reduce system cost

Utility goals under objective	r_{ic} $0<r_{ic}<1$	*F*-value	*F*-critical at 0.05 level	*F*-test result
Reduce system cost	−.108	0.659	5.9874	H_0 cannot be rejected

Table 11.14 Final intraclass correlation coefficient and *F*-value of utility goals

Utility goals under objective	r_{ic} $0<r_{ic}<1$	*F*-value	*F*-critical at 0.05 level	*F*-test result
Promote regional development	0.031	1.148	3.8853	H_0 cannot be rejected
Increase operating flexibility and reliability	0.617	8.574	3.8853	Reject H_0
Reduce system cost	0.632	7.000	5.9874	Reject H_0

It should be noted that although table above shows that there was evident group disagreement for the case of "Promote regional development," final results were accepted since one of the experts decided not to change prior judgment with logical reasoning.

11.6.3 Alternatives Level: Expert Panel 3

Expert panel 3 consisted of four experts who were asked to evaluate energy efficiency program alternatives with respect to utility goals under selected utility objectives. Accordingly, evaluation was based on the relative importance of energy efficiency program alternatives in contributing to corresponding utility goal.

11.6.3.1 Expert Panel 3 Results

It is interesting to observe that, "Ductless heat pump program in residential family end use" has the highest, "Light emitting diodes program in grocery end use" has the second highest, and "Variable frequency drive ventilation program in potato shedding use" has the least amount of contributions with respect to all utility goals. The summary of the results is presented in Table 11.15.

Table 11.15 Summary of local contributions of energy efficiency program alternatives to corresponding utility goals

		Local contributions of alternatives on goals C_{ij}^{A-G}		
Objectives (O_k)	Goals (G_{jk})	DHP Program in single family end use	LED Program in grocery end use	VFD Program in potato onion shedding end use
Promote regional development	Create or retain job opportunities	0.63	0.21	0.16
	Keep local industry competitive	0.58	0.24	0.18
	Improve life standards (nonenergy benefits)	0.70	0.26	0.04
Increase operating flexibility and reliability	Reduce need for critical resources	0.76	0.19	0.05
	Increase power system reliability	0.75	0.16	0.08
	Increase transmission and distribution system reliability	0.69	0.25	0.07
Reduce system cost	Reduce operating costs	0.69	0.22	0.08
	Reduce postpone capital investments	0.58	0.29	0.13

11.6.3.2 Analysis of Expert Judgment Inconsistencies

Expert inconsistencies in evaluation of measure alternatives were analyzed with respect to each utility goal. You can see each of the analyses below.

Expert Inconsistencies in Evaluation of Energy Efficiency Programs Under Utility Goal "Create or Retain Job Opportunities"

Experts reflected their judgment without showing significant degree of inconsistency. Accordingly, all expert judgment was accepted. The summary of consistencies with respect to each expert is presented in Table 11.16.

Expert Inconsistencies in Evaluation of Energy Efficiency Programs Under Utility Goal "Keep Local Industry Competitive"

Experts reflected their judgment without showing significant degree of inconsistency. Accordingly, all expert judgment was accepted. The summary of consistencies with respect to each expert is presented in Table 11.17.

Table 11.16 Inconsistencies and relative contributions of alternatives under goal "Create or retain job opportunities"

Expert	DHP program in single family end use	LED Program in grocery end use	VFD program in potato onion shedding end use	Inconsistency
Expert 8	0.58	0.07	0.34	0.005
Expert 9	0.69	0.23	0.08	0.000
Expert 10	0.71	0.29	0.00	0.051
Expert 11	0.54	0.26	0.20	0.005
Mean	0.63	0.21	0.16	

Table 11.17 Inconsistencies and relative contributions of alternatives to goal "Keep local industry competitive"

Expert	DHP program in single family end use	LED program in grocery end use	VFD program in potato onion shedding end use	Inconsistency
Expert 8	0.58	0.07	0.34	0.005
Expert 9	0.69	0.23	0.08	0.000
Expert 10	0.56	0.43	0.01	0.006
Expert 11	0.48	0.24	0.28	0.006
Mean	0.58	0.24	0.18	

Expert Inconsistencies in Evaluation of Energy Efficiency Programs Under Utility Goal "Improve Life Standards"

Experts reflected their judgment without showing significant degree of inconsistency. Accordingly, all expert judgment was accepted. The summary of consistencies with respect to each expert is presented in Table 11.18.

Expert Inconsistencies in Evaluation of Energy Efficiency Programs Under Utility Goal "Reduce Need for Critical Resources"

Experts reflected their judgment without showing significant degree of inconsistency. Accordingly, all expert judgment was accepted. The summary of consistencies with respect to each expert is presented in Table 11.19.

Expert Inconsistencies in Evaluation of Energy Efficiency Programs Under Utility Goal "Increase Power System Reliability"

Experts reflected their judgment without showing significant degree of inconsistency. Accordingly, all expert judgment was accepted. The summary of consistencies with respect to each expert is presented in Table 11.20.

Table 11.18 Inconsistencies and relative contributions of alternatives under goal "Improve life standards"

Expert	DHP program in single family end use	LED program in grocery end use	VFD program in potato onion shedding end use	Inconsistency
Expert 8	0.69	0.23	0.08	0.000
Expert 9	0.69	0.24	0.08	0.000
Expert 10	0.71	0.29	0.00	0.051
Expert 11	0.71	0.29	0.00	0.051
Mean	0.70	0.26	0.04	

Table 11.19 Inconsistencies and relative contributions of alternatives under goal "Reduce need for critical resources"

Expert	DHP program in single family end use	LED program in grocery end use	VFD program in potato onion shedding end use	Inconsistency
Expert 8	0.70	0.19	0.11	0.020
Expert 9	0.73	0.19	0.08	0.000
Expert 10	0.63	0.36	0.00	0.023
Expert 11	0.98	0.01	0.01	0.000
Mean	0.76	0.19	0.05	

Table 11.20 Inconsistencies and relative contributions of alternatives under goal "Increase power system reliability"

Expert	DHP program in single family end use	LED program in grocery end use	VFD program in potato onion shedding end use	Inconsistency
Expert 8	0.67	0.24	0.09	0.007
Expert 9	0.82	0.1	0.08	0.062
Expert 10	0.79	0.21	0	0.088
Expert 11	0.74	0.11	0.16	0.018
Mean	0.75	0.16	0.08	

Table 11.21 Inconsistencies and relative contributions of alternatives under goal "Increase transmission and distribution system reliability"

Expert	DHP program in single family end use	LED program in grocery end use	VFD program in potato onion shedding end use	Inconsistency
Expert 8	0.58	0.30	0.12	0.009
Expert 9	0.69	0.23	0.08	0.000
Expert 10	0.79	0.21	0.00	0.088
Mean	0.69	0.25	0.07	–

Table 11.22 Inconsistencies and relative contributions of alternatives under goal "Reduce operating costs"

Expert	DHP program in single family end use	LED program in grocery end use	VFD program in potato onion shedding end use	Inconsistency
Expert 8	0.56	0.29	0.15	0.001
Expert 9	0.69	0.23	0.08	0.000
Expert 10	0.81	0.18	0.01	0.001
Expert 11	0.71	0.19	0.10	0.097
Mean	0.69	0.22	0.08	

Expert Inconsistencies in Evaluation of Energy Efficiency Programs Under Utility Goal "Increase Transmission and Distribution System Reliability"

Experts reflected their judgment without showing significant degree of inconsistency. Accordingly, all expert judgment was accepted. The summary of consistencies with respect to each expert is presented in Table 11.21.

Expert Inconsistencies in Evaluation of Energy Efficiency Programs Under Utility Goal "Reduce Operating Costs"

Experts reflected their judgment without showing significant degree of inconsistency. Accordingly, all expert judgment was accepted. The summary of consistencies with respect to each expert is presented in Table 11.22.

Expert Inconsistencies in Evaluation of Energy Efficiency Programs Under Utility Goal "Reduce or Postpone Capital Investments"

Experts reflected their judgment without showing significant degree of inconsistency. Accordingly, all expert judgment was accepted. The summary of consistencies with respect to each expert is presented in Table 11.23.

11.6.3.3 Analysis of Group Disagreements

There were three group disagreements observed in expert panel 3, and all of the disagreements occurred on the utility goals under objective "Increase operating flexibility and reliability." For instance, these group disagreements occurred while assessment of alternatives with respect to utility goals "Reduce need for critical resources," "Increase power system reliability," and "Increase transmission and distribution system reliability." Details about the process undertaken to cope with the aforesaid group disagreements are presented below.

Group Disagreements in Evaluation of Energy Efficiency Program Alternatives Under Utility Goal "Reduce Need for Critical Resources"

Results from the first round of expert evaluations revealed that there was a significant degree of group disagreement on relative contributions of energy efficiency program alternatives to "Reduce need for critical resources." Prior results of ICC and *F*-tests are presented in Table 11.24.

Table 11.23 Inconsistencies and relative contributions of alternatives under goal "Reduce or postpone capital investments"

Expert	DHP program in single family end use	LED program in grocery end use	VFD program in potato onion shedding end use	Inconsistency
Expert 8	0.49	0.40	0.11	0.013
Expert 9	0.69	0.23	0.08	0.000
Expert 10	0.79	0.21	0.00	0.088
Expert 11	0.33	0.33	0.33	0.000
Mean	0.58	0.29	0.13	

Table 11.24 ICC and *F*-test of reduce need for critical resources

Energy efficiency program alternatives under utility goal	r_{ic} $0<r_{ic}<1$	*F*-value	*F*-critical at 0.05 level	*F*-test result
Reduce need for critical resources	0.520	3.884	5.1433	H_0 cannot be rejected

As observed from the results of hierarchical clustering analysis (please refer to "Analysis and data validation of alternatives with respect to "Reduce need for critical resources"" in Appendix 6), there were two different opinions; however, agreement between Expert 8 and Expert 11 is greater than the one between Expert 9 and Expert 10. Accordingly, Expert 9 and Expert 10 were asked to reevaluate their judgment. It was observed that Expert 9 did not interpret the meaning of variable "Reduce need for critical resources" comprehensive enough. Expert was provided with the definition of the variable and presented a related case study at BPA. Based on the information, Expert 9 decided to allocate more weights on variable "DHP Program in Single Family End Use" which significantly decreases winter and summer peak loads, contributing to BPA's load balancing objectives. Expert 10 decided to increase relative weight of "DHP Program in Single Family End Use" due to its greater contribution to reducing winter peak load issue at BPA. It should further be stated that Expert 10 is the responsible project manager for LED energy efficiency program at BPA, thus his prior judgment might have been biased due to aforementioned reason. Accordingly, new judgment was incorporated; calculation of relative weights, inconsistency, and group disagreement analyses was conducted.

Group Disagreements in Evaluation of Energy Efficiency Program Alternatives Under Utility Goal "Increase Power System Reliability"

Results from the first round of expert evaluations revealed that there was a significant degree of group disagreement on relative contributions of energy efficiency program alternatives to "Increase power system reliability." Prior results of ICC and *F*-tests are presented in Table 11.25.

As observed from the results of hierarchical clustering analysis (please refer to "Analysis and data validation of alternatives with respect to "Increase power system reliability"" in Appendix 6), there were two groups which reflect different opinions. Accordingly, Group 1 consists of Expert 8 and Expert 9, and Group 2 consisted of Expert 10 and Expert 11. Furthermore, experts in any of the groups had quite similar expert judgment, thus it was hard to pinpoint which of the groups would be treated as the base. Researcher carefully looked at the other responses of each expert and found out that Expert 9 had tendency of rating all technologies equal. This situation was considered as an indication of misjudge and researcher decided to contact Group 1. Expert 8 works as a contractor with BPA and affiliated with WSU energy program. It was observed that Expert 8 had a good understanding of what the alternative measures were capable of in terms of reducing loads, however not very

Table 11.25 ICC and *F*-test of increase power system reliability

Energy efficiency program alternatives under utility goal	r_{ic} $0<r_{ic}<1$	*F*-value	*F*-critical at 0.05 level	*F*-test result
Increase power system reliability	0.223	1.767	5.1433	H_0 cannot be rejected

Table 11.26 ICC and *F*-test of increase transmission and distribution system reliability

Energy efficiency program alternatives under utility goal	r_{ic} $0<r_{ic}<1$	*F*-value	*F*-critical at 0.05 level	*F*-test result
Increase transmission and distribution system reliability	0.356	2.108	6.9444	H_0 cannot be rejected

knowledgeable about BPA-specific generation loads. Having received the feedback, he stated that he acquired related information and modified prior judgment. Expert 9 indicated that she had difficulty in interpreting variable "Increase power system reliability." She had been provided with the detailed explanation of the aforesaid variable. Further information regarding how energy efficiency technologies could impact power system reliability was provided as well. Accordingly, new judgment was incorporated; calculation of relative weights, inconsistency, and group disagreement analyses was conducted.

Group Disagreements in Evaluation of Energy Efficiency Program Alternatives Under Utility Goal "Increase Transmission and Distribution System Reliability"

Results from the first round of expert evaluations revealed that there was a significant degree of group disagreement on relative contributions of energy efficiency program alternatives to "Increase transmission and distribution system reliability." Prior results of ICC and *F*-tests are presented in Table 11.26.

As observed from the results of hierarchical clustering analysis (please refer to "Analysis and data validation of alternatives with respect to "Increase transmission and distribution system reliability"" in Appendix 6), Expert 8 reflected significantly different opinion than the rest of the group. One of the experts stated that he did not have much knowledge on BPA's transmission and distribution loads, so opted to respond the survey for only this part. Expert 8 was contacted and requested to reconsider his relative weights. As mentioned earlier, Expert 8 works as a contractor with BPA and affiliated with WSU energy program. It was observed that Expert 8 had good understanding of what the alternative measures were capable of in terms of reducing loads, however not very knowledgeable about BPA-specific transmission and distribution loads. Having received the feedback, he stated that he acquired related information and modified prior judgment. Accordingly, new judgment was incorporated; calculation of relative weights, inconsistency, and group disagreement analyses was conducted.

11.6.3.4 Final Group Disagreement Analyses Results

Final group disagreement results were calculated based on the modifications made in the second round of expert evaluations. Final results are presented in Table 11.27.

Table 11.27 Final intraclass correlation coefficient and *F*-value of energy efficiency program alternatives

Energy efficiency program alternatives under utility goal	r_{ic} $0<r_{ic}<1$	F-value	F-critical at 0.05 level	F-test result
Create or retain job opportunities	0.830	13.983	5.1433	Reject H_0
Keep local industry competitive	0.686	6.826	5.1433	Reject H_0
Improve life standards (nonenergy benefits)	0.990	274.726	5.1433	Reject H_0
Reduce need for critical resources	0.898	24.368	5.1433	Reject H_0
Increase power system reliability	0.967	79.426	5.1433	Reject H_0
Increase transmission and distribution system reliability	0.946	35.906	6.9444	Reject H_0
Reduce operating costs	0.948	49.451	5.1433	Reject H_0
Reduce postpone capital investments	0.642	5.791	5.1433	Reject H_0

11.6.4 Synthesis of Priorities

Synthesis of priorities was analyzed with respect to three sections which are global contributions of goals to utility mission, global contributions of decision alternatives to the utility objectives, and global contributions of decision alternatives to the mission. You can find related analyses below.

11.6.4.1 Global Contribution of Utility Goals to the Mission

It has been observed that utility goal "Reduce operating costs" (0.288) has the highest global contribution to the mission whereas the second highest weighted utility goal is "Reduce or postpone capital investments" (0.192). It is also worth mentioning that "Increase transmission and distribution system reliability" (0.151) and "Increase power system reliability" (0.155) have relatively very close contribution scores which are slightly greater than utility goal "Reduce need for critical resources" (0.113). Lastly, it is observed that utility goals under objective "Promote regional development" have very low contribution scores compared to the rest of the utility goals. Global contributions of each utility goal to the mission are presented in Table 11.28.

11.6.4.2 Global Contributions of Decision Alternatives to the Utility Objectives

It is clear that alternative "DHP Program in Single Family End Use" is weighted significantly higher than the rest of the decision alternatives. If carefully analyzed, it will be realized that alternative "LED Program in Grocery End Use" dominates alternative "VFD Program in Potato Onion Shedding End Use" with respect to all

Table 11.28 Global contributions of utility goals to the mission

Objectives (O_k)	Local contributions of objectives on mission (C_k^O)	Goals (G_{jk})	Local contributions of goals on objectives $\left(C_{jk}^{G-O}\right)$	Global contributions of goals on mission $\left(C_{jk}^{G-M}\right)$
Promote regional development	0.10	Create or retain job opportunities	0.31	0.031
		Keep local industry competitive	0.36	0.036
		Improve life standards (nonenergy benefits)	0.32	0.032
Increase operating flexibility and reliability	0.42	Reduce need for critical resources	0.27	0.113
		Increase power system reliability	0.37	0.155
		Increase transmission and distribution system reliability	0.36	0.151
Reduce system cost	0.48	Reduce operating costs	0.60	0.288
		Reduce postpone capital investments	0.40	0.192

Table 11.29 Global contributions of decision alternatives to the utility objectives

	Global contributions of alternatives to the objectives C_{ik}^{A-O}		
Objectives (O_k)	DHP program in single family end use	LED program in grocery end use	VFD program in potato onion shedding end use
Promote regional development	0.06	0.02	0.01
Increase operating flexibility and reliability	0.31	0.08	0.03
Reduce system cost	0.31	0.12	0.05

utility objectives. The same relationship can be observed between variable "DHP Program in Single Family End Use" and "LED Program in Grocery End Use." Global contributions of decision alternatives to each utility objective are presented in Table 11.29.

11.6.4.3 Global Contributions of Decision Alternatives to the Mission

Synthesis of the earlier results shows that alternative "DHP Program in Single Family End Use" is ranked as the best alternative with the score of 0.68, alternative "LED Program in Grocery End Use" is ranked as the second alternative with the

Table 11.30 Global contributions of decision alternatives to the mission

Alternatives (A_i)	Global contributions of alternatives to the mission C_i^A
DHP program in single family end use	0.68
LED program in grocery end use	0.23
VFD program in potato onion shedding end use	0.09

score of 0.23, and "VFD Program in Potato Onion Shedding End Use" is ranked as the third with the score of 0.09. Global contributions of decision alternatives to the mission are presented in Table 11.30.

11.6.5 Sensitivity Analysis

In this section, sensitivity analysis of the case study is illustrated using two scenarios. First scenario helps demonstrate how changes in relative importance of a utility objective may impact the ranking order of the optimum solution. This analysis is useful in cases where there is only one decision alternative that can be invested on due to resource limitations. Second scenario helps capture how changes in relative importance of utility objectives may impact the ranking order of all decision alternatives. This case provides insight when there are decision alternatives which are very closely weighted, and decision makers may want to keep a close eye on all decision alternatives. Both sensitivity analyses are conducted by observing impacts of perturbations on utility objective "Reduce system cost." Reduce system cost has been selected for demonstration purposes since it is ranked as the most important utility objective.

Method used for the sensitivity analysis is developed by Chen [7] and the formula used is shown below. Please refer to Question 3D in case you needed more information about the procedures and notations used. You can also see the data used in the analyses in Tables 11.31, 11.32, and 11.33.

$$\lambda \geq P_{k^*}^{O} \lambda^{O}$$

where

$$\lambda^{O} = C_{r+n,k^*}^{A-O} - C_{rk^*}^{A-O} - \left[\sum_{k=1,k\neq k^*}^{K} C_{r+n,k}^{A-O} \times \left(C_k^O \Big/ \sum_{k=1,k\neq k^*}^{K} C_k^O \right) \right]$$

$$+ \left[\sum_{k=1,k\neq k^*}^{K} C_{rk}^{A-O} \times \left(C_k^O \Big/ \sum_{k=1,k\neq k^*}^{K} C_k^O \right) \right]$$

Table 11.31 Contributions of objectives to the mission

C_k^O	Reduce system cost	Increase operating flexibility and reliability	Promote regional development
Contributions of objectives to the mission	0.48	0.42	0.10

Table 11.32 Global contributions of alternatives to the mission

C_i^A	DHP program in single family end use	LED program in grocery end use	VFD program in potato onion shedding end use
Global contributions of alternatives to the mission	0.68	0.23	0.09
Ranking	1	2	3

Table 11.33 Global contributions of alternatives to the objectives

C_{ik}^{A-O}	Global contributions of alternatives to the objectives		
	DHP program in single family end use	LED program in grocery end use	VFD program in potato onion shedding end use
Reduce system cost	0.31	0.12	0.05
Increase operating flexibility and reliability	0.31	0.08	0.03
Promote regional development	0.06	0.02	0.01

11.6.5.1 Case 1: Keep Ranking Order of the Optimum Decision Alternative

This case is going to analyze the sensitivity of the objective "Reduce system cost" while keeping the current ranking order of the best decision alternative.

In order to keep the current ranking of the best decision alternative, the formula presented earlier is solved for all $r=1$ and $n=1,\ldots, I-1$

$$-C_{k^*}^O \le P_{K^*}^O \le 1 - C_{k^*}^O \rightarrow -0.48 \le P_{1^*}^O \le 0.52$$

$r=1$, $n=1$

$$= C_r^A - C_{r+n}^A = C_1^A - C_2^A = 0.68 - 0.23 = 0.45$$

$$\lambda^O = C_{2,1^*}^{A-O} - C_{11^*}^{A-O} - \left[\sum_{k=2}^{3} C_{2k}^{A-O} \times \left(C_k^O \Big/ \sum_{k=2}^{3} C_k^O\right)\right] + \left[\sum_{k=2}^{3} C_{1k}^{A} \times \left(C_k^O \Big/ \sum_{k=2}^{3} C_k^O\right)\right]$$

$$\lambda^O = 0.12 - 0.31 - \left(0.08 \times \frac{0.42}{0.52} + 0.02 \times \frac{0.10}{0.52}\right)$$
$$+ \left(0.31 \times \frac{0.42}{0.52} + 0.06 \times \frac{0.10}{0.52}\right) = 0.00346$$

$$\frac{\lambda}{\lambda^O} = \frac{0.45}{0.00346} = 130.05$$

$$P_{1^*}^O \leq 130.05$$

$r = 1, n = 2$

$$\lambda = C_r^A - C_{r+n}^A = C_1^A - C_3^A = 0.68 - 0.09 = 0.59$$

$$\lambda^O = C_{31^*}^{A-O} - C_{11^*}^{A-O} - \left[\sum_{k=2}^{3} C_{3k}^{A-O} \times \left(C_k^O \Big/ \sum_{k=2}^{3} C_k^O\right)\right] + \left[\sum_{k=2}^{3} C_{1k}^{A-O} \times \left(C_k^O \Big/ \sum_{k=2}^{3} C_k^O\right)\right]$$

$$\lambda^O = 0.05 - 0.31 - \left(0.03 \times \frac{0.42}{0.52} + 0.01 \times \frac{0.10}{0.52}\right)$$
$$+ \left(0.31 \times \frac{0.42}{0.52} + 0.06 \times \frac{0.10}{0.52}\right) = -0.0242$$

$$\frac{\lambda}{\lambda^O} = \frac{0.59}{-0.0242} = -24.38$$

$$P_{1^*}^O \leq -24.38$$

$$-0.48 \leq P_{1^*}^O \leq 0.52$$
$$\begin{Bmatrix} P_{1^*}^O \leq 130.05\ (\text{when } r = 1, n = 1) \\ P_{1^*}^O \leq -24.38\ (\text{when } r = 1, n = 2) \end{Bmatrix}$$

Combining all the inequalities leads to no feasible solution. Accordingly, it can be said that any magnitude of perturbation on "Reduce system cost" is not going to change the ranking order of the best decision alternative. Therefore, allowable range of perturbations on $C_{1^*}^O$ is [−0.48, 0.52] to preserve the current ranking of the best program alternative.

11.6.5.2 Case 2: Keep the Current Ranking Order of All Decision Alternatives

This case is going to analyze the sensitivity of the objective "Reduce system cost" while keeping the current ranking order of all decision alternatives.

In order to keep the ranking of the best decision alternative, the formula presented earlier is solved for all $r = 1, \ldots, \mathrm{I}-1$ and $n = 1$.

$$-C^O_{k^*} \le P^O_{K^*} \le 1 - C^O_{k^*} \rightarrow -0.48 \le P^O_{1^*} \le 0.52$$

$r = 1$, $n = 1$

$$= C^A_r - C^A_{r+n} = C^A_1 - C^A_2 = 0.68 - 0.23 = 0.45$$

$$\lambda^O = \mathrm{C}^{A-O}_{2,1^*} - \mathrm{C}^{A-O}_{11^*} - \left[\sum_{k=2}^{3} C^{A-O}_{2k} \times \left(C^O_k \Big/ \sum_{k=2}^{3} C^O_k \right)\right] + \left[\sum_{k=2}^{3} C^{A-O}_{1k} \times \left(C^O_k \Big/ \sum_{k=2}^{3} C^O_k \right)\right]$$

$$\lambda^O = 0.12 - 0.31 - \left(0.08 \times \frac{0.42}{0.52} + 0.02 \times \frac{0.10}{0.52}\right)$$
$$+ \left(0.31 \times \frac{0.42}{0.52} + 0.06 \times \frac{0.10}{0.52}\right) = 0.00346$$

$$\frac{\lambda}{\lambda^O} = \frac{0.45}{0.00346} = 130.05$$

$$P^O_{1^*} \le 130.05$$

$r = 2$, $n = 1$

$$= C^A_r - C^A_{r+n} = C^A_2 - C^A_3 = 0.23 - 0.09 = 0.14$$

$$\lambda^O = \mathrm{C}^{A-O}_{31^*} - \mathrm{C}^{A-O}_{21^*} - \left[\sum_{k=2}^{3} C^{A-O}_{3k} \times \left(C^O_k \Big/ \sum_{k=2}^{3} C^O_k \right)\right] + \left[\sum_{k=2}^{3} C^{A-O}_{2k} \times \left(C^O_k \Big/ \sum_{k=2}^{3} C^O_k \right)\right]$$

$$\lambda^O = 0.05 - 0.12 - \left(0.03 \times \frac{0.42}{0.52} + 0.01 \times \frac{0.10}{0.52}\right)$$
$$+ \left(0.08 \times \frac{0.42}{0.52} + 0.02 \times \frac{0.10}{0.52}\right) = -0.0276$$

$$\frac{\lambda}{\lambda^O} = \frac{0.14}{-0.0276} = -5.05$$

$$P^O_{1^*} \le -5.05$$

$$\left\{\begin{array}{c} -0.48 \le P_{1^*}^{O} \le 0.52 \\ P_{1^*}^{O} \le 130.05(\text{when}\, r = 1, n = 1) \\ P_{1^*}^{O} \le -5.05(\text{when}\, r = 1, n = 2) \end{array}\right\}$$

Combining all the inequalities leads to no feasible solution. Accordingly, it can be said that any magnitude of perturbation on "Reduce system cost" is not going to change the ranking order of all decision alternatives. Therefore, allowable range of perturbations on $C_{1^*}^{O}$ is [−0.48, 0.52] to preserve the current ranking of the best program alternative.

11.6.5.3 Results of the Sensitivity Analyses

Results from both of the cases could also be obtained by observing that decision alternative "DHP Program in Single Family End Use" which is the best alternative, has always received the highest relative scores with respect to all utility goals. Similarly, decision alternative "LED Program in Grocery End Use" has always been weighted higher than the alternative "VFD Program in Potato Onion Shedding End Use" with respect to each of the utility goals. As a result, it is expected that any change in relative importance of utility objectives is not going to impact neither the best alternative's nor all the alternatives' current ranking orders. Due to special condition of this case, study researcher is not going to conduct sensitivity analyses to observe impacts of perturbations on other utility objectives.

11.6.6 Criteria-Related Validity

Criteria-related validation refers to the degree of effectiveness of a proposed model in predicting real-life phenomenon. In this research, criteria-related validity refers to the degree to which results of the model have sufficient impacts.

For demonstration purposes in this analysis, alternatives have been selected from RTF website. RTF is a region-wide organization whose mission is to verify the accuracy of the potential energy savings obtained from proposed energy efficiency programs. Selected energy efficiency programs have already been deemed and provided reimbursements by BPA. The logic behind this decision was to get expert judgment on the technologies where there was sufficient energy savings data. Researcher wanted to compare the outcome of the proposed model with real data for criteria-related validation purposes. With this kind of approach, experts were able to base their judgment on real data which would increase the accuracy of criteria validation test.

Result of the study was summarized in a report and discussed with supervisory engineer and a program manager in energy efficiency group at BPA. Each participant was given enough time to state their feedback on the usefulness of the model in assessing value of energy efficiency program alternatives. An interesting suggestion was made by one of the experts regarding the alternatives. Alternatives for the proposed assessment model are combination technology and end use domains. This approach is adopted to take both energy efficiency potential and the measure load shape into consideration. However, similar accuracy was stated to be reached by only using technology alternatives separated from end use domains. Logic behind the suggestion was that on the strategic level, relationship between the utility and measure load shape is largely impacted by the load characteristic of a technology rather than the end use. For instance, an example was given for the case of ductless heat pumps. According to the expert, no matter what the end use is, DHP applications have potential savings during summer and its implications are always the same on the strategic level. Researcher wants to acknowledge that aforementioned suggestion has value in simplifying alternative level to a significant degree and allow for assessment of more technologies. This issue has been decided to be discussed with more experts and clarified in the later stages.

Based on the feedback it has been confirmed that proposed research is able to capture different features of the technology and able to integrate their implications to utility goals and objectives. Research acknowledges that demonstration results do not reflect much diversity in terms of showing strong and weak points of each alternative due to special condition derived from alternatives. Accordingly, following expert comments are presented for further insight and support.

11.6.7 DHP Program in Single Family End Use

Ductless heat pump technology is stated to have by far the biggest kWh potential savings and potential to create the biggest additional labor and business development such as supporting distributors, contractors, and installers. It has further been stated heat pumps are least efficient when it is very cold, which is also the peak load for the region. They may not reduce the need for peak capacity at all, and therefore not increase system reliability, etc., as one would hope.

11.6.8 LED Program in Grocery End Use

LED technology is stated to have potential for good, reliable, and predictable energy and peak load savings; however, the magnitude of savings is not as high as DHPs. LED applications are currently expensive, thus cost of annual kWh savings is high as of now. Currently, the color and quality of light is still often not as good as what customers have come to expect from incandescent technology, though they are even with or probably surpassing CFLs.

11.6.9 VFD Program in Potato Onion Shedding End Use

VFD technology in onion shedding end use is stated to be a good measure where it applies, especially in summer/fall period where its use reaches its peak which is also good for the NW. VFD applications are stated to give a high percentage savings and are technically dispatchable. That is, if the utility chose to have some control over these, they could turn them down during peak loads with little detrimental effect for the end user. Overall savings of this measure is small since there are relatively few applications. The savings at any given time is probably less predictable than the LEDs.

11.7 Appendix 1: Content Validation Survey

11.7.1 Survey Instructions

1. Please note that each page contains one question and short definitions of the relevant terms in case you are not familiar with the wording.
2. Please use Likert scale of 1–3 to reflect your judgment, if you have any comments/suggestions please use the specified field under the scoring table.
3. After you are done with all the questions, please save the word document and send the file back to the researcher using the same mail you have received the survey (ibrahimiskin@gmail.com).

Please proceed to the next page for the survey

1. Please rate the importance of the following attributes in defining overall utility mission. If you believe the list below misses any important attributes, please use the comment box below.

   1: Essential, 2: Useful but not essential, 3: Not necessary

Promote regional development	
Minimize environmental impacts	
Increase operating flexibility and reliability	
Reduce system cost	
Minimize adverse effects on public	

Command:

- Promote regional development: Electric utilities need to support regional development by providing accessible service.
- Minimize environmental impacts: Electric utilities need to operate in an environmentally sustainable manner.
- Increase operating flexibility and reliability: Electric utilities need to maintain their operations without compromising reliability.

- Reduce system cost: Electric utilities need to keep rates as low as possible within sound business principles.
- Minimize adverse effects on public: Electric utilities need to operate collaboratively with public entities and minimize adverse effects caused by their operations.

2. Please rate the importance of the following attributes in defining "Promote regional development." If you believe the list below misses any important attributes, please use the comment box below.
   1: Essential, 2: Useful but not essential, 3: Not necessary

Create or retain job opportunities	
Keep local industry competitive	
Improve life standards (nonenergy benefits)	

Command:

- Promote regional development: Electric utilities need to support regional development by providing accessible service.
- Create or retain job opportunities: Electric utilities can support regional development by creating new job opportunities.
- Keep local industry competitive: Electric utilities can support regional development by providing local industry with low rate energy and enabling diffusion of new technologies.
- Improve life standards (nonenergy benefits): Electric utilities can promote regional development by improving life standards of public through introduction of new technologies.

3. Please rate the importance of the following attributes in defining "Minimize environmental impacts." If you believe the list below misses any important attributes, please use the comment box below.
   1: Essential, 2: Useful but not essential, 3: Not necessary

Reduce GHG emissions	
Reduce emission of soil, air, and water contaminants	
Avoid flora and fauna habitat loss	

Command:

- Minimize environmental impacts: Electric utilities need to operate in an environmentally sustainable manner.
- Reduce GHG emissions: Electric utilities can reduce environmental impacts by reducing greenhouse gas emissions.

- Reduce emission of soil, air, and water contaminants: Electric utilities can reduce environmental impacts by reducing emission of soil, air, and water contaminants.
- Avoid flora and fauna loss: Electric utilities can reduce environmental impacts by protecting flora and fauna species.

4. Please rate the importance of the following attributes in defining "Increase operating flexibility and reliability." If you believe the list below misses any important attributes, please use the comment box below.
   1: Essential, 2: Useful but not essential, 3: Not necessary

Reduce need for critical resources	
Increase power system reliability	
Increase transmission and distribution system reliability	

Command:

- Increase operating flexibility and reliability: Electric utilities need to maintain their operations without compromising reliability.
- Reduce need for critical resources: Electric utilities can increase their operating flexibility and reliability by reducing need for critical resources.
- Increase power system reliability: Electric utilities can increase their operating flexibility and reliability by increasing power system reliability.
- Increase transmission and distribution system reliability: Electric utilities can increase their operating flexibility and reliability by increasing transmission and distribution system reliability.

5. Please rate the importance of the following attributes in defining "Reduce system cost." If you believe the list below misses any important attributes, please use the comment box below.
   1: Essential, 2: Useful but not essential, 3: Not necessary

Reduce/postpone capital investments	
Reduce operating costs	

Command:

- Reduce system cost: Electric utilities need to keep rates as low as possible within sound business principles.
- Reduce/postpone capital investments: Electric utilities can reduce system cost by reducing or postponing capital investments.
- Reduce operating costs: Electric utilities can reduce system cost by reducing operating costs.

6. Please rate the importance of the following attributes in defining "Minimize adverse effects on public." If you believe the list below misses any important attributes, please use the comment box below.

   1: Essential, 2: Useful but not essential, 3: Not necessary

Avoid noise and odor	
Avoid visual impacts	
Avoid property damage and impact on lifestyles	

Command:

- Minimize adverse effects on public: Electric utilities need to operate collaboratively with public entities and minimize adverse effects caused by their operations.
- Avoid noise and odor: Electric utilities can reduce adverse effects on public by reducing noise and odor-related disturbances caused by their operations.
- Avoid visual impacts: Electric utilities can reduce adverse effects on public by reducing visual disturbances caused by their operations.
- Avoid property damage and impacts on lifestyles: Electric utilities can reduce adverse effects on public by reducing property damage and lifestyle-related disturbances caused by their operations.

11.8 Appendix 2: Content Validation Analyses

11.8.1 Chi-Square Test Results

Test Statistics

	Chi-Square	df	Asymp. Sig.
Promote regional development	1.750[a]	2	.417
Create or retain job opportunities	1.750[a]	2	.417
Keep local industry competitive	7.000[a]	2	.030
Improve life standards (non-energy benefits)	3.250[a]	2	.197
Minimize environmental impacts	4.000[a]	2	.135
Reduce GHG emissions	4.750[a]	2	.093
Reduce emission of soil, air and water contaminants	4.000[a]	2	.135
Avoid flora and fauna habitat loss	1.750[a]	2	.417
Increase operating flexibility and reliability	4.750[a]	2	.093
Reduce need for critical resources	4.750[a]	2	.093
Increase power system reliability	16.000[a]	2	.000
Increase transmission and distribution system reliability	16.000[a]	2	.000
Reduce system cost	7.000[a]	2	.030
Reduce/postpone capital investments	10.750[a]	2	.005
Reduce operating costs	7.000[a]	2	.030
Minimize adverse effects on public	4.750[a]	2	.093
Avoid noise and odor	6.250[a]	2	.044
Avoid visual impacts	6.250[a]	2	.044
Avoid property damage and impact on life styles	4.000[a]	2	.135

a. 3 cells (100.0%) have expected frequencies less than 5. The minimum expected cell frequency is 2.7.

11.8.1.1 One Sample *T* test for Mean of 1

One-Sample Test

	Test Value = 1					
					95%	
	t	df	Sig. (2-tailed)	Mean Difference	Lower	Upper
Promote regional development	3.000	7	.020	.75000	.1588	1.3412
Create or retain job opportunities	2.376	7	.049	.62500	.0030	1.2470
Keep local industry competitive	4.583	7	.003	.75000	.3630	1.1370
Improve life standards (non-energy benefits)	3.862	7	.006	.87500	.3392	1.4108
Minimize environmental impacts	2.646	7	.033	.50000	.0531	.9469
Reduce GHG emissions	2.049	7	.080	.37500	-.0577	.8077
Reduce emission of soil, air and water contaminants	2.646	7	.033	.50000	.0531	.9469
Avoid flora and fauna habitat loss	3.000	7	.020	.75000	.1588	1.3412
Increase operating flexibility and reliability	2.049	7	.080	.37500	-.0577	.8077
Reduce need for critical resources	2.049	7	.080	.37500	-.0577	.8077
Reduce system cost	1.528	7	.170	.25000	-.1370	.6370
Reduce/postpone capital investments	1.000	7	.351	.12500	-.1706	.4206
Reduce operating costs	1.528	7	.170	.25000	-.1370	.6370
Minimize adverse effects on public	3.416	7	.011	.62500	.1923	1.0577
Avoid noise and odor	5.292	7	.001	1.00000	.5531	1.4469
Avoid visual impacts	5.292	7	.001	1.00000	.5531	1.4469
Avoid property damage and impact on life styles	2.646	7	.033	.50000	.0531	.9469

11.8.1.2 One Sample *T* test for Mean of 2

One-Sample Test

	Test Value = 2					
					95%	
	t	df	Sig. (2-tailed)	Mean Difference	Lower	Upper
Promote regional development	-1.000	7	.351	-.25000	-.8412	.3412
Create or retain job opportunities	-1.426	7	.197	-.37500	-.9970	.2470
Keep local industry competitive	-1.528	7	.170	-.25000	-.6370	.1370
Improve life standards (non-energy benefits)	-.552	7	.598	-.12500	-.6608	.4108
Minimize environmental impacts	-2.646	7	.033	-.50000	-.9469	-.0531
Reduce GHG emissions	-3.416	7	.011	-.62500	-1.0577	-.1923
Reduce emission of soil, air and water contaminants	-2.646	7	.033	-.50000	-.9469	-.0531
Avoid flora and fauna habitat loss	-1.000	7	.351	-.25000	-.8412	.3412
Increase operating flexibility and reliability	-3.416	7	.011	-.62500	-1.0577	-.1923
Reduce need for critical resources	-3.416	7	.011	-.62500	-1.0577	-.1923
Reduce system cost	-4.583	7	.003	-.75000	-1.1370	-.3630
Reduce/postpone capital investments	-7.000	7	.000	-.87500	-1.1706	-.5794
Reduce operating costs	-4.583	7	.003	-.75000	-1.1370	-.3630
Minimize adverse effects on public	-2.049	7	.080	-.37500	-.8077	.0577
Avoid noise and odor	0.000	7	1.000	0.00000	-.4469	.4469
Avoid visual impacts	0.000	7	1.000	0.00000	-.4469	.4469
Avoid property damage and impact on life styles	-2.646	7	.033	-.50000	-.9469	-.0531

11.8.1.3 One Sample *T* test for Mean of 3

One-Sample Test

	Test Value = 3					
					95% Confidence	
	t	df	Sig. (2-tailed)	Mean Difference	Lower	Upper
Promote regional development	-5.000	7	.002	-1.25000	-1.8412	-.6588
Create or retain job opportunities	-5.227	7	.001	-1.37500	-1.9970	-.7530
Keep local industry competitive	-7.638	7	.000	-1.25000	-1.6370	-.8630
Improve life standards (non-energy benefits)	-4.965	7	.002	-1.12500	-1.6608	-.5892
Minimize environmental impacts	-7.937	7	.000	-1.50000	-1.9469	-1.0531
Reduce GHG emissions	-8.881	7	.000	-1.62500	-2.0577	-1.1923
Reduce emission of soil, air and water contaminants	-7.937	7	.000	-1.50000	-1.9469	-1.0531
Avoid flora and fauna habitat loss	-5.000	7	.002	-1.25000	-1.8412	-.6588
Increase operating flexibility and reliability	-8.881	7	.000	-1.62500	-2.0577	-1.1923
Reduce need for critical resources	-8.881	7	.000	-1.62500	-2.0577	-1.1923
Reduce system cost	-10.693	7	.000	-1.75000	-2.1370	-1.3630
Reduce/postpone capital investments	-15.000	7	.000	-1.87500	-2.1706	-1.5794
Reduce operating costs	-10.693	7	.000	-1.75000	-2.1370	-1.3630
Minimize adverse effects on public	-7.514	7	.000	-1.37500	-1.8077	-.9423
Avoid noise and odor	-5.292	7	.001	-1.00000	-1.4469	-.5531
Avoid visual impacts	-5.292	7	.001	-1.00000	-1.4469	-.5531
Avoid property damage and impact on life styles	-7.937	7	.000	-1.50000	-1.9469	-1.0531

11.9 Appendix 3: Pairwise Comparison Survey for Objectives Level

11.9.1 Survey Instructions

11.9.1.1 Sample Question

Please distribute 100 points between the following pairs of utility objectives to reflect your judgment on their relative importance to overall electric utility mission.

Promote regional development			Increase operating flexibility and reliability
Promote regional development			Reduce system costs
Increase operating flexibility and reliability			Reduce system costs

11.9.1.2 Rating Sample

Promote regional development	50	50	Increase operating flexibility and reliability
Promote regional development	75	25	Reduce system costs
Increase operating flexibility and reliability	1	99	Reduce system costs

If you believe objectives "Promote regional development" and "Increase operating flexibility and reliability" have equal importance to overall utility mission, then allocate 50 points on both sides of the rating table.

If you believe "Promote regional development" has three times as much importance as "Reduce system costs," then allocate left-hand side of the table three times as much points as the right-hand side of the table.

If you believe importance of "Increase operating flexibility and reliability" is negligible compared to "Reduce system costs" then use the scores demonstrated earlier.

Please proceed to the next page for the survey

1. Please distribute 100 points between the following pairs of utility objectives to reflect your judgment on their relative importance to overall utility mission.

Promote regional development			Increase operating flexibility and reliability
Promote regional development			Reduce system costs
Increase operating flexibility and reliability			Reduce system costs

Command:

- Promote regional development: Electric utilities need to support regional development by providing accessible service.

- Increase operating flexibility and reliability: Electric utilities need to maintain their operations without compromising reliability.
- Reduce system cost: Electric utilities need to keep rates as low as possible within sound business principles.

11.10 Appendix 4: Pairwise Comparison Survey for Goals Level

11.10.1 Survey Instructions

11.10.1.1 Sample Question

Please distribute 100 points between the following pairs of utility goals to reflect your judgment on their relative contributions to "Promote regional development"

Create or retain job opportunities			Keep local industry competitive
Create or retain job opportunities			Improve life standards (nonenergy benefits)
Keep local industry competitive			Improve life standards (nonenergy benefits)

11.10.2 Rating Sample

Create or retain job opportunities	50	50	Keep local industry competitive
Create or retain job opportunities	75	25	Improve life standards (nonenergy benefits)
Keep local industry competitive	1	99	Improve life standards (nonenergy benefits)

If you believe "Create or retain job opportunities" and "Keep local industry competitive" have equal contributions in fulfilling objective "Promote regional development," then allocate 50 points on both sides of the rating table.

If you believe "Create or retain job opportunities" has three times as much contribution as "Improve life standards (nonenergy benefits)" in fulfilling objective, then allocate left-hand side of the table three times as much points as the right-hand side of the table.

If you believe contribution of "Keep local industry competitive" is negligible compared to "Improve life standards (nonenergy benefits)" in fulfilling the objective, then use the scores demonstrated earlier.

Please proceed to the next page for the survey

1. Please distribute 100 points between the following pairs attributes to reflect your judgment on their relative contributions to "Promote regional development"

Create or retain job opportunities			Keep local industry competitive
Create or retain job opportunities			Improve life standards (nonenergy benefits)
Keep local industry competitive			Improve life standards (nonenergy benefits)

Command:

- Promote regional development: Electric utilities need to support regional development by providing accessible service.
- Create or retain job opportunities: Electric utilities can support regional development by creating new job opportunities.
- Keep local industry competitive: Electric utilities can support regional development by providing local industry with low rate energy and enabling diffusion of new technologies.
- Improve life standards (nonenergy benefits): Electric utilities can promote regional development by improving life standards of public through introduction of new technologies.

2. Please distribute 100 points between the following pairs of attributes to reflect your judgment on their relative contributions to "Increase operating flexibility and reliability"

Reduce need for critical resources			Increase power system reliability
Reduce need for critical resources			Increase transmission and distribution system reliability
Increase power system reliability			Increase transmission and distribution system reliability

Command:

Operations without compromising reliability.

- Reduce need for critical resources: Electric utilities can increase their operating flexibility and reliability by reducing need for critical resources.
- Increase power system reliability: Electric utilities can increase their operating flexibility and reliability by increasing power system reliability.
- Increase transmission and distribution system reliability: Electric utilities can increase their operating flexibility and reliability by increasing transmission and distribution system reliability.

3. Please distribute 100 points between the following attributes to reflect your judgment on their relative contributions to "Reduce system cost"

Reduce/postpone capital investments			Reduce operating costs

Command:

Sound business principles.

- Reduce/postpone capital investments: Electric utilities can reduce system cost by reducing or postponing capital investments.
- Reduce operating costs: Electric utilities can reduce system cost by reducing operating costs.

11.11 Appendix 5: Pairwise Comparison Survey for Alternatives Level

11.11.1 Survey Instructions

11.11.1.1 Sample Question

Please distribute 100 points between the following pairs of program alternatives to reflect your judgment on their relative potentials to "Increase power system reliability"

- Alternative 1: An energy efficiency program for promoting ductless heat pump technology for replacing electric baseboard heating technology in single family residential end use.
- Alternative 2: An energy efficiency program for promoting light emitting diodes technology for replacing compact fluorescent technology in grocery commercial end use.
- Alternative 3: An energy efficiency program for promoting variable frequency drive ventilation fan systems for replacing nonvariable frequency drive ventilation fan systems in agricultural potato onion shedding end use.

Alternative 1			Alternative 2
Alternative 1			Alternative 3
Alternative 2			Alternative 3

11.11.2 Rating Sample

Please note that, magnitude of each alternative's contribution is a function of utility load profile, corresponding technologies' end use load profile, and potential market size for diffusion. Please respond the following questions for BPA's case.

Alternative 1	50	50	Alternative 2
Alternative 1	75	25	Alternative 3
Alternative 2	1	99	Alternative 3

If you believe "Alternative 1" and "Alternative 2" have equal contributions to "Increase power system reliability" then allocate 50 points on both sides of the rating table.

If you believe alternative "Alternative 1" has three times as much contribution as "Alternative 3" then allocate left-hand side of the table three times as much points as the right-hand side of the table.

If you believe the contribution of alternative "Alternative 2" is negligible compared to "Alternative 3" then use the scores demonstrated earlier.

11.11.3 Survey

- Alternative 1: An energy efficiency program for promoting ductless heat pump technology for replacing electric baseboard heating technology in single family residential end use.
- Alternative 2: An energy efficiency program for promoting light emitting diodes technology for replacing compact fluorescent technology in grocery commercial end use.
- Alternative 3: An energy efficiency program for promoting variable frequency drive ventilation fan systems technology for replacing nonvariable frequency drive ventilation fan systems technology in agricultural potato onion shedding end use.

1. Please distribute 100 points between the following pairs of program alternatives to reflect your judgment on their relative potentials to "Create or retain job opportunities"

Alternative 1			Alternative 2
Alternative 1			Alternative 3
Alternative 2			Alternative 3

Create or retain job opportunities: Development and delivery of energy efficiency programs create new job opportunities throughout the industry supply chain such as manufacturers, retailers, designers, contractors, etc.

2. Please distribute 100 points between the following pairs of program alternatives to reflect your judgment on their relative potentials to "Keep local industry competitive"

Alternative 1			Alternative 2
Alternative 1			Alternative 3
Alternative 2			Alternative 3

Keep local industry competitive: Energy efficiency programs can enable rapid diffusion of some of the new manufacturing technologies by eliminating implementation and operation-related concerns through demonstration projects.

3. Please distribute 100 points between the following pairs of program alternatives to reflect your judgment on their relative potentials to "Improve life standards (non energy benefits)"

Alternative 1			Alternative 2
Alternative 1			Alternative 3
Alternative 2			Alternative 3

Improve life standards (nonenergy benefits): New technology alternatives provide not only energy savings, but also improve life standards of public through newly added functions embedded in new products.

4. Please distribute 100 points between the following pairs of program alternatives to reflect your judgment on their relative potentials to "Reduce need for critical resources"

Alternative 1			Alternative 2
Alternative 1			Alternative 3
Alternative 2			Alternative 3

Reduce need for critical resources: Energy efficiency programs can reduce/alter loads to help with allocation of critical resources between competing bodies. A relevant example might be a utility case where regulations of environment, fish, and wildlife require a steady water flow for wildlife habitat and this situation causes power generation difficulty in managing available water reservoir.

5. Please distribute 100 points between the following pairs of program alternatives to reflect your judgment on their relative potentials to "Increase power system reliability"

Alternative 1			Alternative 2
Alternative 1			Alternative 3
Alternative 2			Alternative 3

Increase power system reliability: Energy efficiency programs can reduce/alter loads on critical parts of power generation systems in order to cope with some of the challenges such as increasing population, seasonality in magnitude peak loads, generation variability in renewable energy alternatives, and aging infrastructure.

6. Please distribute 100 points between the following pairs of program alternatives to reflect your judgment on their relative potentials to "Increase transmission and distribution system reliability"

Alternative 1			Alternative 2
Alternative 1			Alternative 3
Alternative 2			Alternative 3

Increase transmission and distribution system reliability: Energy efficiency programs can reduce/alter loads on critical parts of power transmission systems in order to cope with some of the challenges such as increasing population, seasonality in load and peak demands, and aging transmission and distribution systems.

7. Please distribute 100 points between the following pairs of program alternatives to reflect your judgment on their relative potentials to "Reduce/postpone capital investments"

Alternative 1			Alternative 2
Alternative 1			Alternative 3
Alternative 2			Alternative 3

Reduce/postpone capital investments: Energy efficiency programs can be utilized to reduce magnitude of peak loads and help reduce or postpone capital investments which require high upfront capital requirements.

8. Please distribute 100 points between the following pairs of program alternatives to reflect your judgment on their relative potentials to "Reduce operating costs"

Alternative 1			Alternative 2
Alternative 1			Alternative 3
Alternative 2			Alternative 3

Reduce operating costs: Energy efficiency programs can be utilized to reduce need for peak load generation units whose marginal cost of energy generation is higher than base load generation units.

11.12 Appendix 6: Analysis and Data Validation of Expert Judgment

11.12.1 Analysis and Data Validation of Utility Objectives with Respect to Mission

Inconsistencies and relative contributions of objectives to mission

Relative Weights

Project Title: Mission

Users	1	2	3	Incn
Person 1	0.16	0.38	0.46	0.014
Person 2	0.16	0.49	0.35	0.001
Person 3	0.14	0.50	0.36	0.004
Person 4	0.09	0.24	0.67	0.007
Person 5	0.09	0.27	0.64	0.015
Person 6	0.00	0.67	0.32	0.036
Person 7	0.05	0.41	0.54	0.005
Mean	0.10	0.42	0.48	0.122
Min	0.00	0.24	0.32	
Max	0.16	0.67	0.67	
Std Dev	0.06	0.15	0.14	

Group disagreements on utility objectives under mission

Intraclass Correlation Coefficient

	Intraclass Correlation[b]	95% Confidence Interval		F Test with True Value 0			
		Lower Bound	Upper Bound	Value	df1	df2	Sig
Single Measures	.717[a]	.245	.991	12.830	2	12	.001
Average Measures	.947	.695	.999	12.830	2	12	.001

Two-way random effects model where both people effects and measures effects are random.

a. The estimator is the same, whether the interaction effect is present or not.

b. Type A intraclass correlation coefficients using an absolute agreement definition.

Hierarchical clustering analysis of utility objectives under mission

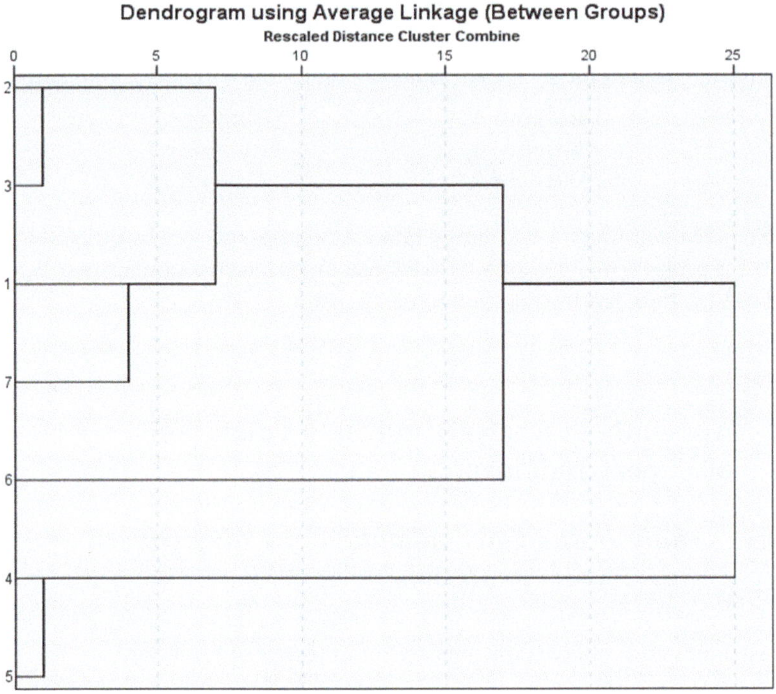

11.13 Analysis and Data Validation of Utility Goals with Respect to "Promote Regional Development"

11.13.1 First Round

Inconsistencies and relative contributions of utility goals under "Promote regional development"

Relative Weights

Project Title: Promote regional development

Users	1	2	3	Incn
Person 1	0.31	0.38	0.31	0.012
Person 2	0.27	0.42	0.31	0.002
Person 3	0.33	0.33	0.33	0.000
Person 4	0.67	0.17	0.17	0.000
Person 5	0.25	0.38	0.38	0.000
Person 6	0.43	0.43	0.14	0.000
Person 7	0.12	0.31	0.57	0.056
Mean	0.34	0.34	0.32	0.139
Min	0.12	0.17	0.14	
Max	0.67	0.43	0.57	
Std Dev	0.17	0.09	0.14	

Group disagreements on utility goals under "Promote regional development"

Intraclass Correlation Coefficient

	Intraclass Correlation[b]	95% Confidence Interval		F Test with True Value 0			
		Lower Bound	Upper Bound	Value	df1	df2	Sig
Single Measures	-.252[a]	-.268	.233	.061	2	12	.941
Average Measures	3.453	3.076	.680	.061	2	12	.941

Two-way random effects model where both people effects and measures effects are random.

a. The estimator is the same, whether the interaction effect is present or not.

b. Type A intraclass correlation coefficients using an absolute agreement definition.

Hierarchical clustering analysis of utility goals under "Promote regional development"

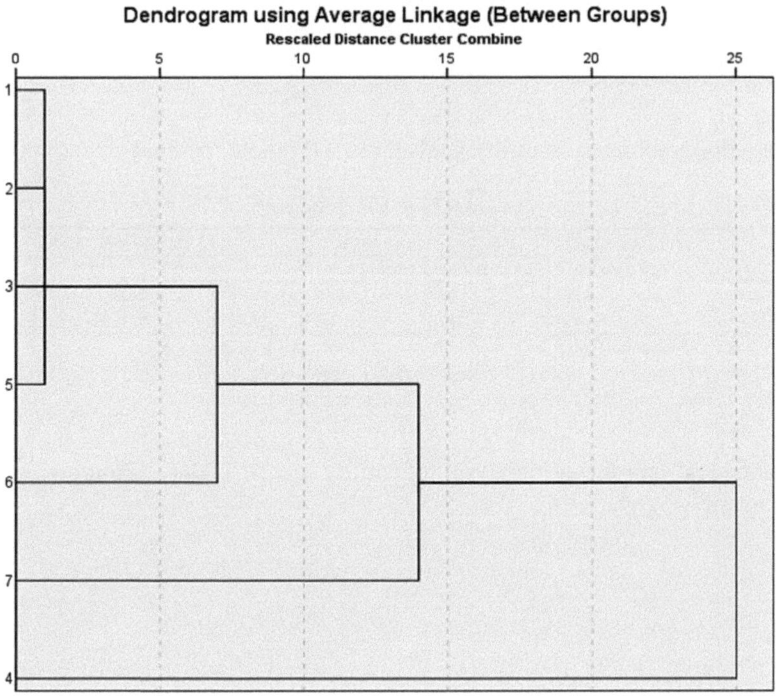

11.13.2 Second Round

Inconsistencies and relative contributions of utility goals under "Promote regional development"

Relative Weights

Project Title: Promote regional development

Users	1	2	3	Incn
Person 1	0.31	0.38	0.31	0.012
Person 2	0.27	0.42	0.31	0.002
Person 3	0.33	0.33	0.33	0.000
Person 4	0.38	0.31	0.31	0.000
Person 5	0.25	0.38	0.38	0.000
Person 6	0.39	0.39	0.21	0.000
Person 7	0.27	0.33	0.40	0.000
Mean	0.32	0.36	0.32	0.053
Min	0.25	0.31	0.21	
Max	0.39	0.42	0.40	
Std Dev	0.06	0.04	0.06	

Group disagreements on utility goals under "Promote regional development"

Intraclass Correlation Coefficient

	Intraclass Correlation[b]	95% Confidence Interval		F Test with True Value 0			
		Lower Bound	Upper Bound	Value	df1	df2	Sig
Single Measures	.031[a]	-.199	.904	1.148	2	12	.350
Average Measures	.182	7.262	.985	1.148	2	12	.350

Two-way random effects model where both people effects and measures effects are random.

a. The estimator is the same, whether the interaction effect is present or not.

b. Type A intraclass correlation coefficients using an absolute agreement definition.

Hierarchical clustering analysis of utility goals under "Promote regional development"

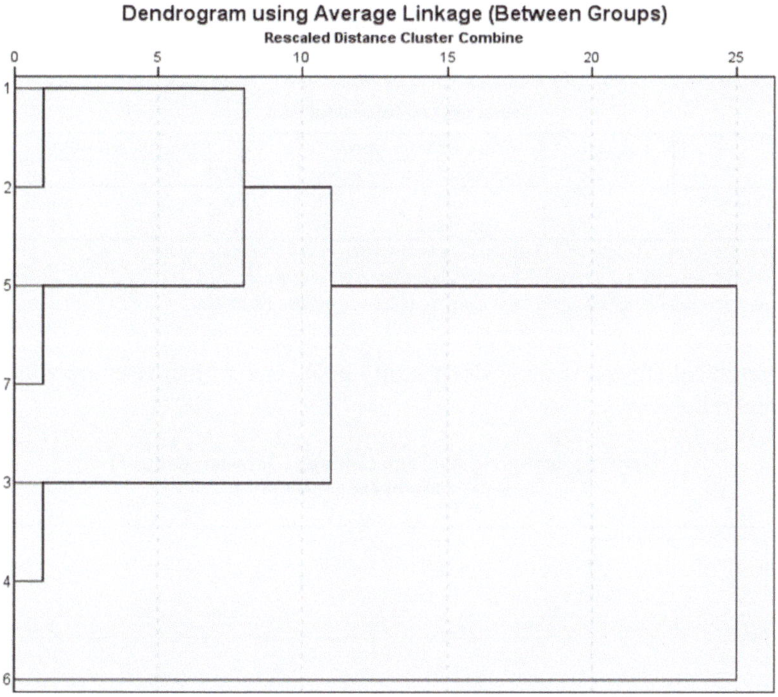

11.13.3 *Analysis and Data Validation of Utility Goals with Respect to "Increase Operating Flexibility and Reliability"*

11.13.3.1 First Round

Inconsistencies and relative contributions of utility goals under "Increase operating flexibility and reliability"

Relative Weights

Project Title: Increase operating flex. and rel

Users	1	2	3	Incn
Person 1	0.36	0.31	0.33	0.005
Person 2	0.43	0.25	0.33	0.002
Person 3	0.25	0.38	0.38	0.000
Person 4	0.11	0.44	0.44	0.000
Person 5	0.25	0.45	0.30	0.012
Person 6	0.33	0.33	0.33	0.000
Person 7	0.21	0.39	0.39	0.000
Mean	0.28	0.37	0.36	0.079
Min	0.11	0.25	0.30	
Max	0.43	0.45	0.44	
Std Dev	0.10	0.07	0.05	

Group disagreements on utility goals under "Increase operating flexibility and reliability"

Intraclass Correlation Coefficient

	Intraclass Correlation[b]	95% Confidence Interval		F Test with True Value 0			
		Lower Bound	Upper Bound	Value	df1	df2	Sig
Single Measures	.139[a]	-.163	.936	1.757	2	12	.214
Average Measures	.531	-51.919	.990	1.757	2	12	.214

Two-way random effects model where both people effects and measures effects are random.

a. The estimator is the same, whether the interaction effect is present or not.

b. Type A intraclass correlation coefficients using an absolute agreement definition.

Hierarchical clustering analysis of utility goals under "Increase operating flexibility and reliability"

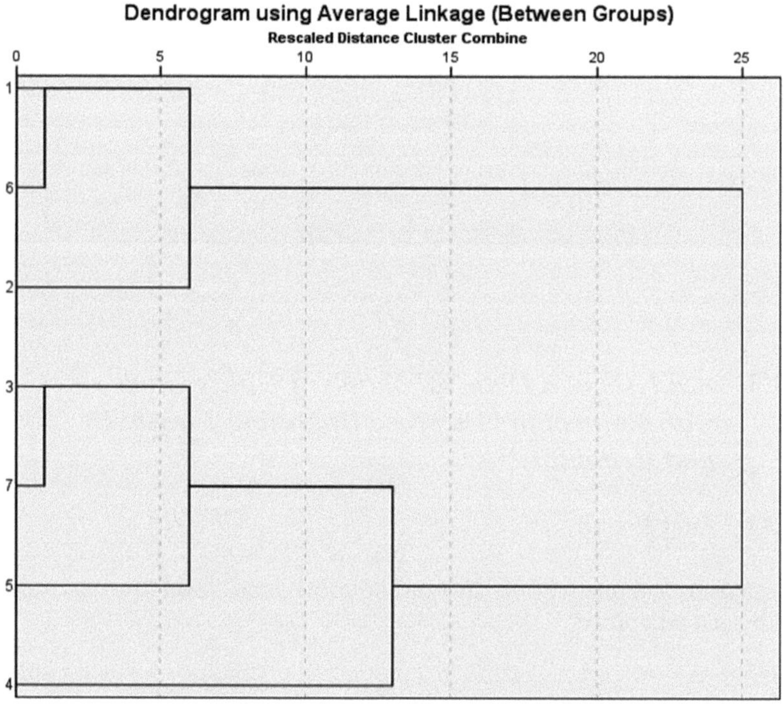

11.13.3.2 Second Round

Inconsistencies and relative contributions of utility goals under "Increase operating flexibility and reliability"

Relative Weights

Project Title: Increase operating flex. and rel

Users	1	2	3	Incn
Person 1	0.27	0.35	0.38	0.001
Person 2	0.33	0.33	0.33	0.000
Person 3	0.25	0.38	0.38	0.000
Person 4	0.25	0.38	0.38	0.000
Person 5	0.25	0.45	0.30	0.012
Person 6	0.33	0.33	0.33	0.000
Person 7	0.21	0.39	0.39	0.000
Mean	0.27	0.37	0.36	0.041
Min	0.21	0.33	0.30	
Max	0.33	0.45	0.39	
Std Dev	0.05	0.04	0.03	

Group disagreements on utility goals under "Increase operating flexibility and reliability"

Intraclass Correlation Coefficient

	Intraclass Correlation[b]	95% Confidence Interval		F Test with True Value 0			
		Lower Bound	Upper Bound	Value	df1	df2	Sig
Single Measures	.617[a]	.128	.986	8.574	2	12	.005
Average Measures	.919	.506	.998	8.574	2	12	.005

Two-way random effects model where both people effects and measures effects are random.

a. The estimator is the same, whether the interaction effect is present or not.

b. Type A intraclass correlation coefficients using an absolute agreement definition.

Hierarchical clustering analysis of utility goals under "Increase operating flexibility and reliability"

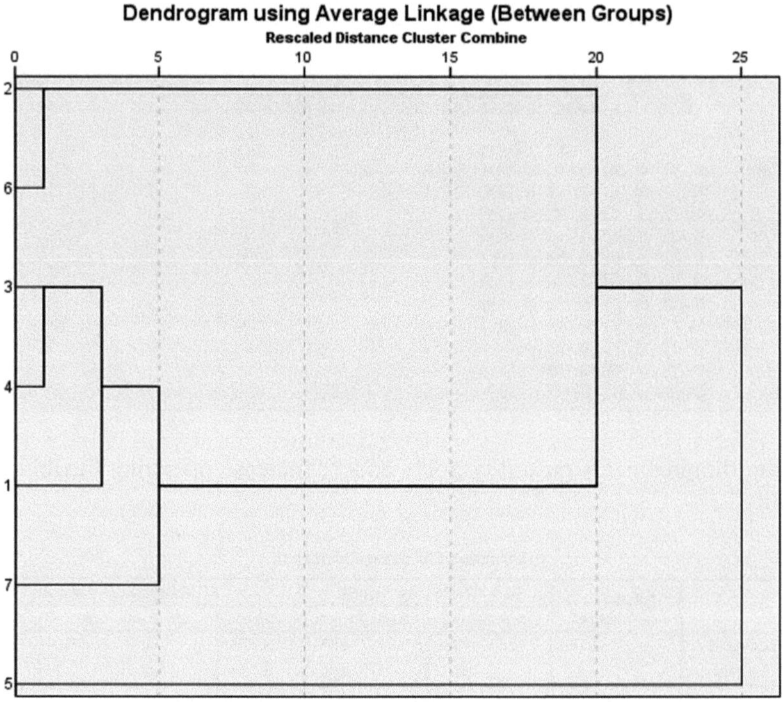

11.13.4 *Analysis and Data Validation of Utility Goals with Respect to "Reduce System Cost"*

11.13.4.1 First Round

Relative contributions of utility goals under "Reduce system cost"

	Reduce operating costs	Reduce postpone capital investments
Expert 1	0.5	0.5
Expert 2	0.6	0.4
Expert 3	0.7	0.3
Expert 4	0.5	0.5
Expert 5	0.7	0.3
Expert 6	0.5	0.5
Expert 7	0.3	0.7
Aggregated	0.54	0.46

Group disagreements on utility goals under "Reduce system cost"

Intraclass Correlation Coefficient

	Intraclass Correlation[b]	95% Confidence Interval		F Test with True Value 0			
		Lower Bound	Upper Bound	Value	df1	df2	Sig
Single Measures	-.108[a]	-.359	.994	.659	1	6	.448
Average Measures	-2.154	2.176	.999	.659	1	6	.448

Two-way random effects model where both people effects and measures effects are random.

a. The estimator is the same, whether the interaction effect is present or not.

b. Type A intraclass correlation coefficients using an absolute agreement definition.

Hierarchical clustering analysis of utility goals under "Reduce system cost"

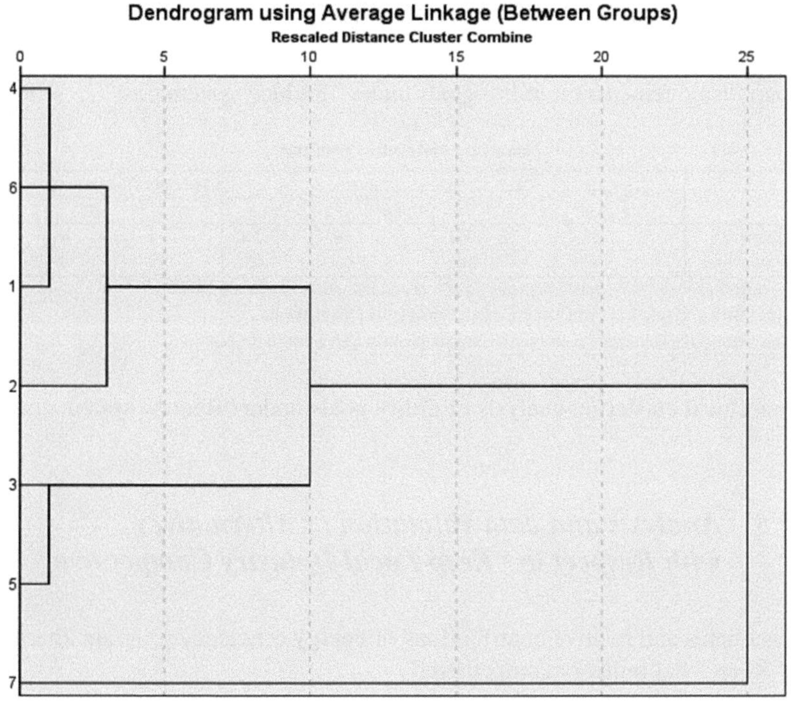

11.13.4.2 Second Round

Relative contributions of utility goals under "Reduce system cost"

	Reduce operating costs	Reduce postpone capital investments
Expert 1	0.5	0.5
Expert 2	0.6	0.4
Expert 3	0.7	0.3
Expert 4	0.5	0.5
Expert 5	0.7	0.3
Expert 6	0.5	0.5
Expert 7	0.7	0.3
Aggregated	0.60	0.40

Group disagreements on utility goals under "Reduce system cost"

Intraclass Correlation Coefficient

	Intraclass Correlation[b]	95% Confidence Interval		F Test with True Value 0			
		Lower Bound	Upper Bound	Value	df1	df2	Sig
Single Measures	.632[a]	-.062	.999	7.000	1	6	.038
Average Measures	.923	-.699	1.000	7.000	1	6	.038

Two-way random effects model where both people effects and measures effects are random.

a. The estimator is the same, whether the interaction effect is present or not.

b. Type A intraclass correlation coefficients using an absolute agreement definition.

Hierarchical clustering analysis of utility goals under "Reduce system cost"

11.13.5 Analysis and data Validation of Alternatives with Respect to "Keep Local Industry Competitive"

Inconsistencies and relative contributions of energy efficiency program alternatives under "Keep local industry competitive"

Relative Weights

Project Title: Keep local industry competitive

Users	1	2	3	Incn
Person 1	0.58	0.07	0.34	0.005
Person 2	0.69	0.23	0.08	0.000
Person 3	0.56	0.43	0.01	0.006
Person 4	0.48	0.24	0.28	0.006
Mean	0.58	0.24	0.18	0.133
Min	0.48	0.07	0.01	
Max	0.69	0.43	0.34	
Std Dev	0.09	0.14	0.16	

Group disagreements on energy efficiency program alternatives under "Keep local industry competitive"

Intraclass Correlation Coefficient

	Intraclass Correlation[b]	95% Confidence Interval		F Test with True Value 0			
		Lower Bound	Upper Bound	Value	df1	df2	Sig
Single Measures	.686[a]	-.023	.990	6.826	2	6	.028
Average Measures	.897	-.098	.998	6.826	2	6	.028

Two-way random effects model where both people effects and measures effects are random.

a. The estimator is the same, whether the interaction effect is present or not.

b. Type A intraclass correlation coefficients using an absolute agreement definition.

Hierarchical clustering analysis of energy efficiency program alternatives under "Keep local industry competitive"

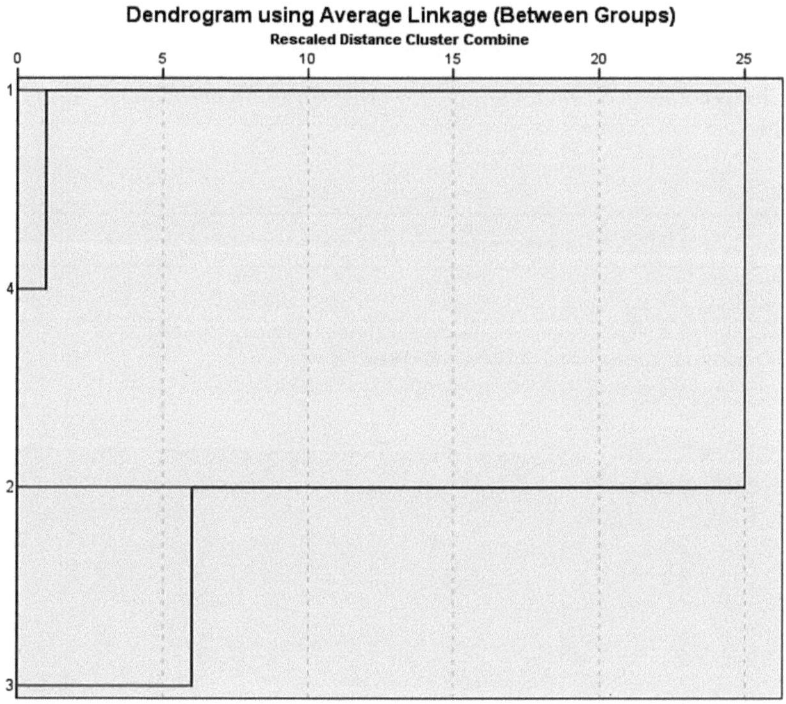

11.13.6 Analysis and Data Validation of Alternatives with Respect to "Improve Life Standards"

Inconsistencies and relative contributions of energy efficiency program alternatives under "Improve life standards"

Relative Weights

Project Title: Improve life standards

Users	1	2	3	Incn
Person 1	0.69	0.23	0.08	0.000
Person 2	0.69	0.24	0.08	0.000
Person 3	0.71	0.29	0.00	0.051
Person 4	0.71	0.29	0.00	0.051
Mean	0.70	0.26	0.04	0.031
Min	0.69	0.23	0.00	
Max	0.71	0.29	0.08	
Std Dev	0.01	0.03	0.04	

Group disagreements on energy efficiency program alternatives under "Improve life standards"

Intraclass Correlation Coefficient

	Intraclass Correlation[b]	95% Confidence Interval		F Test with True Value 0			
		Lower Bound	Upper Bound	Value	df1	df2	Sig
Single Measures	.990[a]	.933	1.000	274.726	2	6	.000
Average Measures	.998	.982	1.000	274.726	2	6	.000

Two-way random effects model where both people effects and measures effects are random.

a. The estimator is the same, whether the interaction effect is present or not.

b. Type A intraclass correlation coefficients using an absolute agreement definition.

Hierarchical clustering analysis of energy efficiency program alternatives under "Improve life standards"

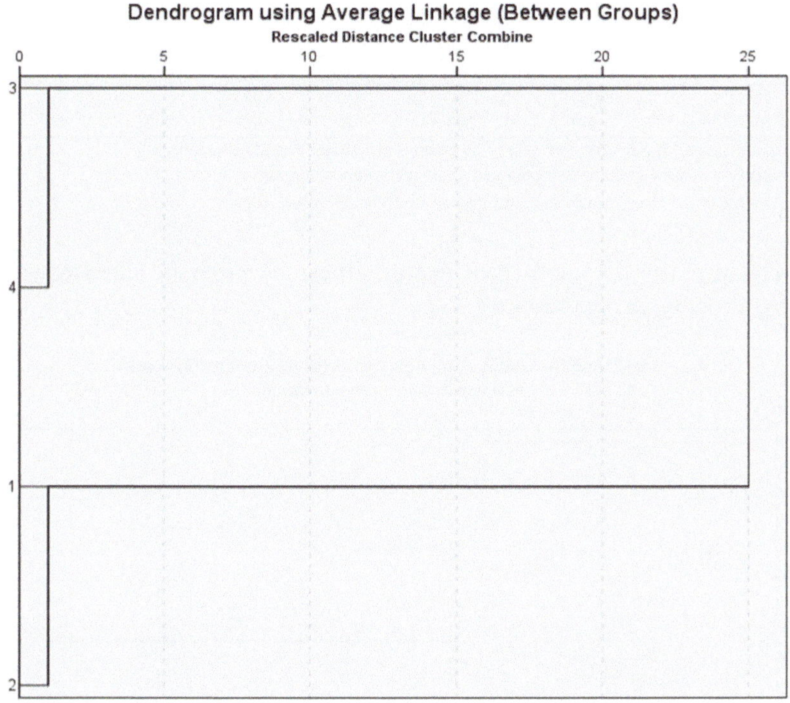

11.13.7 Analysis and Data Validation of Alternatives with Respect to "Create or Retain Job Opportunities"

Inconsistencies and relative contributions of energy efficiency program alternatives under "Create or retain job opportunities"

Relative Weights

Project Title: Create or retain job opportunities

Users	1	2	3	Incn
Person 1	0.58	0.07	0.34	0.005
Person 2	0.69	0.23	0.08	0.000
Person 3	0.71	0.29	0.00	0.051
Person 4	0.54	0.26	0.20	0.005
Mean	0.63	0.21	0.16	0.112
Min	0.54	0.07	0.00	
Max	0.71	0.29	0.34	
Std Dev	0.08	0.10	0.15	

Group disagreements on energy efficiency program alternatives under "Create or retain job opportunities"

Intraclass Correlation Coefficient

	Intraclass Correlation[b]	95% Confidence Interval		F Test with True Value 0			
		Lower Bound	Upper Bound	Value	df1	df2	Sig
Single Measures	.830[a]	.258	.995	13.983	2	6	.006
Average Measures	.951	.582	.999	13.983	2	6	.006

Two-way random effects model where both people effects and measures effects are random.

a. The estimator is the same, whether the interaction effect is present or not.

b. Type A intraclass correlation coefficients using an absolute agreement definition.

Hierarchical clustering analysis of energy efficiency program alternatives under "Create or retain job opportunities"

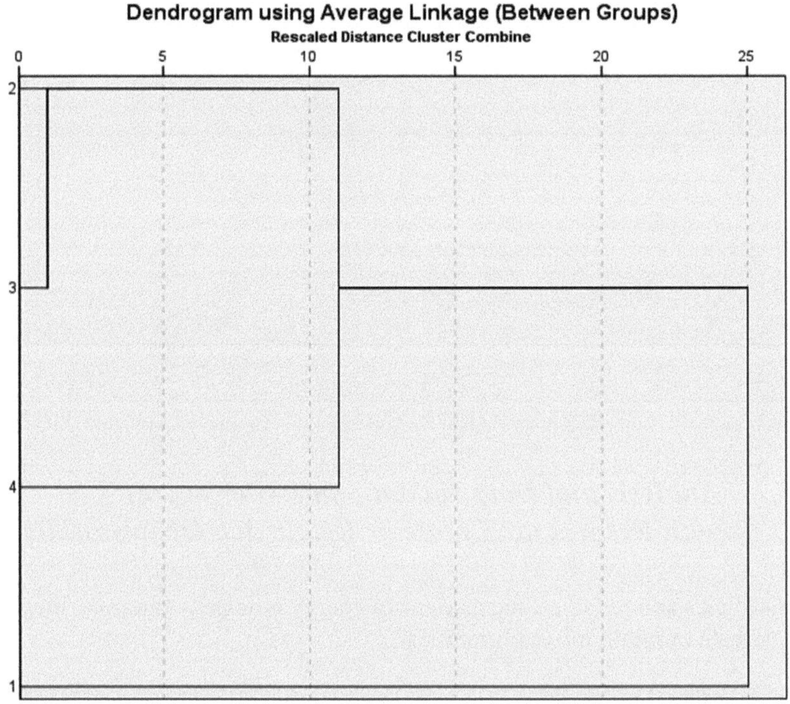

11.13.8 Analysis and Data Validation of Alternatives with Respect to "Reduce Need for Critical Resources"

11.13.8.1 Round 1

Inconsistencies and relative contributions of energy efficiency program alternatives under "Reduce need for critical resources"

Relative Weights

Project Title: Reduce need for critical resources

Users	1	2	3	Incn
Person 1	0.70	0.19	0.11	0.002
Person 2	0.33	0.33	0.33	0.000
Person 3	0.50	0.50	0.01	0.000
Person 4	0.98	0.01	0.01	0.000
Mean	0.63	0.26	0.12	0.219
Min	0.33	0.01	0.01	
Max	0.98	0.50	0.33	
Std Dev	0.28	0.21	0.15	

Group disagreements on energy efficiency program alternatives under "Reduce need for critical resources"

Intraclass Correlation Coefficient

	Intraclass Correlation[b]	95% Confidence Interval		F Test with True Value 0			
		Lower Bound	Upper Bound	Value	df1	df2	Sig
Single Measures	.520[a]	-.211	.983	3.884	2	6	.083
Average Measures	.812	-2.303	.996	3.884	2	6	.083

Two-way random effects model where both people effects and measures effects are random.

a. The estimator is the same, whether the interaction effect is present or not.

b. Type A intraclass correlation coefficients using an absolute agreement definition.

Hierarchical clustering analysis of energy efficiency program alternatives under "Reduce need for critical resources"

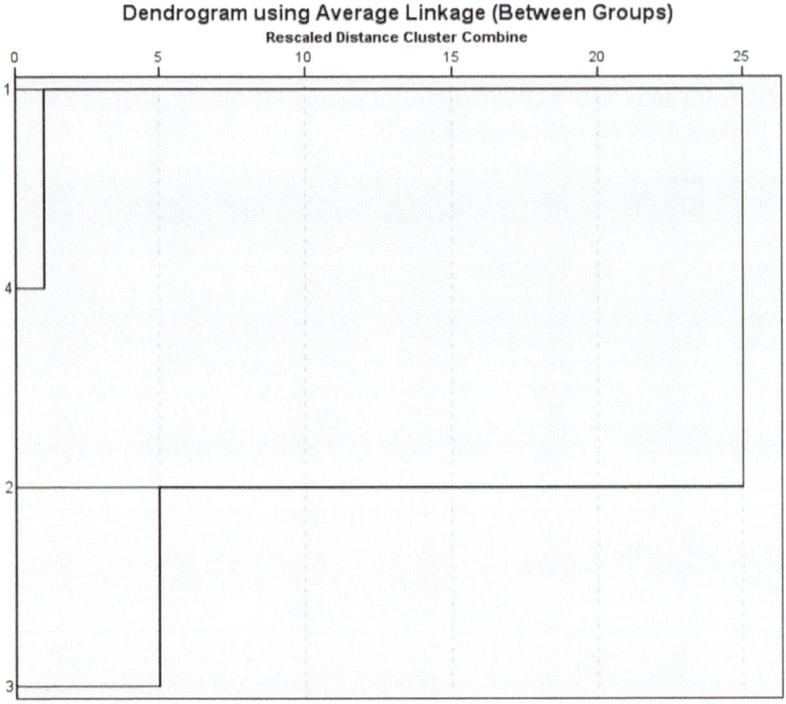

11.13.8.2 Round 2

Inconsistencies and relative contributions of energy efficiency program alternatives under "Reduce need for critical resources"

Relative Weights

Project Title: Reduce need for critical resources

Users	1	2	3	Incn
Person 1	0.70	0.19	0.11	0.002
Person 2	0.73	0.19	0.08	0.000
Person 3	0.63	0.36	0.00	0.023
Person 4	0.98	0.01	0.01	0.000
Mean	0.76	0.19	0.05	0.124
Min	0.63	0.01	0.00	
Max	0.98	0.36	0.11	
Std Dev	0.15	0.14	0.05	

Group disagreements on energy efficiency program alternatives under "Reduce need for critical resources"

Intraclass Correlation Coefficient

	Intraclass Correlation[b]	95% Confidence Interval		F Test with True Value 0			
		Lower Bound	Upper Bound	Value	df1	df2	Sig
Single Measures	.898[a]	.469	.997	24.368	2	6	.001
Average Measures	.972	.780	.999	24.368	2	6	.001

Two-way random effects model where both people effects and measures effects are random.

a. The estimator is the same, whether the interaction effect is present or not.

b. Type A intraclass correlation coefficients using an absolute agreement definition.

Hierarchical clustering analysis of energy efficiency program alternatives under "Reduce need for critical resources"

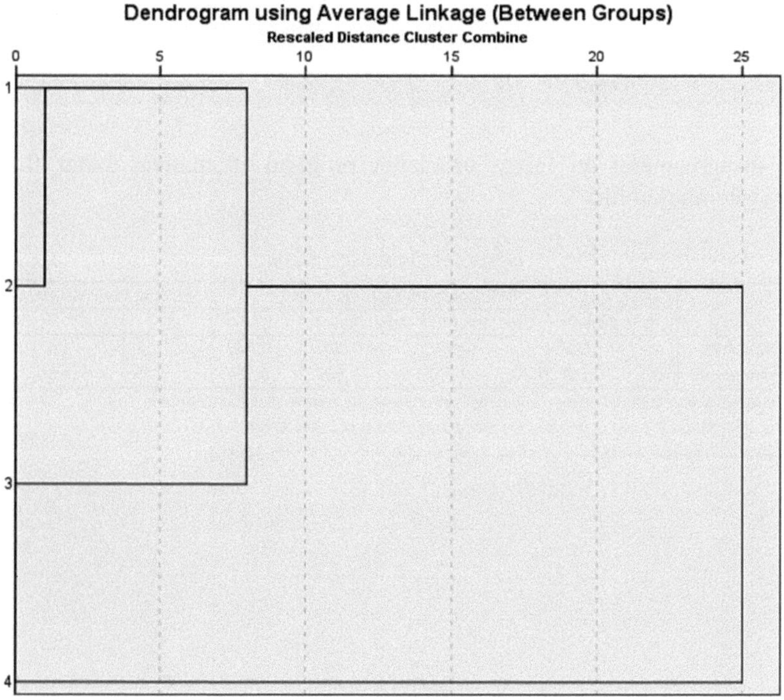

11.13.9 Analysis and Data Validation of Alternatives with Respect to "Increase Power System Reliability"

11.13.9.1 Round 1

Inconsistencies and relative contributions of energy efficiency program alternatives under "Increase power system reliability"

Relative Weights

Project Title: Increase power system reliability

Users	1	2	3	Incn
Person 1	0.23	0.43	0.34	0.014
Person 2	0.33	0.33	0.33	0.000
Person 3	0.79	0.21	0.00	0.088
Person 4	0.76	0.10	0.14	0.018
Mean	0.53	0.27	0.21	0.208
Min	0.23	0.10	0.00	
Max	0.79	0.43	0.34	
Std Dev	0.29	0.14	0.16	

Group disagreements on energy efficiency program alternatives under "Increase power system reliability"

Intraclass Correlation Coefficient

	Intraclass Correlation[b]	95% Confidence Interval		F Test with True Value 0			
		Lower Bound	Upper Bound	Value	df1	df2	Sig
Single Measures	.223[a]	-.396	.963	1.767	2	6	.249
Average Measures	.535	8.431	.990	1.767	2	6	.249

Two-way random effects model where both people effects and measures effects are random.

a. The estimator is the same, whether the interaction effect is present or not.

b. Type A intraclass correlation coefficients using an absolute agreement definition.

Hierarchical clustering analysis of energy efficiency program alternatives under "Increase power system reliability"

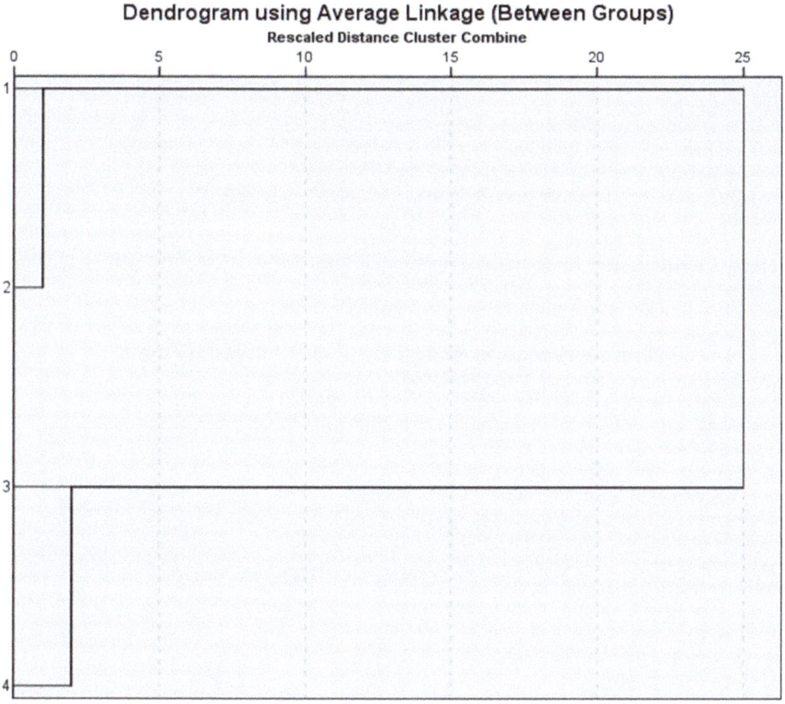

11.13.9.2 Round 2

Inconsistencies and relative contributions of energy efficiency program alternatives under "Increase power system reliability"

Relative Weights

Project Title: Increase power system reliability

Users	1	2	3	Incn
Person 1	0.67	0.24	0.09	0.007
Person 2	0.82	0.10	0.08	0.062
Person 3	0.79	0.21	0.00	0.088
Person 4	0.76	0.10	0.14	0.018
Mean	0.76	0.16	0.08	0.065
Min	0.67	0.10	0.00	
Max	0.82	0.24	0.14	
Std Dev	0.06	0.07	0.06	

Group disagreements on energy efficiency program alternatives under "Increase power system reliability"

Intraclass Correlation Coefficient

	Intraclass Correlation[b]	95% Confidence Interval		F Test with True Value 0			
		Lower Bound	Upper Bound	Value	df1	df2	Sig
Single Measures	.967[a]	.789	.999	79.426	2	6	.000
Average Measures	.992	.937	1.000	79.426	2	6	.000

Two-way random effects model where both people effects and measures effects are random.

a. The estimator is the same, whether the interaction effect is present or not.

b. Type A intraclass correlation coefficients using an absolute agreement definition.

Hierarchical clustering analysis of energy efficiency program alternatives under "Increase power system reliability"

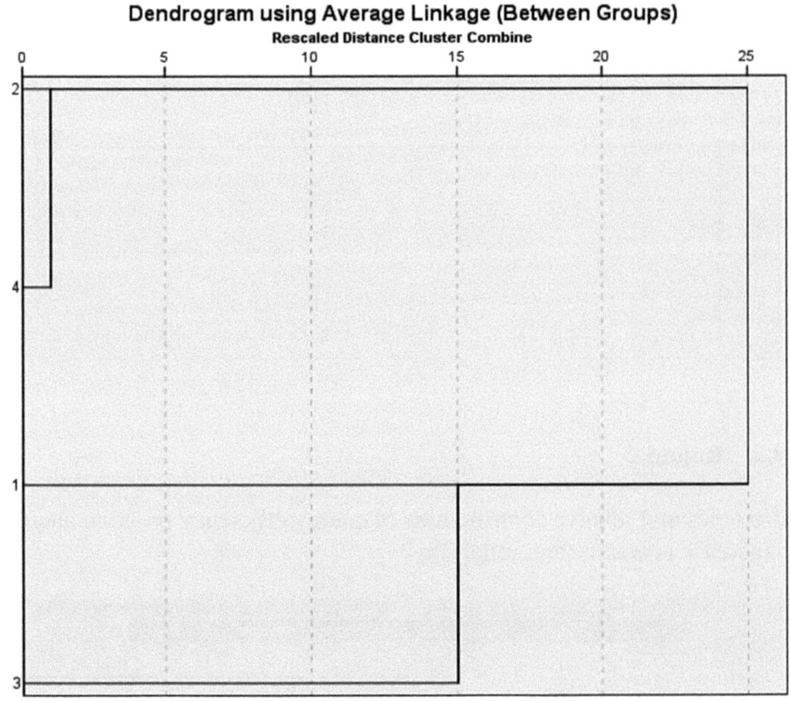

11.13.10 Analysis and Data Validation of Alternatives with Respect to "Increase Transmission and Distribution System Reliability"

11.13.10.1 Round 1

Inconsistencies and relative contributions of energy efficiency program alternatives under "Increase transmission and distribution system reliability"

Relative Weights

Project Title: Increase trans. and dist. sys. rel

Users	1	2	3	Incn
Person 1	0.23	0.43	0.34	0.014
Person 2	0.69	0.23	0.08	0.000
Person 3	0.79	0.21	0.00	0.088
Mean	0.57	0.29	0.14	0.213
Min	0.23	0.21	0.00	
Max	0.79	0.43	0.34	
Std Dev	0.30	0.12	0.18	

Group disagreements on energy efficiency program alternatives under "Increase transmission and distribution system reliability"

Intraclass Correlation Coefficient

	Intraclass Correlation[b]	95% Confidence Interval		F Test with True Value 0			
		Lower Bound	Upper Bound	Value	df1	df2	Sig
Single Measures	.356[a]	-.670	.976	2.108	2	4	.237
Average Measures	.624	5.923	.992	2.108	2	4	.237

Two-way random effects model where both people effects and measures effects are random.

a. The estimator is the same, whether the interaction effect is present or not.

b. Type A intraclass correlation coefficients using an absolute agreement definition.

Hierarchical clustering analysis of energy efficiency program alternatives under "Increase transmission and distribution system reliability"

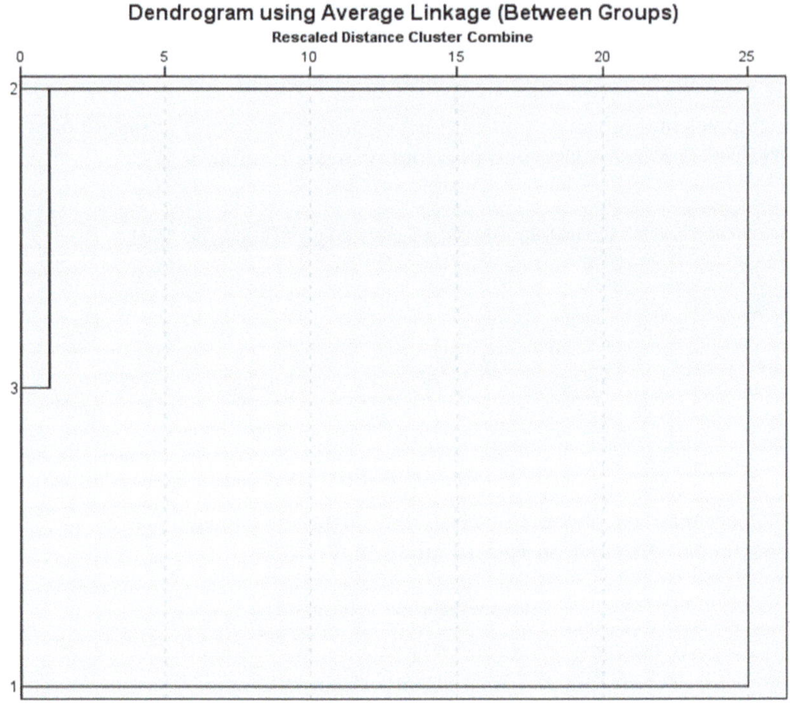

11.13.10.2 Round 2

Inconsistencies and relative contributions of energy efficiency program alternatives under "Increase transmission and distribution system reliability"

Relative Weights

Project Title: Increase trans. and dist. sys. rel

Users	1	2	3	Incn
Person 1	0.58	0.30	0.12	0.009
Person 2	0.69	0.23	0.08	0.000
Person 3	0.79	0.21	0.00	0.088
Mean	0.69	0.25	0.07	0.074
Min	0.58	0.21	0.00	
Max	0.79	0.30	0.12	
Std Dev	0.10	0.05	0.06	

Group disagreements on energy efficiency program alternatives under "Increase transmission and distribution system reliability"

Intraclass Correlation Coefficient

	Intraclass Correlation[b]	95% Confidence Interval		F Test with True Value 0			
		Lower Bound	Upper Bound	Value	df1	df2	Sig
Single Measures	.946[a]	.543	.999	35.906	2	4	.003
Average Measures	.981	.781	1.000	35.906	2	4	.003

Two-way random effects model where both people effects and measures effects are random.

a. The estimator is the same, whether the interaction effect is present or not.

b. Type A intraclass correlation coefficients using an absolute agreement definition.

Hierarchical clustering analysis of energy efficiency program alternatives under "Increase transmission and distribution system reliability"

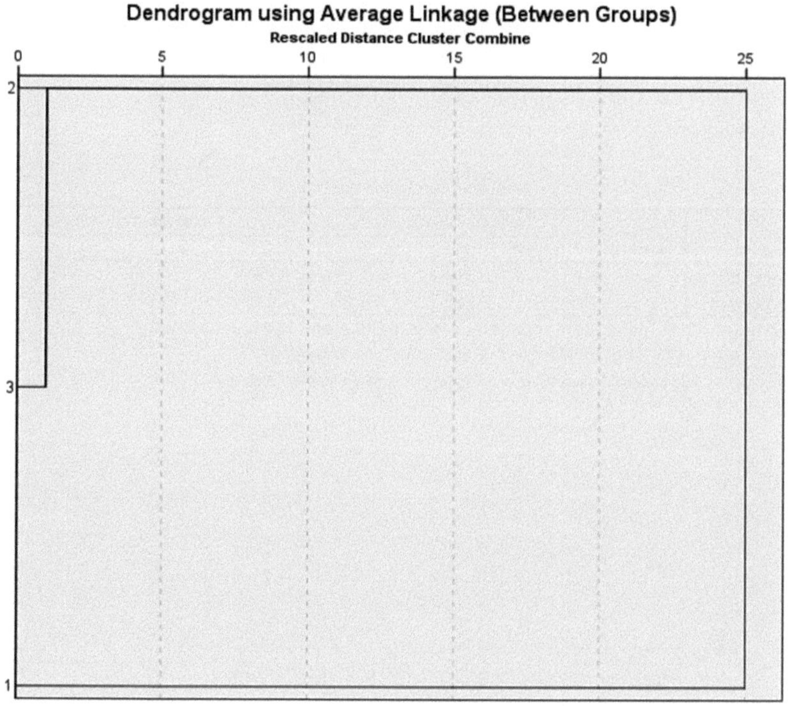

11.13.11 Analysis and Data Validation of Alternatives with Respect to "Reduce Operating Costs"

Inconsistencies and relative contributions of energy efficiency program alternatives under "Reduce operating costs"

Relative Weights

Project Title: Reduce operating costs

Users	1	2	3	Incn
Person 1	0.56	0.29	0.15	0.001
Person 2	0.69	0.23	0.08	0.000
Person 3	0.81	0.18	0.01	0.001
Person 4	0.71	0.19	0.10	0.097
Mean	0.69	0.22	0.08	0.073
Min	0.56	0.18	0.01	
Max	0.81	0.29	0.15	
Std Dev	0.10	0.05	0.06	

Group disagreements on energy efficiency program alternatives under "Reduce operating costs"

Intraclass Correlation Coefficient

	Intraclass Correlation[b]	95% Confidence Interval		F Test with True Value 0			
		Lower Bound	Upper Bound	Value	df1	df2	Sig
Single Measures	.948[a]	.685	.999	49.451	2	6	.000
Average Measures	.986	.897	1.000	49.451	2	6	.000

Two-way random effects model where both people effects and measures effects are random.

a. The estimator is the same, whether the interaction effect is present or not.

b. Type A intraclass correlation coefficients using an absolute agreement definition.

Hierarchical clustering analysis of energy efficiency program alternatives under "Reduce operating costs"

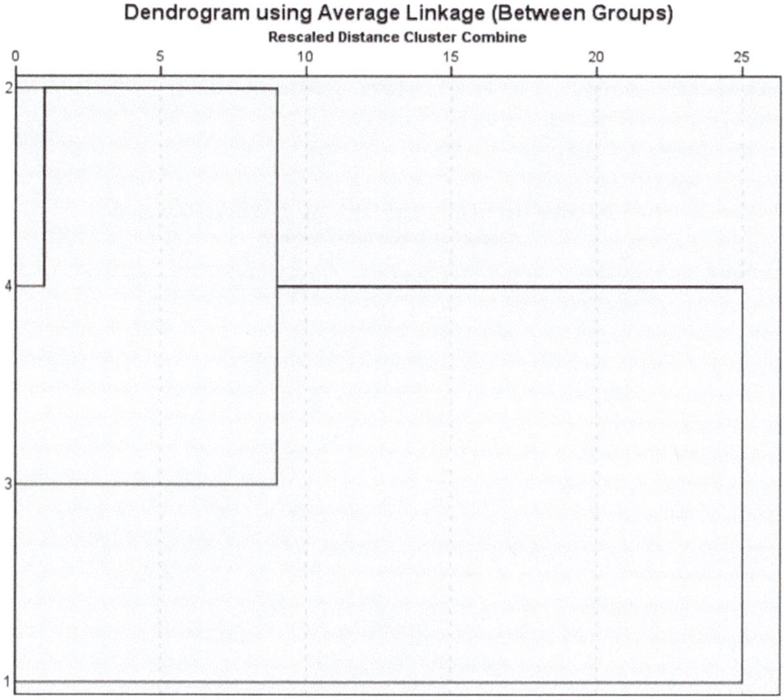

11.13.12 *Analysis and Data Validation of Alternatives with Respect to "Reduce Postpone Capital Investments"*

Inconsistencies and relative contributions of energy efficiency program alternatives under "Reduce or postpone capital investments"

Relative Weights

Project Title: Reduce postpone capital investments

Users	1	2	3	Incn
Person 1	0.49	0.40	0.11	0.013
Person 2	0.69	0.23	0.08	0.000
Person 3	0.79	0.21	0.00	0.088
Person 4	0.33	0.33	0.33	0.000
Mean	0.58	0.29	0.13	0.152
Min	0.33	0.21	0.00	
Max	0.79	0.40	0.33	
Std Dev	0.20	0.09	0.14	

Intraclass Correlation Coefficient

	Intraclass Correlation[b]	95% Confidence Interval		F Test with True Value 0			
		Lower Bound	Upper Bound	Value	df1	df2	Sig
Single Measures	.642[a]	-.082	.988	5.791	2	6	.040
Average Measures	.878	-.435	.997	5.791	2	6	.040

Two-way random effects model where both people effects and measures effects are random.

a. The estimator is the same, whether the interaction effect is present or not.

b. Type A intraclass correlation coefficients using an absolute agreement definition.

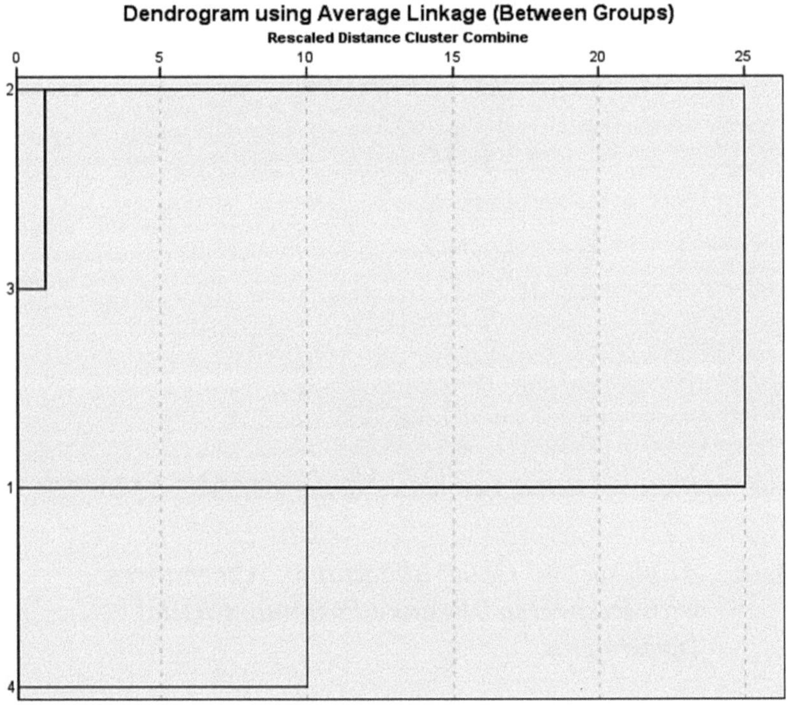

Group disagreements on energy efficiency program alternatives under "Reduce postpone capital investments"

Hierarchical clustering analysis of energy efficiency program alternatives under "Reduce postpone capital investments"

References

1. N.H. Afgan, P.A. Pilavachi, M.G. Carvalho, Multi-criteria evaluation of natural gas resources. Energy Policy **35**(1), 704–713 (2007)
2. N.H. Afgan, M.G. Carvalho, P.A. Pilavachi, N. Matins, Evaluation of natural gas supply options for Southeast and Central Europe: part 2. Multi-criteria assessment. Energy Convers. Manag. **49**(8), 2345–2353 (2008)
3. A.I. Chatzimouratidis, P.A. Pilavachi, Objective and subjective evaluation of power plants and their non-radioactive emissions using the analytic hierarchy process. Energy Policy **35**(8), 4027–4038 (2007)
4. A. Chatzimouratidis, P.A. Pilavachi, Technological, economic and sustainability evaluation of power plants using the analytic hierarchy process. Energy Policy **37**(3), 778–787 (2008). doi:10.1016/j.enpol.2008.10.009
5. A.I. Chatzimouratidis, P.A. Pilavachi, Sensitivity analysis of the evaluation of power plants impact on the living standard using the analytic hierarchy process. Energy Convers. Manag. **49**(12), 3599–3611 (2008)
6. A.I. Chatzimouratidis, P.A. Pilavachi, Multicriteria evaluation of power plants impact on the living standard using the analytic hierarchy process. Energy Policy **36**(3), 1074–1089 (2008)
7. H. Chen, Sensitivity analysis for hierarchical decision models. Doctoral dissertation, Portland State University, 2007
8. C.W. Gellings, W.M. Smith, Integrating demand-side management into utility planning. Proc. IEEE **77**(6), 908–918 (1989)
9. B.F. Hobbs, G.T. Horn, Building public confidence in energy planning: a multimethod MCDM approach to demand-side planning at BC gas. Energy Policy **25**(3), 357–375 (1997)
10. B.F. Hobbs, P.M. Meier, Multicriteria methods for resource planning: an experimental comparison. IEEE Trans. Power Syst. **9**(4), 1811–1817 (1994)
11. I. Iskin, Literature review on adoption of energy efficient technologies from a demand side management perspective: taxonomy of adoption drivers, barriers and policy tools. Technology Management in Energy Smart World (PICMET 2011) (2011), pp. 1–16
12. I. Iskin, J. Callahan, M. Vowles, R Fedie, Variable capacity heat pump program development roadmap. Technology Management in Energy Smart World (PICMET 2011) (2011), pp. 1–12
13. R.L. Keeney, T.L. McDaniels, Value-focused thinking about strategic decisions at BC Hydro. Interfaces **22**, 94–109 (1992)
14. R.L. Keeney, T.L. McDaniels, Identifying and structuring values to guide integrated resource planning at BC Gas. Oper. Res. **47**(5), 651–662 (1999)
15. C.H. Lawshe, A quantitative approach to content validity. Pers. Psychol. **28**(4), 563–575 (1975)
16. D.K. Lee, S.Y. Park, S.U. Park, Development of assessment model for demand-side management investment programs in Korea. Energy Policy **35**, 5585–5590 (2007)
17. S.K. Lee, Y.J. Yoon, J.W. Kim, A study on making a long-term improvement in the national energy efficiency and GHG control plans by the AHP approach. Energy Policy **35**(5), 2862–2868 (2007)
18. D. Mills, I. Lowe, L.A. Vlacic, Hierarchical decision support system for integrated resource planning. J. Decis. Syst. **7**(1), 21–38 (1998)
19. N. Nagesha, P. Balachandra, Barriers to energy efficiency in small industry clusters: multi-criteria-based prioritization using the analytic hierarchy process. Energy **31**(12), 1633–1647 (2006)
20. M.A. Papalexandrou, P.A. Pilavachi, A.I. Chatzimouratidis, Evaluation of liquid bio-fuels using the analytic hierarchy process. Process Saf. Environ. Prot. **86**(5), 360–374 (2008)
21. P.A. Pilavachi, C.P. Roumpeas, S. Minett, N.H. Afgan, Multi-criteria evaluation for CHP system options. Energy Convers. Manag. **47**(20), 3519–3529 (2006)
22. P.A. Pilavachi, S.D. Stephanidis, V.A. Pappas, N.H. Afgan, Multi-criteria evaluation of hydrogen and natural gas fuelled power plant technologies. Appl. Therm. Eng. **29**(11–12), 2228–2234 (2009)

23. P.A. Pilavachi, A.I. Chatzipanagi, A.I. Spyropoulou, Evaluation of hydrogen production methods using the analytic hierarchy process. Int. J. Hydrog. Energy **34**(13), 5294–5303 (2009)
24. R. Ramanathan, L.S. Ganesh, Energy alternatives for lighting in households: an evaluation using integrated goal programming-AHP model. Energy **20**(1), 63–72 (1995)
25. V. Ravi, R. Shankar, Analysis of interactions among the barriers of reverse logistics. Technol. Forecast. Soc. Chang. **72**(8), 1011–1029 (2005)
26. A.K.N. Reddy, Barriers t o improvement in energy efficiency. Energy Policy **19**(10), 953–961 (1991)
27. A.H. Sanstad, R.B. Howarth, "Normal" markets, market imperfections and energy efficiency. Energy Policy **22**(10), 811–818 (1994)
28. J. Sathaye, A.J. Gadgil, Aggressive cost-effective electricity conservation: novel approaches. Energy Policy **20**(2), 163–172 (1992)
29. S. Vashishtha, M. Ramachandran, Multicriteria evaluation of demand side management (DSM) implementation strategies in the Indian power sector. Energy **31**, 2210–2225 (2006)
30. G. Wang, Y. Wang, T. Zhao, Analysis of interactions among the barriers to energy saving in China. Energy Policy **36**(6), 1879–1889 (2008)

Chapter 12
Evaluating Consumer Preferences for Clothes Dryers

Kevin C. van Blommestein and Tugrul U. Daim

Abstract Focus on encouraging the adoption of residential energy efficient appliances over recent years has resulted in major energy savings. Even though clothes dryers account for approximately 4 % of residential energy use in the USA, there is still no energy efficient clothes dryer in the market. There has been a lot of focus recently on the development of energy efficient clothes dryers; however, there is limited research on consumer preferences when purchasing a clothes dryer. These preferences should be taken into account when developing the new clothes dryer technology that may aid in encouraging adoption. This study utilizes the Hierarchical Decision Model (HDM) to capture consumer preferences and to capture the perception of manufacturers on what they think the consumer preferences are. Energy efficiency experts are used to quantify the sub-criteria for each technology and the resulting technology preferences are determined. The two highest consumer preferences were the purchase and installation cost and the operating lifetime. The results for the consumers and the manufacturers mainly align, except for the importance of potential savings, drying cycle time, and operating lifetime. It is determined that the most preferred technology is moisture and temperature sensors for clothes dryers; however, the overall scores for each technology were relatively close to one another.

K.C. van Blommestein (✉)
Portland State University, SW 4th Avenue 50-02 1900, Portland, OR 97207, USA
e-mail: kevin9@pdx.edu

T.U. Daim
Department of Engineering and Technology Management, Portland State University, 1900 SW 4th Avenue 50-02, Portland, OR 97201, USA
e-mail: tugrul@etm.pdx.edu

T.U. Daim et al. (eds.), *Policies and Programs for Sustainable Energy Innovations*, Innovation, Technology, and Knowledge Management,
DOI 10.1007/978-3-319-16033-7_12

12.1 Introduction

Based on the 2009 Residential Energy Consumption Survey (RECS), approximately 80 % of households have clothes dryers, of which approximately 80 % of the clothes dryers are electric [1]. A breakdown of the US market is illustrated in Fig. 12.1. Due to the large clothes dryer market and the current inefficient clothes dryer options in the USA, focus has shifted on improving clothes dryer efficiencies. This is emphasized by the fact that more than 4 % of residential energy use in the USA is from clothes dryers [2]. The Energy Star market and industry scoping report [3] and the U.S. Department of Energy (DOE) technical support document for residential clothes dryers [4], evaluate potential saving options for clothes dryers, which include alternatives for the drum, motor, controls, heat recovery, heat generation, and several other performance improvements.

Due to the large potential savings from heat generation, majority of efficiency improvements have been focused on this area. The Energy Star market and industry scoping report [3] and the DOE Roadmap for Next-Generation Appliances [5] lists alternative methods of heat generation, specially heat pump, microwave, modulating, and indirect heating. The heat pump has the maximum estimated savings of the heat generating technologies (20–60 %). However, even though the estimated savings are high for this technology, drying times are generally longer and the purchase cost is higher. Therefore, a tradeoff needs to be made by the consumer between these factors.

The purpose of this study is to determine the preferences of consumers when purchasing a new clothes dryer. By determining what criteria are most important to consumers before the energy efficient product enters the market, it is possible for manufacturers to make adjustments to their products that may result in increased adoption by the consumer. The overall objective of this study is to aid energy efficiency program administrators in increasing the adoption of energy efficient

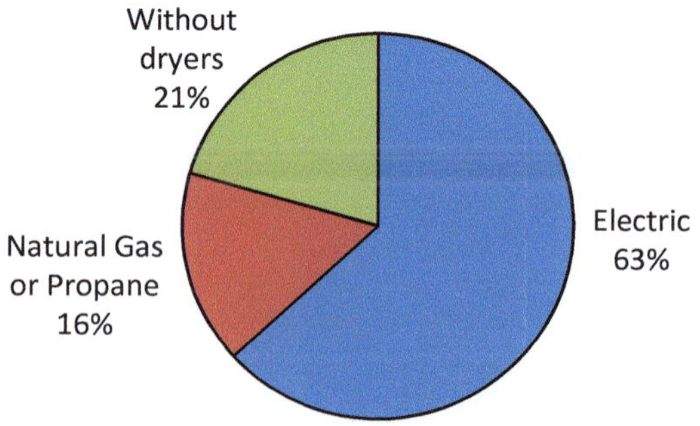

Fig. 12.1 US clothes dryer market share [1]

clothes dryers, thereby increasing overall energy savings. Depending on consumer preferences, these energy efficient program administrators can adjust their policies accordingly to align with these preferences.

12.2 Methods for Determining Consumer Preferences

Extensive research has been placed in trying to understand consumer preferences. Adaptive conjoint analysis and choice-based conjoint analysis have become the most prevalent methods, specifically related to evaluating new product concepts [6]. Even though not as extensively used, Analytic Hierarchy Process (AHP) and Hierarchical Decision Model (HDM) have been used to evaluate mobile service adoption factors [7], energy efficient device adoption [8], remote health monitoring [9], hybrid cars [10], to name a few. Mulye [11] compares AHP to conjoint analysis and mentions that one of the benefits of AHP is the ability to conduct consistency checks and sensitivity analysis, while the main limitation is the low flexibility of the model. In comparison, the benefit of conjoint analysis is the high level of flexibility, while the limitation is the lack of consistency checks and sensitivity analysis. Scholz et al. [12] compare the results of a modified version of AHP, referred to as Paired Comparison-based Preference Measurement (PCPM), with adaptive conjoint analysis, and determined that PCPM yielded better results and allows for more precise preferences. In summary, there are many approaches to determining consumer preferences, each with their own benefits and limitations.

12.3 Methodology

Figure 12.2 illustrates the methodology followed for the evaluation of consumer preferences for clothes dryers. The first step of the methodology was to identify energy efficient experts, clothes dryer consumers, and clothes dryer manufacturers. The second step was to identify the relevant criteria, sub-criteria, and technologies with the aid of the identified experts, consumers, and manufacturers. Once these criteria, sub-criteria, and technologies were identified an HDM was created. A pairwise comparison survey was then sent to the consumers and manufacturers in order to determine weights for the criteria and sub-criteria, and a Likert scale survey was sent to the experts to quantify the sub-criteria for each technology. The reason why experts were required to quantify the sub-criteria for each technology was because not all technologies were available in the market, and therefore these values were unknown. These results were then evaluated, and any incomplete or incorrectly completed surveys were sent back to the relevant respondent for rectification. The end results were then discussed together with suggestions on how the study could be updated and improved. The following sections will discuss each of these steps.

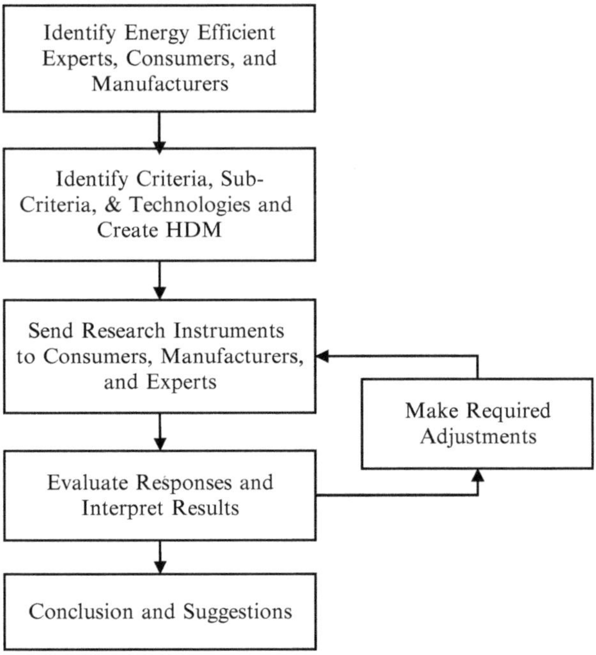

Fig. 12.2 Research methodology

12.4 Consumer, Manufacturer, and Expert Identification

12.4.1 Consumers and Manufacturers

The energy efficiency program administrator requesting this research maintained a database of energy efficient clothes washer consumers. These consumers were chosen and contacted due to the similarity between purchasing a clothes washer and purchasing a clothes dryer. The clothes dryer manufacturer contact details however were not as readily available and required communication through external contacts.

12.4.2 Energy Efficiency Experts

The energy efficiency program administrator maintained a small list of energy efficiency expert contact details; however, additional contacts were required. One method of identifying experts in a chosen field is by finding journal articles and conference proceedings associated with the topic, and identify the author(s) with his/her/their contact details. The question is, how do you know which authors to choose when a list of hundreds or thousands of authors is identified? You could simply choose the authors with the most number of publications; however, these

Table 12.1 Journal articles and conference proceedings search criteria (general search)

Keywords	(clothes OR tumble) AND (dr*) AND ((energy NEAR efficient) OR (energy NEAR efficiency) OR (energy NEAR saving))
Country	USA
Year	2000–2014
Language	English

publications could all be focused on one specific area of the technology, and does not take into account the quality of the articles. Additionally, it does not take into account the position of an author in the network of authors, neither does it include the relationship among these authors.

Social Network Analysis (SNA) is a method that can be used to understand and analyze relationships among nodes (e.g., authors) in a social network. Social networks consist of nodes with edges representing the relationship among these nodes. Some metrics that can be obtained from SNA are [13]:

1. Degree Centrality—the number of direct connections a node (actor) has in the social network.
2. Betweenness Centrality—the number of times a node (actor) occurs on the shortest path between other nodes. This represents the influence the node has on the exchange of information in the network.
3. Closeness Centrality—the sum of the shortest distances from a specific node to every other node in the network. This represents how quickly the node can interact with other nodes.

By searching the Compendex database using keywords associated with clothes dryers, adjacency matrices were created and networks constructed. The search criteria used to identify the appropriate journal articles and conference proceedings is shown in Table 12.1. The focus of the search was to identify experts (authors) with a general understanding of most, if not all, recent developments in energy efficient clothes dryer technologies.

It was discovered from the results of the SNA that the network was mainly disconnected and therefore the closeness centrality values had little meaning. Additionally, since degree centrality measures the number of direct connections, if there were a large number of authors for the same article, they would be connected to one another and have high degree centralities. Since the intention was not to discover authors that only focus in one specific area, a high degree centrality had little meaning. However, by using a local cluster coefficient, it can be determined whether the connecting authors are also connected to one another. A lower local cluster coefficient means that some connected authors are not connected to one another, and therefore are not all coauthors of the same article. Therefore, a high-degree centrality with a low cluster coefficient would be preferred. The betweenness centrality was a useful metric for determining the importance of an author in the network.

As a result of an iterative search process, nine articles were identified based on the search criteria mentioned above. All articles identified were relevant to the topic at hand.

Twenty-nine US authors were identified in total. Due to the small disconnected network, SNA may have been unnecessary for this specific application, however for topics with much higher interests, SNA would be more appropriate.

12.5 Hierarchical Decision Model

In order to evaluate the decision of a consumer when purchasing a new clothes dryer, the HDM in Fig. 12.3 was created. The purpose of the HDM was to quantify consumer preferences based on the consumer's decision-making process. The HDM consists of the objective at the top of the hierarchy, followed by the criteria, sub-criteria, and finally the technologies.

12.5.1 Criteria and Sub-criteria

An initial set of criteria and sub-criteria were decided upon after discussions with the energy efficiency program administrator and a few potential consumers. Additionally, an initial concept test of the study was conducted, and the comments received were assessed and the relevant updates were made. The model was evaluated and accepted and the following criteria and sub-criteria were deemed as relevant:

The main criteria were

1. Performance—Relates to the effectiveness of the clothes dryer. This includes the time it takes the clothes dryer to complete one cycle, whether the clothes are completely dry after one cycle, and the level of wear and tear on the clothing.

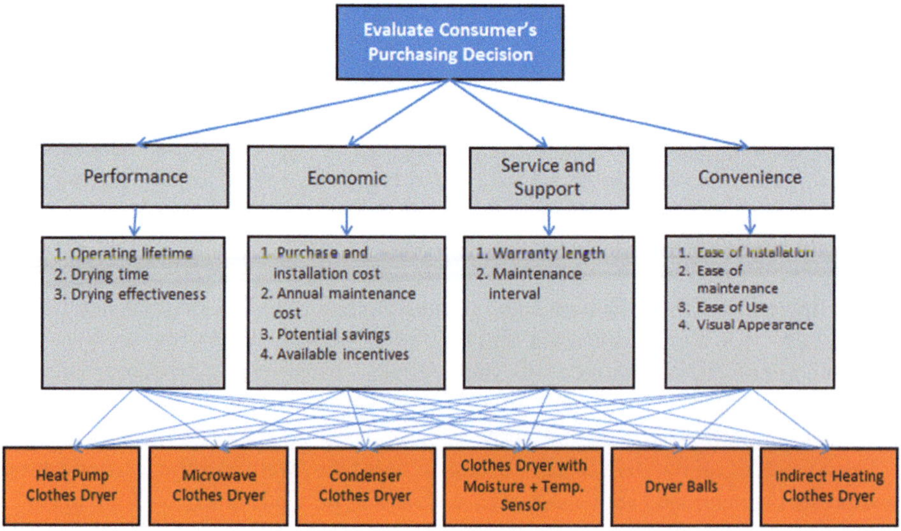

Fig. 12.3 HDM model for evaluating consumer preference

2. Cost—Relates to the purchase and installation cost of the clothes dryer, the annual maintenance cost, the potential electricity savings compared to your existing clothes dryer, and the incentives offered by utilities and energy efficiency program administrators.
3. Reliability—Relates to the service interval, the length of warranty, and the operating lifetime of the clothes dryer.
4. Convenience—Relates to the ease of installing, operating, and maintaining the clothes dryer. It also includes the visual appearance of the clothes dryer (i.e., does it match the clothes washer).

The sub-criteria under performance were

1. Operating Lifetime—The life expectancy of the clothes dryer.
2. Drying Time—The time the clothes dryer takes to complete one cycle.
3. Drying Effectiveness—Whether the clothes are completely dry after one cycle.

The sub-criteria under economic were

1. Purchase and Installation Cost—The fixed purchase and installation cost for a new clothes dryer.
2. Annual Maintenance Cost—The annual cost associated with maintaining the clothes dryer.
3. Potential Savings—The monetary savings obtained per annum after installing the clothes dryer.
4. Available Incentives—The incentives/rebates offered when purchasing the clothes dryer.

The sub-criteria under “service and support” were

1. Warranty Length—The duration of the warranty for the clothes dryer (in years).
2. Maintenance Interval—The expected duration between routine maintenance activities on the clothes dryer.

The sub-criteria under convenience were

1. Ease of Installation—How easy is it for the installation contractor or user to install and set up the clothes dryer.
2. Ease of Maintenance—How easy is it for a contractor or user to maintain the clothes dryer at the maintenance intervals.
3. Ease of Use—How easy is it to operate the clothes dryer effectively.
4. Visual Appearance—Must the clothes dryer matche the clothes washer.

12.5.2 Technologies

Three relevant clothes dryer technologies were indentified from the DOE Roadmap for Next-Generation Appliances, specifically the heat pump, microwave, and indirect heating clothes dryer [5]. An indirect heating clothes dryer obtains heat indirectly from a home hydronic heater system. The dryer control sensors (moisture, temperature) and condenser clothes dryer were identified from the Energy Star Market & Industry Scoping Report [3]. Dryer balls were identified as an alternative technology due to the lack of extensive knowledge about this technology. Dryer balls are wool, plastic, or rubber balls used to soften and separate clothing while drying, thereby reducing the drying time.

12.5.3 Research Instrument

The initial intention of the study was to only evaluate the purchasing decision of the consumer; however, this was extended to also evaluate manufacturers and to use energy efficient experts to quantify the technology level of the HDM. The responses and intended outcomes for each of the groups of respondents were:

1. Consumers completed the pairwise comparisons for the HDM in order to understand what their preferences were when purchasing a new clothes dryer. The research instrument used for pairwise comparisons is shown under Appendix 1.
2. Manufacturers completed the pairwise comparisons for the HDM in order to understand what they think consumer preferences are when purchasing a new clothes dryer. The research instrument used for pairwise comparisons is shown under Appendix 1.
3. Energy efficiency experts completed the technology survey, shown under Appendix 2, which was used to quantify each sub-criteria for each technology. Due to the fact that most of the technologies are not yet available in the US market, expert opinion was required in order to quantify the sub-criterion values for each technology. As an example, energy efficiency experts are asked to specify what the purchase and installation cost of each technology may be at time of release, since this information is not yet available.

12.5.4 Concept Test

In order to test the HDM, the instrument used for the pairwise comparisons was sent out to five Portland State University (PSU) students in the Engineering and Technology Management (ETM) department for completion. The results were satisfactory and the suggested comments were integrated into the research instrument. The main suggestions: were rewording some of the explanations of the sub-criteria and to make the second element of each pairwise comparison auto-complete.

12.6 Results

The research instrument was sent out to 12 consumers, 2 manufacturers, and 8 energy efficiency experts. From these contacts only 3 consumers, 1 manufacturer, and 5 energy efficiency experts responded completely and correctly.

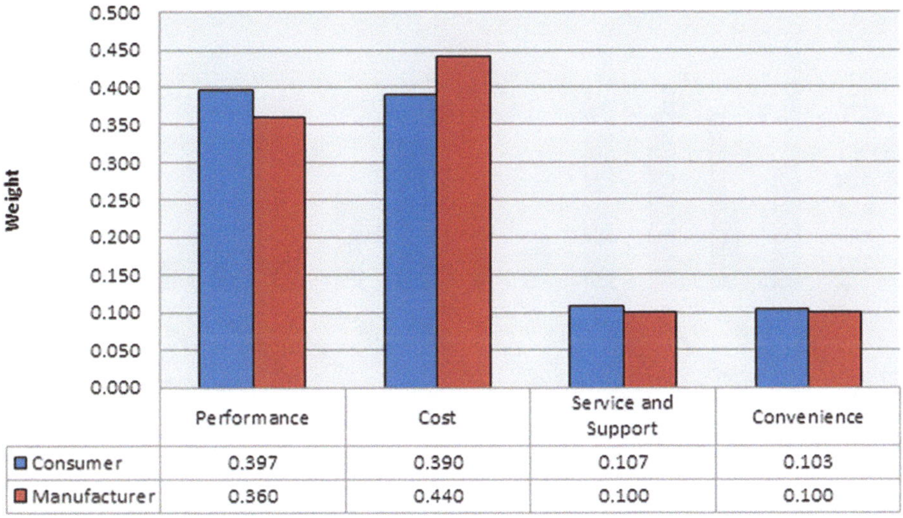

Fig. 12.4 Main criteria weights

The results can be seen in Figs. 12.4, 12.5, and 12.6. It can be noted that the weights from the consumers and manufacturer mainly align, with both Performance and Cost seen as the most important criteria. However, for the sub-criteria there are significant differences, specifically for Operating Lifetime, Drying Cycle Time, and Potential Savings. Consumers see Operating Lifetime as a much higher priority compared to manufacturers, as opposed to Drying Cycle Time and Potential Savings which are seen as more important to the manufacturer.

The overall scores for the technologies are shown in Fig. 12.6. It can be noted that the clothes dryer with moisture and temperature sensor has the highest score, although all scores are relatively close to one another. Again the consumer and manufacturer responses mainly align. It is understandable that the indirect heating clothes dryer has the lowest score on average, due to higher purchase and installation costs. Additionally, the clothes dryer with sensors had the lowest purchase and installation cost (excluding the dryer balls). The dryer balls were removed from the results due to lack of knowledge from the experts about the technology; however, this did not change the validity of the study since the technologies are the bottom level of the HDM.

12.7 Results Interpretation and Analysis

The intention of obtaining responses from both the consumers and manufacturers was to determine whether there was any misalignment between the understanding of consumer needs by the manufacturer and the actual consumer needs.

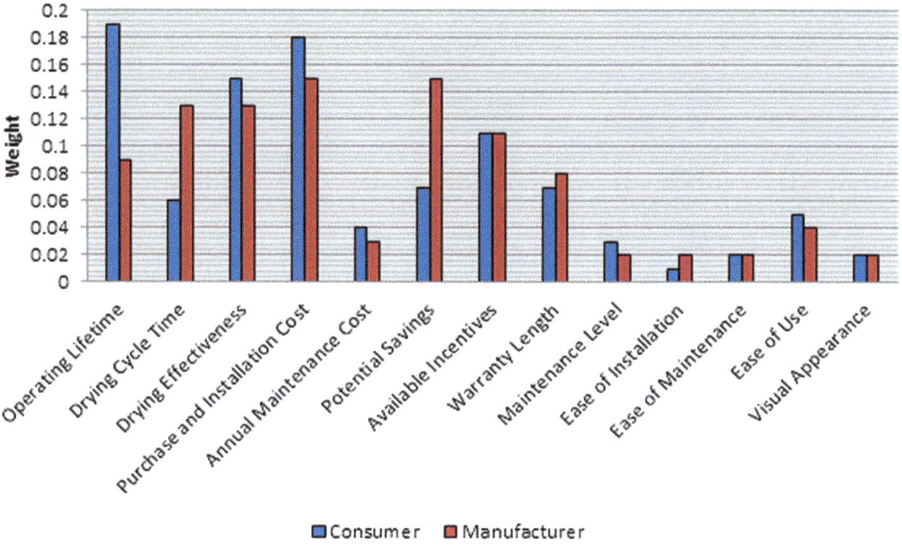

Fig. 12.5 Purchasing sub-criteria weights

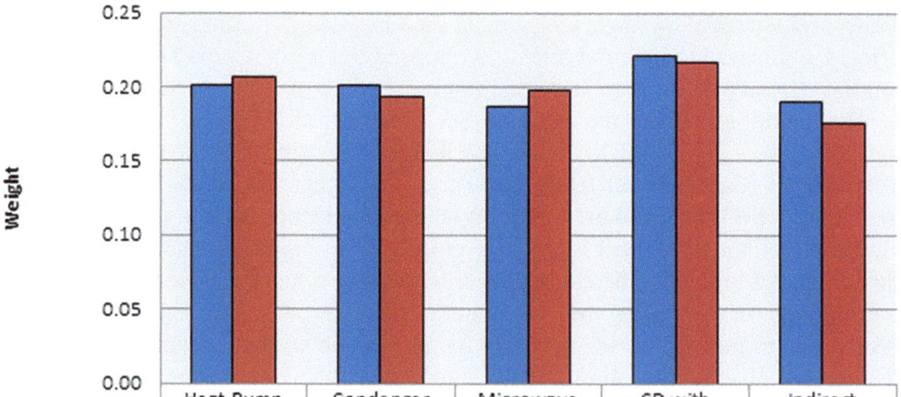

	Heat Pump CD	Condenser CD	Microwave CD	CD with sensors	Indirect Heating CD
Consumer	0.20	0.20	0.19	0.22	0.19
Manufacturer	0.21	0.19	0.20	0.22	0.18

Fig. 12.6 Technology scores

It was determined that there are some major misalignments between the sub-criteria, specifically associated with operating lifetime, drying cycle time, and potential savings. These misalignments could be a valuable insight for the manufacturers, so they can align their interests towards consumer needs before their products reach the market. However, since only one manufacturer responded, this may not be an accurate representation of all manufacturers. By increasing the number of manufacturer responses, there may be more of an alignment of sub-criteria.

TIAX previously conducted consumer feedback focus group studies for the DOE [14] and determined that the top three purchase criteria for clothes dryers were reliability, features and functions, and price. Additionally, consumers were not willing to pay premium prices if there were longer drying times and were not willing to pay more for certain increases in energy savings. Most of these findings align with the results that were obtained from the HDM study. Both operating lifetime and purchase and installation cost were the two highest sub-criteria, which aligns with two of the top three criteria determined from the TIAX study. Potential savings was also not seen as a critical sub-criterion from the HDM study, which again aligns with the TIAX study. The only disagreement between the two studies was that drying cycle time was not seen as a critical sub-criterion to the consumer in the HDM study, as opposed to what was discovered in the TIAX study.

12.8 Conclusion

Even with the limited amount of responses from consumers, manufacturers, and energy efficient experts, interesting results were found. It was clearly shown that the single manufacturer that responded did not have a complete understanding of their clothes dryer market. These misalignments should be taken into consideration by the manufacturer, however after a more extensive study of the market. The three consumer responses gave a basic idea of what the consumer preferences could be however no conclusions can be made with such a limit sample set. It was encouraging however to see that the results aligned with a much more extensive previous study done by TIAX.

Additional insight can be obtained from the energy efficiency expert responses. With the exclusion of dryer balls, the main differences between the technologies were drying cycle time, purchase and installation cost, and potential savings. Where the heat pump technology has a longer drying time and higher purchase and installation cost, it has a higher potential savings. Where the clothes dryer with sensors has a lower drying time and purchase and installation cost, it has a lower potential saving. There is a tradeoff between the different technologies that is clearly taken into account in this study. However, the focus of the energy efficiency program administrator is to reduce energy consumption. If potential savings is not of major concern to the consumer, how is the program administrator suppose to encourage the adoption of energy efficient clothes dryers?

At the moment, the main energy efficient technology of focus is the heat pump clothes dryer. The reason why this is not shown as the most preferred technology from the results of the study is mainly due to the high purchase and installation cost and the long drying time. By using the same purchase and installation cost of the clothes dryer with sensors for the heat pump clothes dryer, and the same drying cycle time, the heat pump clothes dryer becomes the most preferred alternative (however only slightly). The energy efficiency program administrator can influence only one of these criteria, purchase and installation cost, by offering incentives at the time of purchase. However, reducing the drying cycle time would require input from other players involved in the development of the technology.

12.9 Recommendations and Future Work

As part of the research instrument, respondents were given the opportunity to suggest any changes and to mention what may be missing from the criteria and sub-criteria. It was determine that "clothing wear and tear" was a critical missing sub-criterion. This could potentially be added under the performance criterion. Additionally, not all efficiency experts were familiar with dryer balls and indirect heating clothes dryers; therefore, the responses may not have been accurate. The description of each technology and possibly a link to an existing technology should be provided as part of the survey sent to the experts. Finally, there were very few responses and the study is most likely not an accurate representation of the actual market. By increasing the number of responses to a number that is deemed as an acceptable representation of the overall market would be required.

Appendix 1: Clothes Dryer Consumer Preference Study Research Instrument

Clothes Dryer Purchasing Decision Survey

The purpose of this survey is to evaluate different criteria and sub-criteria (factors) that you as the manufacturer believe is important to a consumer when they decide to purchase a new clothes dryer for their home.

12.9.1 Introduction

The following are the steps to follow in order to complete this survey:

- The comparisons in this survey are done by assigning 10 points between two criteria. As an example, the following is a comparison between Cost and Performance of a new clothes dryer:

8	Cost
2	Performance

- Since Cost is seen as 4 times more important than Performance, 8 points are assigned to Cost and 2 points to Performance. This can be verified by: Cost = 4 x Performance (8 = 4 × 2).
- If Cost and Performance are seen as equally important, assign 5 points to both the criteria.
- If only Cost is seen as important, assign 9 points to Cost and 1 point to Performance. 10 points cannot be assigned to only one criterion.

Q1. Which company are you representing in this survey?

12.9.2 Section 1: Main Criteria Comparison

The criteria to be compared in this section of the survey are described below:

1. **Performance**—Relates to the effectiveness of the clothes dryer. This includes the time it takes the clothes dryer to complete one cycle, whether the clothes are completely dry after one cycle, and the level of wear and tear on the clothing.
2. **Cost**—Relates to the purchase and installation cost of the clothes dryer, the annual maintenance cost, the potential electricity savings compared to your existing clothes dryer, and the incentives offered by utilities and energy efficiency program administrators.
3. **Reliability**—Relates to the service interval, the length of warranty, and the operating lifetime of the clothes dryer.
4. **Convenience**—Relates to the ease of installing, operating, maintaining the clothes dryer. It also includes the visual appearance of the clothes dryer (i.e., does it match the clothes washer).

Q2. Assign 10 points among Performance and Cost:

	Performance
	Cost

Q3. Assign 10 points among Performance and Reliability:

	Performance
	Reliability

Q4. Assign 10 points among Performance and Convenience:

	Performance
	Convenience

Q5. Assign 10 points among Cost and Reliability:

	Cost
	Reliability

Q6. Assign 10 points among Cost and Convenience:

	Cost
	Convenience

Q7. Assign 10 points among Reliability and Convenience:

	Reliability
	Convenience

12.9.3 Section 2: Performance Sub-criteria Comparison

The objective of this section is to compare the sub-criteria (factors) for Performance. The comparisons should be completed in the same manner as was previously done. The sub-criteria are as follows:

1. **Clothing Wear and Tear**—The level of wear and tear on the clothing per cycle.
2. **Drying Cycle Time**—The time it takes the clothes dryer to complete one cycle.
3. **Drying Effectiveness**—Whether the clothes are completely dry after one cycle.

Q8. Assign 10 points among "Clothing Wear and Tear" and Drying Cycle Time:

	Clothing Wear and Tear
	Drying Cycle Time

Q9. Assign 10 points among "Clothing Wear and Tear" and Drying Effectiveness:

	Clothing Wear and Tear
	Drying Effectiveness

Q10. Assign 10 points among Drying Cycle Time and Drying Effectiveness:

	Drying Cycle Time
	Drying Effectiveness

12.9.4 Section 3: Cost Sub-criteria Comparison

The objective of this section is to compare the sub-criteria (factors) for Cost. The comparisons should be completed in the same manner as was previously done. The sub-criteria are as follows:

1. **Purchase and Installation Cost**—The fixed purchase and installation cost for a new clothes dryer.
2. **Annual Maintenance Cost**—The annual cost associated with maintaining the clothes dryer.
3. **Potential Savings**—The monetary savings obtained per annum after installing the clothes dryer.
4. **Available Incentives**—The incentives/rebates offered when purchasing the clothes dryer.

Q11. Assign 10 points among "Purchase and Installation Cost" and Annual Maintenance Cost:

	Purchase and Installation Cost
	Annual Maintenance Cost

Q12. Assign 10 points among "Purchase and Installation Cost" and Potential Savings:

	Purchase and Installation Cost
	Potential Savings

Q13. Assign 10 points among "Purchase and Installation Cost" and Available Incentives:

	Purchase and Installation Cost
	Available Incentives

Q14. Assign 10 points among Annual Maintenance Cost and Potential Savings:

	Annual Maintenance Cost
	Potential Savings

Q15. Assign 10 points among Annual Maintenance Cost and Available Incentives:

	Annual Maintenance Cost
	Available Incentives

Q16. Assign 10 points among Potential Savings and Available Incentives:

	Potential Savings
	Available Incentives

12.9.5 Section 4: Reliability Sub-criteria Comparison

The objective of this section is to compare the sub-criteria for Reliability. The comparisons should be completed in the same manner as was previously done. The sub-criteria are as follows:

1. **Operating Lifetime**—The life expectancy of the clothes dryer.
2. **Warranty Length**—The duration of the warranty for the clothes dryer (in years).
3. **Maintenance Interval**—The expected duration between routine maintenance activities on the clothes dryer.

Q17. Assign 10 points among Operating Lifetime and Warranty Length:

	Operating Lifetime
	Warranty Length

Q18. Assign 10 points among Operating Lifetime and Maintenance Interval:

	Operating Lifetime
	Maintenance Interval

Q19. Assign 10 points among Warranty Length and Maintenance Interval:

	Warranty Length
	Maintenance Interval

12.9.6 Section 5: Convenience Sub-criteria Comparison

The objective of this section is to compare the sub-criteria (factors) for Convenience. The comparisons should be completed in the same manner as was done previously. The sub-criteria are as follows:

1. Ease of Installation—How easy is it for the installation contractor or user to install and set up the clothes dryer.
2. Ease of Maintenance—How easy is it for a contractor or user to maintain the clothes dryer at the maintenance intervals.
3. Ease of Use—How easy is it to operate the clothes dryer effectively.
4. Visual Appearance—Must the clothes dryer matches the clothes washer.

Q20. Assign 10 points among Ease of Installation and Ease of Maintenance:

	Ease of Installation
	Ease of Maintenance

Q21. Assign 10 points among Ease of Installation and Ease of Use:

	Ease of Installation
	Ease of Use

Q22. Assign 10 points among Ease of Installation and Visual Appearance:

	Ease of Installation
	Visual Appearance

Q23. Assign 10 points among Ease of Maintenance and Ease of Use:

	Ease of Maintenance
	Ease of Use

Q24. Assign 10 points among Ease of Maintenance and Visual Appearance:

	Ease of Maintenance
	Visual Appearance

Q25. Assign 10 points among Ease of Use and Visual Appearance:

	Ease of Use
	Visual Appearance

12.9.7 General Questions

Q26. Are there any additional comments you would like to add?

Appendix 2: Clothes Dryer Consumer Preference Study Technology Survey

Clothes Dryer Technology Survey

The purpose of this survey is to quantify criteria for clothes dryer technologies. The result of this survey will be integrated into a consumer purchasing decision model to determine consumer preferences.

There are six clothes dryer categories that will be compared throughout this survey. The intention is to compare the best known dryer technology under each of the following categories:

1. Heat Pump Clothes Dryer
2. Condenser Clothes Dryer
3. Microwave/Radio Frequency (RF) Clothes Dryer
4. Clothes Dryer with Moisture and Temperature Sensor
5. Indirect Heating Clothes Dryer—obtains heat indirectly from a home hydronic heater system
6. Dryer Balls—wool, plastic, or rubber balls used to soften and separate clothing while drying, thereby reducing drying time

Section 1: Performance Sub-criteria

For this section, the following sub-criteria for Performance will be assessed:

1. **Clothing Wear and Tear**—The level of wear and tear on the clothing per cycle.
2. **Drying Cycle Time**—The time it takes the clothes dryer to complete one cycle.
3. **Drying Effectiveness**—Whether the clothes are completely dry after one cycle.

Q1. **Clothing Wear and Tear**—The level of wear and tear on the clothing per cycle.

	Low	Medium	High	Don't know
Heat Pump Clothes Dryer	❍	❍	❍	❍
Condenser Clothes Dryer	❍	❍	❍	❍
Microwave/RF Clothes Dryer	❍	❍	❍	❍
Clothes Dryer with Moisture and Temperature Sensor	❍	❍	❍	❍
Indirect Heating Clothes Dryer	❍	❍	❍	❍
Dryer Balls	❍	❍	❍	❍

Q2. **Drying Cycle Time**—The time it takes the clothes dryer to complete one cycle:

	Less than 30 min	Between 30 min and 1 h	Between 1 and 2 h	Between 2 and 3 h	More than 3 h	Don't know
Heat Pump Clothes Dryer	❍	❍	❍	❍	❍	❍
Condenser Clothes Dryer	❍	❍	❍	❍	❍	❍
Microwave/RF Clothes Dryer	❍	❍	❍	❍	❍	❍
Clothes Dryer with Moisture and Temperature Sensor	❍	❍	❍	❍	❍	❍
Indirect Heating Clothes Dryer	❍	❍	❍	❍	❍	❍
Dryer Balls	❍	❍	❍	❍	❍	❍

Q3. **Drying Effectiveness**—Whether the clothes are completely dry after one cycle:

	Completely wet	Mostly wet	Slightly wet	Completely dry	Don't know
Heat Pump Clothes Dryer	❍	❍	❍	❍	❍
Condenser Clothes Dryer	❍	❍	❍	❍	❍
Microwave/RF Clothes Dryer	❍	❍	❍	❍	❍
Clothes Dryer with Moisture and Temperature Sensor	❍	❍	❍	❍	❍
Indirect Heating Clothes Dryer	❍	❍	❍	❍	❍
Dryer Balls	❍	❍	❍	❍	❍

Section 2: Cost Sub-criteria

For this section, the following sub-criteria for Cost will be assessed:

1. **Purchase and Installation Cost**—The fixed purchase and installation cost for the dryer technology.

2. **Annual Maintenance Cost**—The annual cost associated with maintaining the dryer technology (percentage of purchase cost).
3. **Potential Savings**—The monetary savings obtained per annum after installing the dryer technology (percentage of purchase cost).
4. **Available Incentives**—The incentives/rebates offered when purchasing the dryer technology.

Q4. **Purchase and Installation Cost**—The fixed purchase and installation cost for the dryer technology:

	Less than $500	Between $500 and $1,000	Between $1,000 and $1,500	Between 1,500 and $2,000	Greater than $2,000	Don't know
Heat Pump Clothes Dryer	❍	❍	❍	❍	❍	❍
Condenser Clothes Dryer	❍	❍	❍	❍	❍	❍
Microwave/RF Clothes Dryer	❍	❍	❍	❍	❍	❍
Clothes Dryer with Moisture and Temperature Sensor	❍	❍	❍	❍	❍	❍
Indirect Heating Clothes Dryer	❍	❍	❍	❍	❍	❍
Dryer Balls	❍	❍	❍	❍	❍	❍

Q5. **Annual Maintenance Cost**—The annual cost associated with maintaining the dryer technology (percentage of purchase cost):

	Less than 10 %	Between 10 % and 20 %	Between 20 % and 30 %	Between 30 % and 40 %	Greater than 40 %	Don't know
Heat Pump Clothes Dryer	❍	❍	❍	❍	❍	❍
Condenser Clothes Dryer	❍	❍	❍	❍	❍	❍
Microwave/RF Clothes Dryer	❍	❍	❍	❍	❍	❍
Clothes Dryer with Moisture and Temperature Sensor	❍	❍	❍	❍	❍	❍
Indirect Heating Clothes Dryer	❍	❍	❍	❍	❍	❍
Dryer Balls	❍	❍	❍	❍	❍	❍

Q6. **Potential Savings**—The monetary savings obtained per annum after installing the dryer technology (percentage of purchase cost):

	Less than 1 %	Between 1 % and 5 %	Between 5 % and 10 %	Between 10 % and 15 %	Greater than 15 %	Don't know
Heat Pump Clothes Dryer	❍	❍	❍	❍	❍	❍
Condenser Clothes Dryer	❍	❍	❍	❍	❍	❍
Microwave/RF Clothes Dryer	❍	❍	❍	❍	❍	❍
Clothes Dryer with Moisture and Temperature Sensor	❍	❍	❍	❍	❍	❍
Indirect Heating Clothes Dryer	❍	❍	❍	❍	❍	❍
Dryer Balls	❍	❍	❍	❍	❍	❍

Q7. **Available Incentives**—Do you think a cash incentive from a utility or an energy efficiency program administrator would increase the market share for the dryer technology?

	Yes	No	Don't know
Heat Pump Clothes Dryer	❍	❍	❍
Condenser Clothes Dryer	❍	❍	❍
Microwave/RF Clothes Dryer	❍	❍	❍
Clothes Dryer with Moisture and Temperature Sensor	❍	❍	❍
Indirect Heating Clothes Dryer	❍	❍	❍
Dryer Balls	❍	❍	❍

12.9.8 Section 3: Reliability Sub-criteria

For this section, the following sub-criteria for Reliability will be assessed:

1. **Operating Lifetime**—The life expectancy of the dryer technology,
2. **Warranty Length**—The duration of the warranty for the dryer technology,
3. **Maintenance Interval**—The expected duration between routine maintenance activities on the dryer technology.

Q8. **Operating Lifetime**—The life expectancy of the dryer technology,

	Less than 5 years	Between 5 and 10 years	Between 10 and 15 years	Between 15 and 20 years	Greater than 20 years	Don't know
Heat Pump Clothes Dryer	❍	❍	❍	❍	❍	❍
Condenser Clothes Dryer	❍	❍	❍	❍	❍	❍
Microwave/RF Clothes Dryer	❍	❍	❍	❍	❍	❍
Clothes Dryer with Moisture and Temperature Sensor	❍	❍	❍	❍	❍	❍
Indirect Heating Clothes Dryer	❍	❍	❍	❍	❍	❍
Dryer Balls	❍	❍	❍	❍	❍	❍

Q9. **Warranty Length**—The duration of the warranty for the dryer technology:

	Less than 1 year	Between 1 and 2 years	Between 2 and 3 years	Between 3 and 4 years	Greater than 4 years	Don't know
Heat Pump Clothes Dryer	❍	❍	❍	❍	❍	❍
Condenser Clothes Dryer	❍	❍	❍	❍	❍	❍
Microwave/RF Clothes Dryer	❍	❍	❍	❍	❍	❍
Clothes Dryer with Moisture and Temperature Sensor	❍	❍	❍	❍	❍	❍
Indirect Heating Clothes Dryer	❍	❍	❍	❍	❍	❍
Dryer Balls	❍	❍	❍	❍	❍	❍

Q10. **Maintenance Interval**—The expected duration between routine maintenance activities on the dryer technology:

	Less than 1 year	Between 1 and 2 years	Between 2 and 3 years	Between 3 and 4 years	Greater than 4 years	Don't know
Heat Pump Clothes Dryer	❍	❍	❍	❍	❍	❍
Condenser Clothes Dryer	❍	❍	❍	❍	❍	❍
Microwave/RF Clothes Dryer	❍	❍	❍	❍	❍	❍
Clothes Dryer with Moisture and Temperature Sensor	❍	❍	❍	❍	❍	❍
Indirect Heating Clothes Dryer	❍	❍	❍	❍	❍	❍
Dryer Balls	❍	❍	❍	❍	❍	❍

12.9.9 Section 4: Convenience Sub-criteria

For this section, the following sub-criteria for Convenience will be assessed

1. **Ease of Installation**—How easy is it for the installation contractor or user to install and set up the dryer technology,
2. **Ease of Maintenance**—How easy is it for a contractor or user to maintain the dryer technology at the maintenance intervals,
3. **Ease of Use**—How easy is it to operate the dryer technology effectively.

Q11. **Ease of Installation**—How easy is it for the installation contractor or user to install and set up the dryer technology?

	Difficult	Somewhat difficult	Neutral	Somewhat easy	Easy	Don't know
Heat Pump Clothes Dryer	❍	❍	❍	❍	❍	❍
Condenser Clothes Dryer	❍	❍	❍	❍	❍	❍
Microwave/RF Clothes Dryer	❍	❍	❍	❍	❍	❍
Clothes Dryer with Moisture and Temperature Sensor	❍	❍	❍	❍	❍	❍
Indirect Heating Clothes Dryer	❍	❍	❍	❍	❍	❍
Dryer Balls	❍	❍	❍	❍	❍	❍

Q12. **Ease of Maintenance**—How easy is it for a contractor or user to maintain the dryer technology at the maintenance intervals?

	Difficult	Somewhat difficult	Neutral	Somewhat easy	Easy	Don't know
Heat Pump Clothes Dryer	❍	❍	❍	❍	❍	❍
Condenser Clothes Dryer	❍	❍	❍	❍	❍	❍
Microwave/RF Clothes Dryer	❍	❍	❍	❍	❍	❍
Clothes Dryer with Moisture and Temperature Sensor	❍	❍	❍	❍	❍	❍
Indirect Heating Clothes Dryer	❍	❍	❍	❍	❍	❍
Dryer Balls	❍	❍	❍	❍	❍	❍

Q13. **Ease of Use**—How easy is it to operate the dryer technology effectively?

	Difficult	Somewhat difficult	Neutral	Somewhat easy	Easy	Don't know
Heat Pump Clothes Dryer	❍	❍	❍	❍	❍	❍
Condenser Clothes Dryer	❍	❍	❍	❍	❍	❍
Microwave/RF Clothes Dryer	❍	❍	❍	❍	❍	❍

	Difficult	Somewhat difficult	Neutral	Somewhat easy	Easy	Don't know
Clothes Dryer with Moisture and Temperature Sensor	❍	❍	❍	❍	❍	❍
Indirect Heating Clothes Dryer	❍	❍	❍	❍	❍	❍
Dryer Balls	❍	❍	❍	❍	❍	❍

12.9.10 Section 5: General Questions

Q14. Are there any sub-criteria that you think were not taken into account?

Q15. Are there any dryer technologies that you think were not taken into account?

Q16. Are there any additional comments that you would like to mention?

References

1. U.S. Energy Information Organization, (Residential Energy Consumption Survey (RECS), 2009), http://www.eia.gov/consumption/residential/. Accessed: 06 Oct 2013
2. U.S. Energy Information Administration, (Residential Energy Consumption Survey (RECS)), http://www.eia.gov/consumption/residential/data/2009/. Accessed 3 Dec 2013
3. Energy Star, ENERGY STAR Market & Industry Scoping Report: Residential Clothes Dryers: November 2011 (2011)
4. U.S. Department of Energy, Technical Support Document: Energy Efficiency Program for Consumer Products and Commercial and Industrial Equipment: Residential Clothes Dryers and Room Air Conditioner. no. (2011)

5. U.S. Department of Energy: Energy Efficiency and Renewable Energy, Roadmap for Next-Generation Appliances. no. (2012)
6. R. Decker, M. Trusov, Estimating aggregate consumer preferences from online product reviews. Int. J. Res. Mark. **27**(4), 293–307 (2010)
7. S. Nikou, J. Mezei, Evaluation of mobile services and substantial adoption factors with Analytic Hierarchy Process (AHP). Telecomm. Policy **37**(10), 915–929 (2013)
8. K.C. van Blommestein, T.U. Daim, Residential energy efficient device adoption in South Africa. Sustain. Energy Technol. Assess. **1**, 13–27 (2013)
9. N. Basoglu, T.U. Daim, U. Topacan, Determining patient preferences for remote monitoring. J. Med. Syst. **36**(3), 1389–1401 (2012)
10. D. Fenwick, T.U. Daim, Choosing a hybrid car using a hierarchical decision model. Int. J. Sustain. Soc **3**(3), 243 (2011)
11. R. Mulye, An empirical comparison of three variants of the AHP and two variants of conjoint analysis. J. Behav. Decis. Mak. **11**(4), 263–280 (1998)
12. S.W. Scholz, M. Meissner, R. Decker, measuring consumer preferences for complex products: a compositional approach based on paired comparisons. J. Mark. Res. **47**(4), 685–698 (2010)
13. D. Knoke, S. Yang, *Social Network Analysis*, vol. 2 (Sage, Los Angeles, 2008)
14. P. Pescatore, P. Carbone, *High Efficiency High Performance Clothes Dryer* (TIAX, Cambridge, 2005)

Chapter 13
Technology Intelligence on Clothes Dryers

Kevin C. van Blommestein and Tugrul U. Daim

Abstract In the US in 2009, almost 80 % of households had clothes dryers, of which approximately 80 % of the clothes dryers were electric (ENERGY STAR Market & Industry Scoping Report: Residential Clothes Dryers, 2011). Due to the large market of clothes dryers and the current inefficient dryer options in the US, focus has shifted on improving clothes dryer efficiency. The Energy Star Market and Industry Scoping Report (ENERGY STAR Market & Industry Scoping Report: Residential Clothes Dryers, 2011) evaluated potential savings options for clothes dryers, which included the drum, motor, dryer control, heat recovery, heat generation, and several other options. These potential clothes dryer feature upgrades are discussed under the clothes dryer adoption study section of this report.

13.1 Landscape Analysis

Figure 13.1 represents a perceptual map of the clothes dryer technologies that were under review by the US Department of Energy (DOE) [1]. The y-axis represents the purchase and installation cost ($2008) and the x-axis represents the Efficiency (EF). The bottom right area of the map represents the preferred position. It can be noted that majority of the vented heat pump clothes dryer technologies are close to this preferred position on the map, with a low purchase and installation cost and high efficiency. At the time of this study, the updated test procedure for residential clothes dryers was not yet finalized.

K.C. van Blommestein
Portland State University, 1900 SW 4th Avenue 50-02, Portland, OR 97207, USA
e-mail: kevin9@pdx.edu

T.U. Daim (✉)
Department of Engineering and Technology Management, Portland State University,
1900 SW 4th Avenue 50-02, Portland, OR 97201, USA
e-mail: tugrul@etm.pdx.edu

T.U. Daim et al. (eds.), *Policies and Programs for Sustainable Energy Innovations*, Innovation, Technology, and Knowledge Management,
DOI 10.1007/978-3-319-16033-7_13

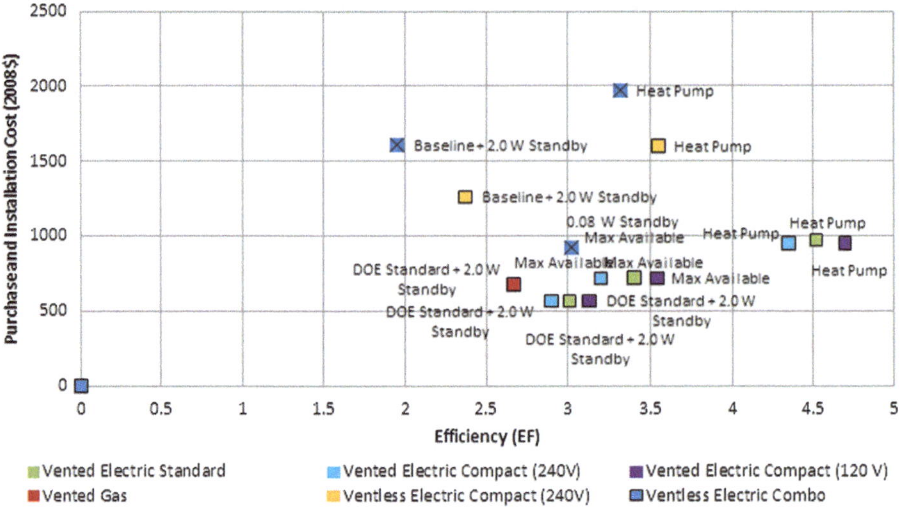

Fig. 13.1 Clothes dryer perceptual map [1]

13.2 Patent Analysis

Due to the advancements in Europe relating the adoption of energy efficient clothes dryers, a comparative patent analysis study was conducted to evaluate the difference between Europe and the US. Figure 13.2 illustrates the cumulative patent counts for heat pump clothes dryers in both Europe and the US, together with important events that took place from 1980 to 2011. It can be noted that the patent count for Europe started to increase around 4 years before the US. Additionally, the European Union (EU) Energy Label was introduced for clothes dryers in 1995; however, there has been no Energy Star label introduced yet in the US. Over recent years however, developments in both the US and Europe are increasing relatively at the same rate.

Figure 13.3 illustrates the top assignees of patents in both Europe and the US.

A logistic growth curve was used to forecast the development of the heat pump clothes dryer technology in the US, as shown in Fig. 13.4. The maximum cumulative patent count (800) was established by determining the cumulative patent count of an existing saturated technology (resistive heat element clothes dryer). However, even though the patent count for this saturated technology has reduced substantially over recent years, patents are still being published. A value of 800 + 20 % and 800 − 20 % was therefore also included as upper and lower bounds for future patent counts. This method of forecasting gives an idea to the potential development of a technology; however, it relies heavily on the assumption that the heat pump clothes dryer technology will have the same interest over time as the saturated technology, which is not always the case. Additionally, the development may occur at a faster or slower rate than what is depicted by the logistic growth curve.

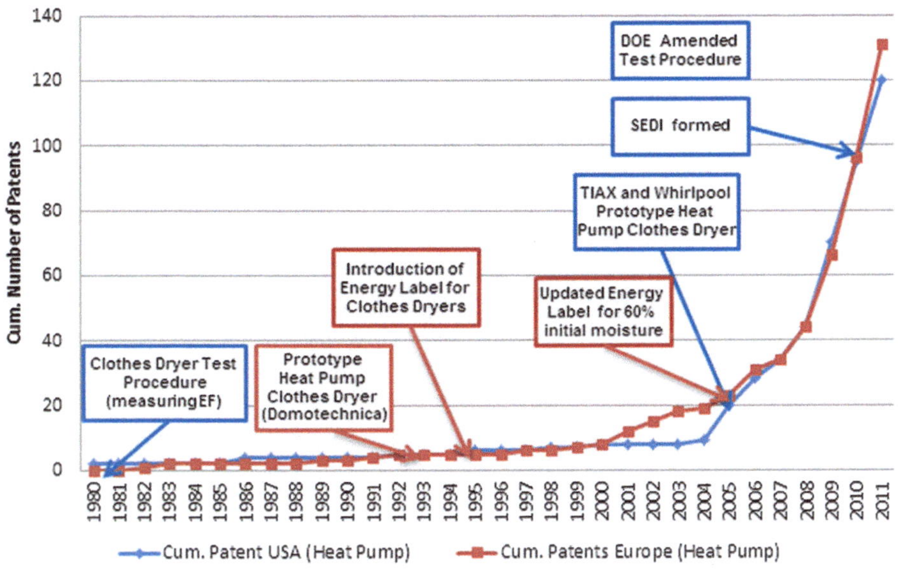

Fig. 13.2 Cumulative patent count (heat pump clothes dryer)

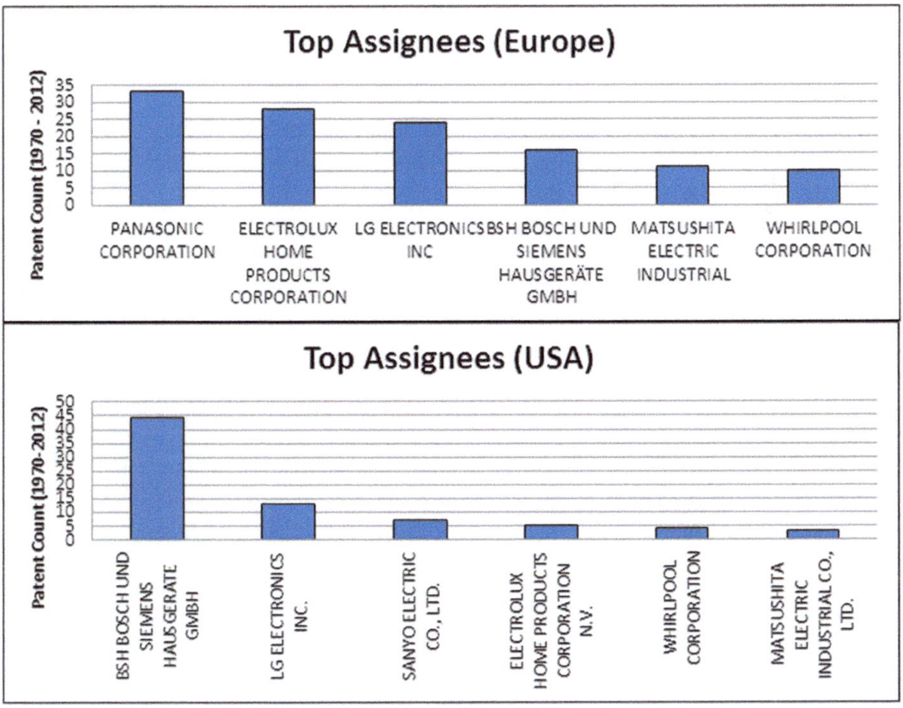

Fig. 13.3 Top patent assignees (heat pump clothes dryer)

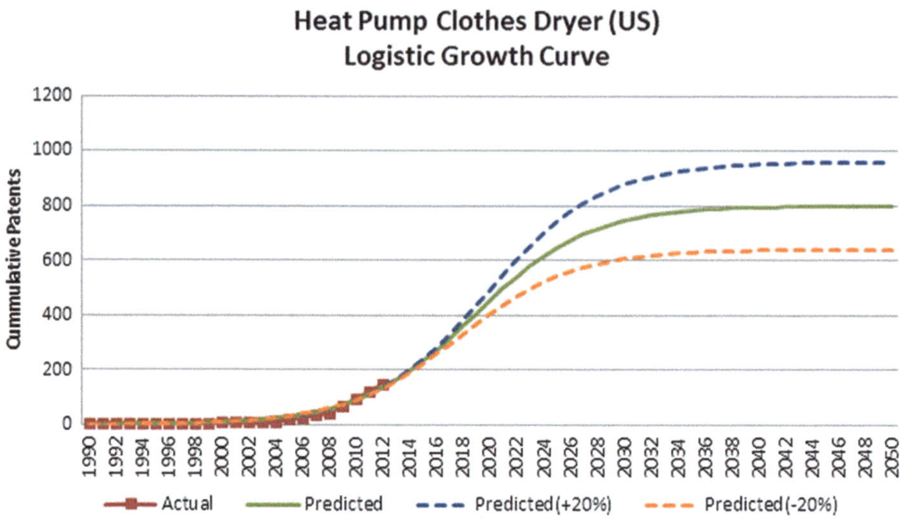

Fig. 13.4 Logistic growth curve forecast (heat pump clothes dryer)

13.3 System Dynamics Model (Concept)

13.3.1 Introduction

The purpose of the study was to model the adoption of both energy efficient and nonenergy efficient clothes washers. The initial intention was to model both of these adoptions using the standard Bass diffusion model introduced by Sterman and Sterman [2], with a few minor changes for a dynamic market size. It was then realized that the adoption is different for energy efficient and nonenergy efficient clothes washers. Nonenergy efficient clothes washers have been in the residential sector for many years and it could be assumed that the adoption is close to saturation and will occur at a constant adoption rate. For energy efficient clothes washers, the Bass diffusion model is more applicable since it could be seen as a new technology introduced into the market. The following potential adopters of clothes washers were identified:

- Purchases of new homes being built
- Home owners currently not owning a clothes washer and purchasing one
- Replacement of existing clothes washers

For each of these situations, a decision needs to be made whether to purchase an energy efficient clothes washer or not. A clothes washer labeled by Energy Star (ES) was used to categorize efficient and nonefficient washers. In order to accommodate the decision between ES and Non-ES (NES) clothes washers, the model was created with three stocks, one for potential adopters, and the other two for NES and ES product adopters. To make things easier a simple model was first developed with only the option to adopt a clothes washer, not taking into account whether it was ES or not. The time period of the model was set from 1990 to 2011. The first ES clothes washer was released in the US in 1997 which was setup in the model as a step function.

13.3.1.1 Reference Behavior Pattern

Data was gathered from multiple sources in order to create the Reference Behavior Pattern (RBP). To understand the adoption of clothes washers in the US, the following data was collected:

- Total clothes washer shipments in the US from 1990 to 2011 [3]
- Percentage of shipments that were ES labeled in the US from 1990 to 2011 [4]
- Total number of new housing in the US from 1990 to 2011 [3, 5]
- Total number of households in the US with and without clothes washers from 1990 to 2011 [6]

Figure 13.5 illustrates the shipment of clothes washers in the US per year, including the number of shipments for ES and NES clothes washers. As previously mentioned, the first ES clothes washer was released in the US in 1997 as shown in the figure. This RBP will be used to verify and validate the flows into the ES and NES stocks. Figure 13.6 illustrates the number of households in the US with and without clothes washers. This will be used to verify and validate the total number of adopters (stocks) from 1990 to 2011. Figure 13.7 illustrates the number of newly constructed single and multifamily housing as well as mobile home replacements. This will be used to verify and validate the growth rate flow increasing the size of potential adopters.

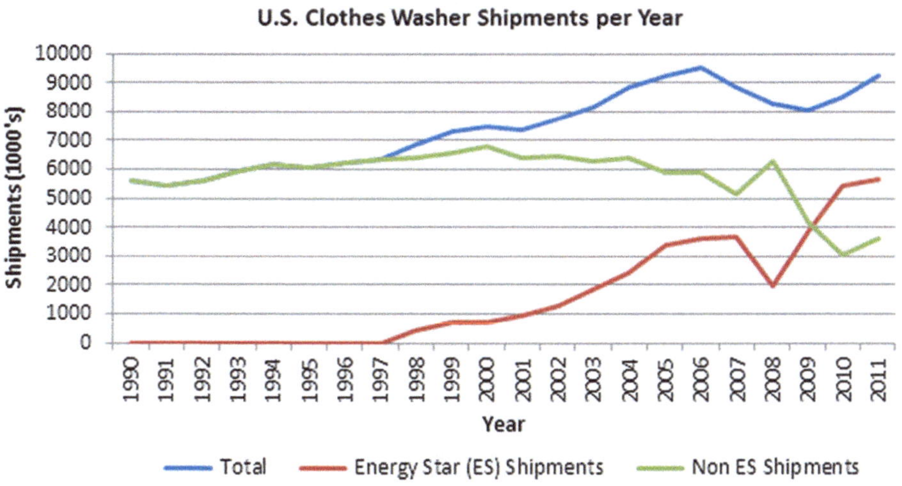

Fig. 13.5 Shipments of clothes washers per year (flows)

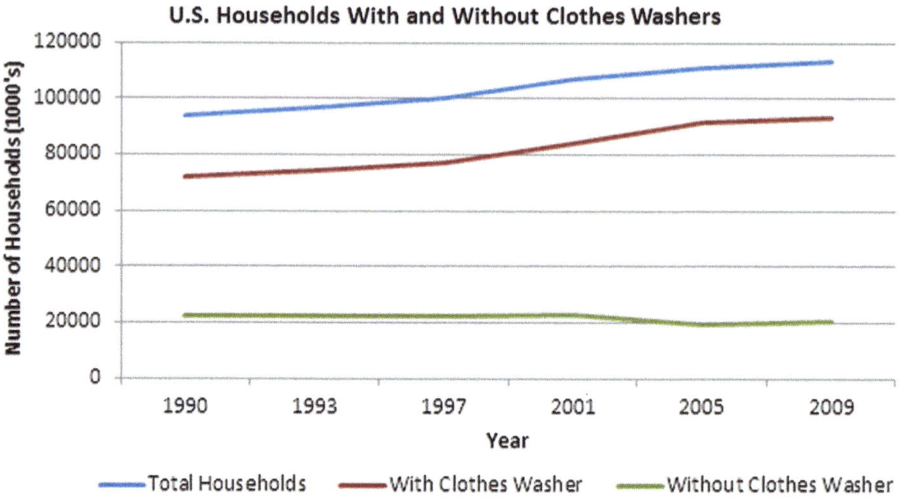

Fig. 13.6 US households owning or not owning clothes washers (Stocks)

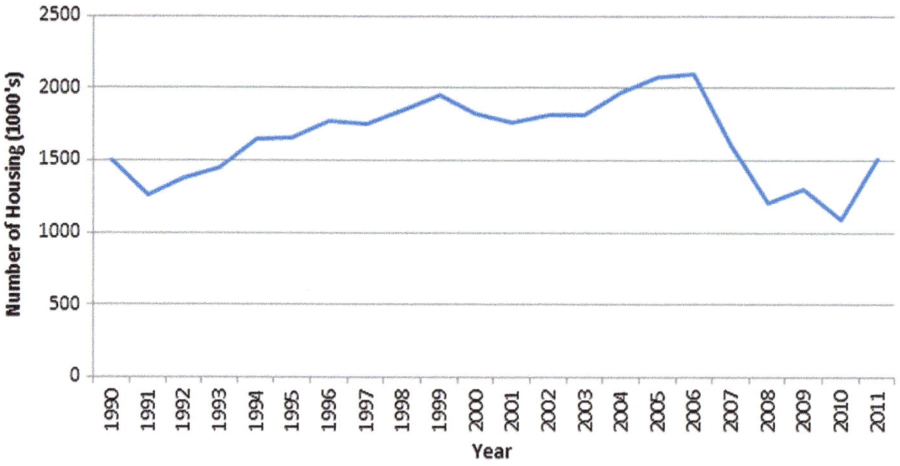

Fig. 13.7 New housing in the US (flow)

13.3.2 Initial Model

Figure 13.8 shows the initial model for the adoption of clothes washers with two feedback loops. The first balancing feedback loop is the adoption of clothes washers and the second balancing feedback loop is the replacement of clothes washers. The reinforcing feedback loop from the Bass model was not taken into account in this

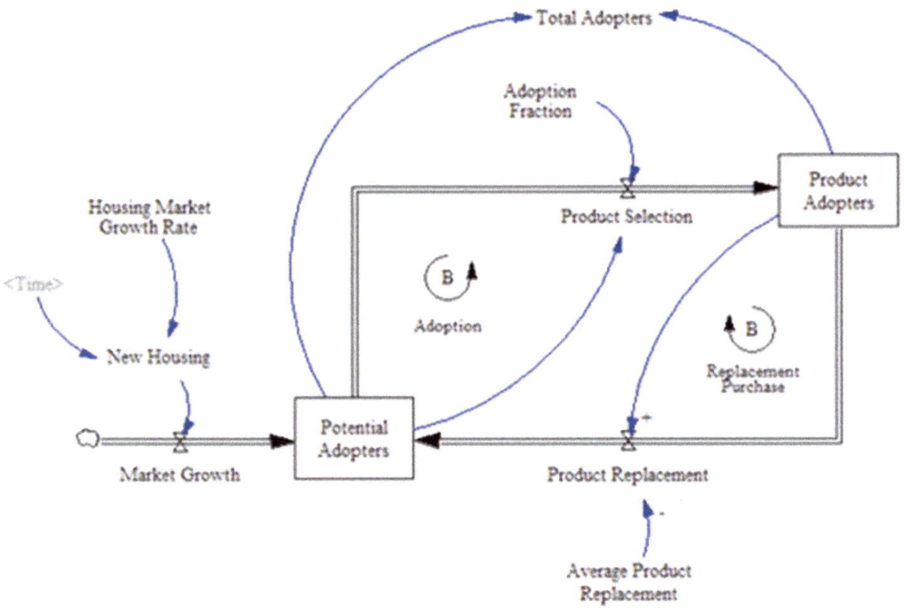

Fig. 13.8 Simple clothes washer adoption model

model since the adoption is based on the replacement of clothes washers and the requirements for new housing. Word of Mouth (WOM) is seen to have little impact on this adoption. The flows and stocks within the model are the following:

- Potential Adopters—These are the households that have not yet adopted a clothes washer. The initial value for this stock was set to 22.3 million, which is the households without clothes washers in 1990.
- Product Adopters—These are the households that have already adopted a clothes washer. The initial value for this stock was set to 71.7 million, which is the households with clothes washers in 1990. The RBP for this stock was shown in Fig. 13.7.
- Market Growth—The market growth is based on the new housing per year (single and multifamily housing and mobile home replacements). Since we expect this value to increase annually, a housing market growth rate is also used. Since we are creating a model from 1990 to 2011, the 10 years before 1990 were used to determine an initial growth rate of 2.73 %. This, however, fluctuates with many other factors and may not be the best approach but will be improved in this exercise. The initial value used in the equation for the flow is 1.496 million, which is the number of new housing in 1990. The RBP for this flow was shown in Fig. 13.8.
- Product Selection—These are households selecting (adopting) the clothes washer, which is a fraction of the potential adopters. The initial adoption fraction was calculated by assuming that all replacement households and new housing

will require clothes washers. The initial adoption fraction was therefore calculated to be 27 % ((initial new housing + initial product replacement)/initial potential adopters). The RBP for this flow is the total number of shipments of clothes washers in the US which was shown in Fig. 13.6.

- Product Replacement—These are households replacing their clothes washers after an average product replacement time. The initial value for the replacement time was set to the expected lifespan of a clothes washer of 16 years.

13.3.2.1 Initial Model Results

The model was too calibrated to match the actual adoption as follows and the results are shown in Fig. 13.8:

- A variable was added that would reduce the number of New Housing during the recession with a value between 0 and 1. The variable decreases the number of new houses to half from 2006 to 2008, remains at a half until 2010, and then starts to increase back up to 1 using the following function:

  1-RAMP (0.25, 2006, 2008) + RAMP (0.25, 2010, 2012)

- The housing market growth rate was reduced to 2 %
- The adoption fraction was reduced to 25 %
- The average product replacement was changed to 13 years

As can be seen in Fig. 13.9, the New Housing output aligns better with the RBP and the same can be said for Product Selection and Product Adopters. The issue remains with the Total Adopters increasing quicker than the RBP, as previously explained.

13.3.3 Final Model: ES and Non-ES Clothes Washer Adoption

As discussed in the introduction, the intention of this study was to separate ES and NES Product Adopters. It was assumed that the total adoption rate of clothes washers was the same as the simple model; however, it is now distributed between two options. This was created by a coflow between NES Product Selection and ES Product Selection. The NES Production Selection was calculated using the Adoption Fraction of 0.25 from the simple model, multiplied by the number of potential adopters minus the ES Product Selection. This simple model was used for the adoption of NES clothes washers since they have been in the market for a long time. Since ES clothes washers were new to the market in 1997, advertising was required to initially encourage the adoption of the product, followed by adopters encouraging other adopters by WOM. This would be a classic case of the Bass model. The selection of an ES clothes washer was therefore setup with the adoption from advertising and the adoption from WOM, as shown in Chap. 9 of Sterman [2]. The overall

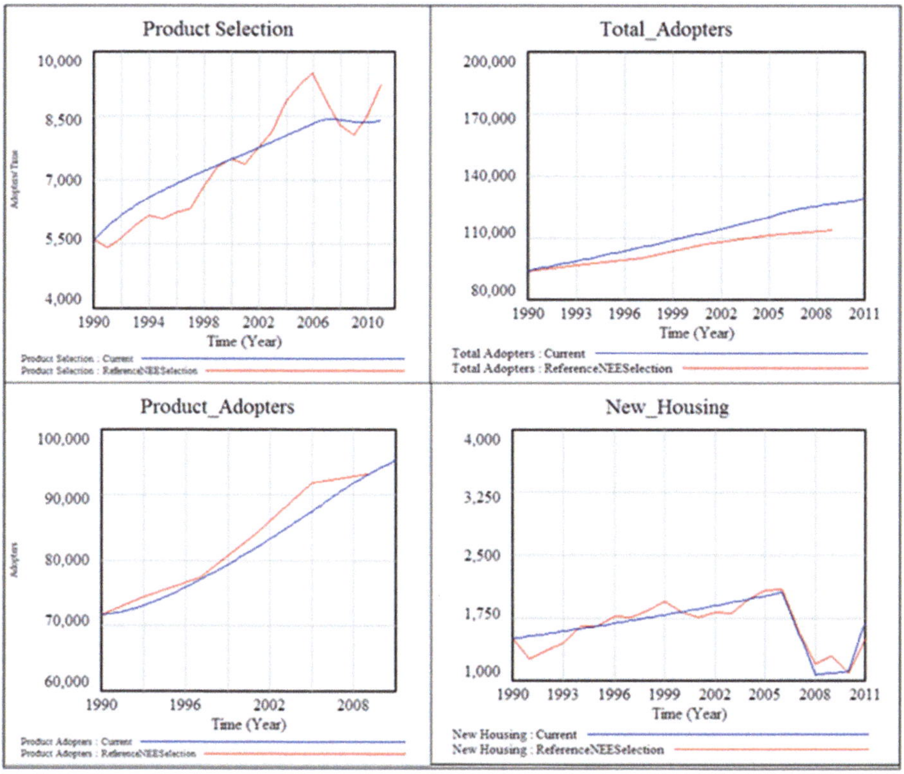

Fig. 13.9 Updated result for simple adoption model

model is shown in Fig. 13.9. There are six balancing feedback loops and one reinforcing loop, where one of the balancing loops is not shown due to the use of a shadow variable. This loop is as follows:

ES Product Selection increases → NES Product Selection decreases → NES Product Adopters decreases → NES Product Replacement decreases → Potential Adopters decreases → ES Product Selection decreases.

The model was also setup with two binary parameters, ES Option Available and NES Option Available. The ES Option Available is set as a step function at 1997 when the first ES clothes washer was released. The NES Option Available is always set to one but can be changed to see the effect of a possible standard preventing the purchase of NES products (Fig. 13.10).

The same values were used for the NES clothes washer selection as was determined for the simple model. The only values that were required for the updated model were the advertising Effect (p) and WOM Effect (q) for the ES clothes washer selection. A common approach for selecting these values is based on previous adoption of an earlier model of the same or similar technology. Based on clothes washer adoption from 1923 to 1971 the p value that was initially used was 0.016 and the

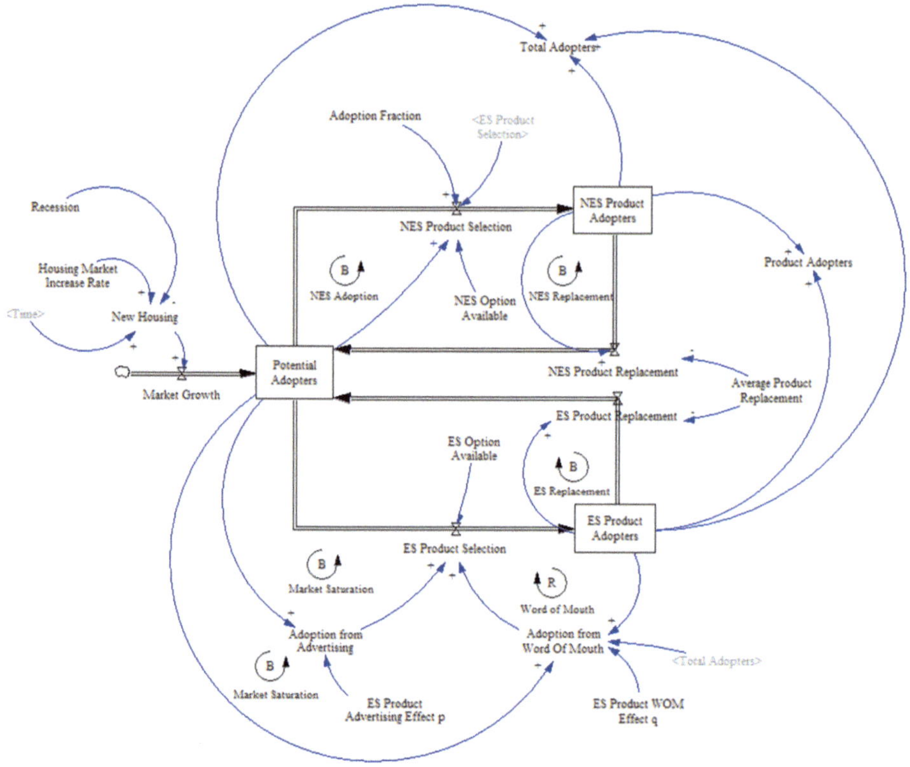

Fig. 13.10 Final adoption model

q value was 0.049. Figure 13.11 shows the result of using these values. It can be seen that New Housing has remained the same since it is affected by exogenous values. The number of Product Adopters has also remained relatively the same. The issue now is that ES Product Selection initially increases when it becomes available; however, the adoption does not increase as in the RBP as time progresses. The opposite occurs for the NES Product Selection, since it is dependent on ES Product Selection. As previously mentioned there are two values affecting ES Product Selection, namely, Advertising Effect (p) and WOM Effect (q). Advertising Effect will adjust the initial jump in adoption when the product is released and WOM Effect will adjust the slope from the time of release onwards.

By adjusting the Advertising Effect (p) and WOM Effect (q), the model outcome matched the RBP more closely. The outcome is shown in Fig. 13.12, where the jump in NES Product Selection and ES Product Selection is due to the step function used to represent the introduction of the ES clothes washer. A p value of 0.025 and q value of 0.65 were eventually selected, where both of these values fall within the boundaries of 0 and 1.

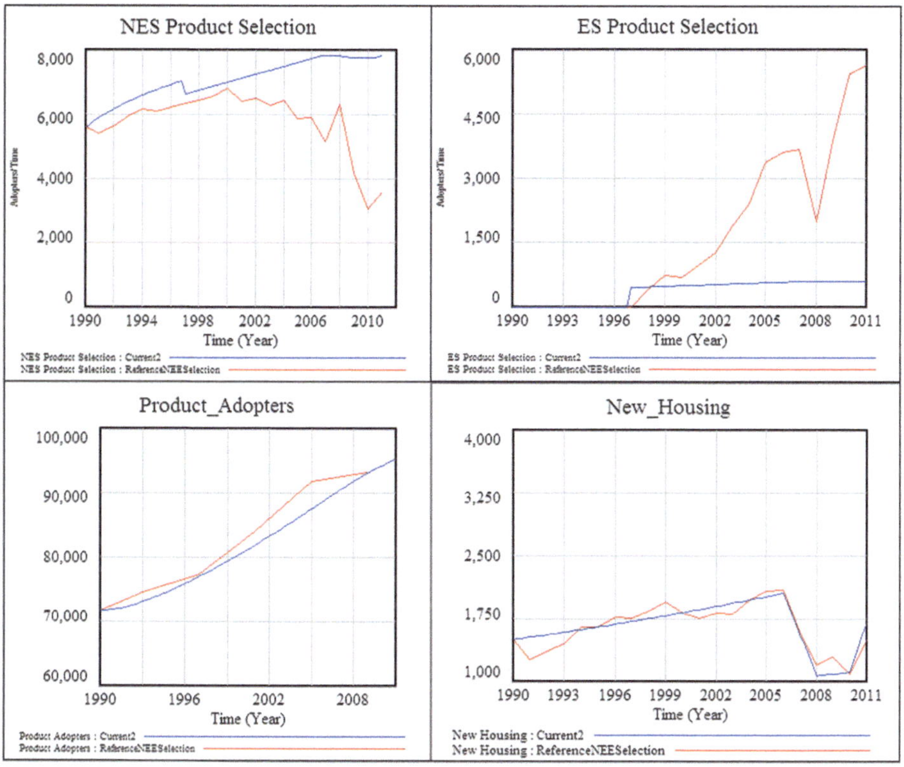

Fig. 13.11 Initial result for final adoption model

13.3.4 Verification

To further verify the model, the effect on the number of ES Product Adopters was determined by varying the value of variables between −20 % and +20 % of the base value. The results are shown in Table 13.1 together with a tornado diagram in Fig. 13.13. The tornado diagram illustrates that the largest change when adjusting a variables value is for NES Adoption Fraction. When reducing NES Adoption Fraction by 20 % the number of ES Product Adopters increases. This makes sense since we are reducing the number of NES Product Adopters which would result in an increase in the adoption of ES Products. The opposite occurs for ES Product WOM and Advertising Effect, where an increase in these values results in an increase in the number of ES Product Adopters.

Since one half of the model has already been tested individually the other part of the model will be tested separately by changing the variable NES Option Available to 0. This prevents any flow into the NES Product Adopters stock, so the only flow will be from both replacements into Potential Adopters and from Potential Adopters

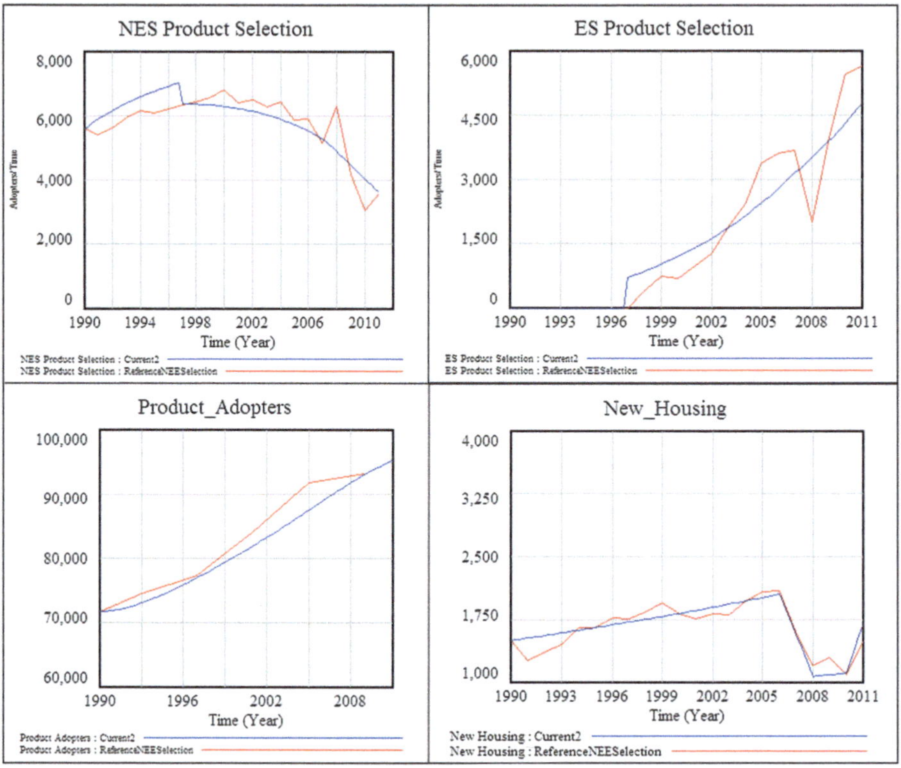

Fig. 13.12 Updated result for final adoption model

Table 13.1 Varying variables from −20 % to +20 % with ES product adopters as the response

	−20 %	−10 %	Base	+10 %	+20 %
Housing market increase rate	22,863.9	23,054.5	23,249.8	23,449.9	23,655.0
NES adoption fraction	34,772.8	28,062.9	23,249.8	19,679.8	16,956.0
ES product advertising effect	18,599.8	20,924.8	23,249.8	25,574.8	27,899.8
ES product WOM effect	17,535.6	20,143.3	23,249.8	26,960.9	31,406.3
Average product replacement	27,865.5	25,363.4	23,249.8	21,461.2	19,939.2

into ES Product Adopters. Figure 13.14 illustrates the outcome of the model. It can be seen that NES Product Selection is always 0 which is what we expected. The number of Total Adopters starts to decrease due to replacements of NES clothes washers and when the ES clothes washer is introduced the number of adopters starts to increase again. The shape of the ES Product Selection is what we would expect from the Bass model.

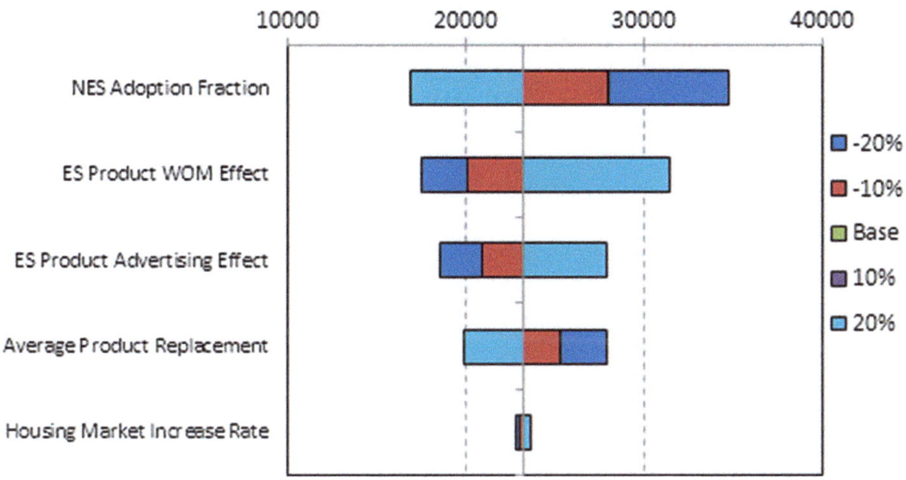

Fig. 13.13 Tornado diagram showing the variance when adjusting variables

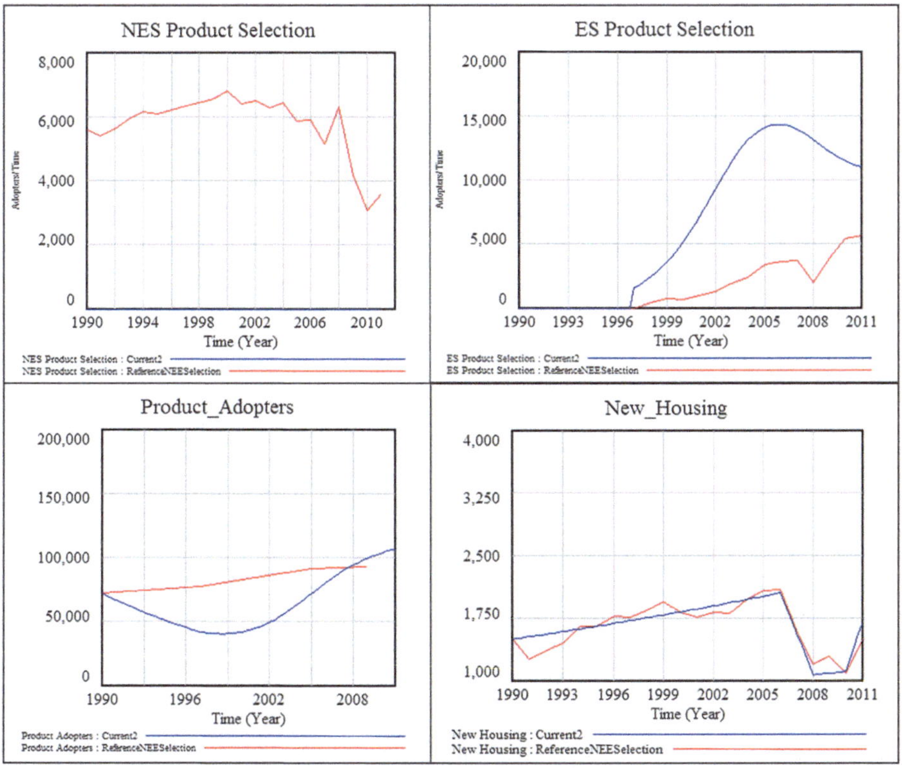

Fig. 13.14 Testing one half of the model for ES adoption

13.4 Conclusion

The adoption of clothes washers could have easily been done by implementing the standard Bass diffusion model; however, it does not cater for a growth in the market size, which is the case in the housing market. It also does not cater for the adoption of the same product with two different technologies. In order to recreate this adoption, the model was developed to cater for adopting an ES or NES clothes washer. By calibrating the model, the outcome was a close match to the RBP except for the total number of households. It was therefore assumed that households could adopt more than one clothes washer. The model tries to incorporate the well-established Bass model for the adoption of the new technology (ES clothes washer) with the adoption of an established technology (NES clothes washer). It was assumed that the effect of WOM was no longer applicable for the NES clothes washer and therefore only one balancing loop was included for this adoption. The model was verified by turning off sections of the model and by determining the effect of certain variables on the number of ES Product Adopters.

13.4.1 Future Work

The model created is a simplified representation of the actual system. There are many other factors that would affect the adoption decision which should be added to the model to further understand their effect. It would be interesting to analyze whether introducing larger incentives for these energy efficient devices would result in increased adoption. This model could also be expanded to study a whole range of different energy efficient devices. It would also be beneficial to look at different perspectives such as technical, economical, organizational, etc., that could affect the adoption.

References

1. US Department of Energy: Energy Efficiency and Renewable Energy, LCC for Clothes Dryers (2008)
2. J. Sterman, J.D. Sterman, *Business Dynamics: Systems Thinking and Modeling for a Complex World with CD-ROM* (New. York, McGraw-Hill/Irwin, 2000)
3. U.S. Department of Energy, Clothes Washer, Chapter 9. Shipment Analysis (2013), http://www1.eere.energy.gov/buildings/appliance_standards/residential/pdfs/rcw_dfr_tsd_ch9.pdf. Accessed 03 Dec 2013

4. Energy Star, Downloads (2013), http://www.energystar.gov/ia/partners/downloads/. Accessed 03 Dec 2013
5. U.S. Department of Energy, Clothes Dryer Chapter 9. Shipment Analysis (2013), https://www1.eere.energy.gov/buildings/appliance_standards/residential/pdfs/clothesdryers_tsd/clothesdryers_roomac_preanalysis_ch9.pdf. Accessed 3 Dec 2013
6. U.S. Energy Information Administration, Residential Energy Consumption Survey (RECS) (2013), http://www.eia.gov/consumption/residential/data/2009/. Accessed 3 Dec 2013

Chapter 14
Furnace Fan Motor Technology Assessment

Kevin C. van Blommestein and Tugrul U. Daim

Abstract Understanding the barriers associated with the lack of adoption of energy efficient technologies in the residential sector has always been an area of major focus. In order to encourage the adoption of these technologies, it is important to first understand what is currently needed, what the current status to meeting these needs are, and then determining the gaps between the current status and needs. These gaps can be used to determine what is possibly restricting the adoption of the technology, and solutions can be determined that can potentially close these gaps. One such energy efficient technology that has been an area of focus is energy efficient furnace fan motors. This study looks at the current technical, organizational, and personal gaps associated with this technology and determines the positive and negative influences restricting adoption. Solutions to the gaps are then identified and are specified as guides to help change negative influences to positive influences. The second section of this study is a bibliometric analysis used to evaluate the current R&D status of the three main technologies currently available in the market. Both journal articles and patents are associated with the respective stages of R&D to determine the specialization of different countries with respect to the technology and to determine the progress over time of the technology in each of the first three R&D stages. The purpose of this study is therefore to evaluate the technology from the R&D perspective as well as the market perspective. Finally, a link between the current R&D status and the technical requirements of the market is identified.

K.C. van Blommestein (✉)
Portland State University, SW 4th Avenue 50-02 1900, Portland, OR 97207, USA
e-mail: kevin9@pdx.edu

T.U. Daim
Dept. of Engineering and Technology, Portland State University,
Mana 1900 SW 4th Avenue 50-02, Portland, OR 97201, USA
e-mail: tugrul@etm.pdx.edu

T.U. Daim et al. (eds.), *Policies and Programs for Sustainable Energy Innovations*, Innovation, Technology, and Knowledge Management,
DOI 10.1007/978-3-319-16033-7_14

14.1 Introduction

The intention of this study was to assess the current and future energy efficient furnace fan motor technologies and factors surrounding the adoption of these technologies. Since the furnace fan motor is only part of the heating system, the entire system was to be taken into account for parts of the assessment.

The efficiency of a furnace is measured according to the following two values [1]:

- The Annual Fuel Utilization Efficiency (AFUE)—the proportion between the amount of fuel converted to space heat to the amount of fuel entering the furnace,
- **Furnace Fan Efficiency** (**e**%)—the ratio of the electrical consumption of the furnace fan motor in heating mode to the total energy consumption of the furnace.

The focus of energy efficient furnace fan motors is to improve the furnace fan efficiency. Energy Star specifies that the furnace fan efficiency should be less than or equal to 2.0 % [1]. In the residential sector, these fans can be found in central HVAC systems; gas-fired, oil-fired, electric furnaces; air handlers; and modular fan coils.

This study focused on furnace fan motor technologies and was broken down into the following two sections:

- Gap Analysis—A gap analysis was conducted for energy efficient furnace fan motors to identify technical, organizational, and personal gaps as well as potential solutions to these gaps. Once these solutions were identified, an influence diagram was created to determine positive and negative influences on the adoption of this technology. Research was then conducted to understand what technologies were currently under review by the relevant organizations.
- Bibliometric Analysis—In order to establish some idea of the current developments relating to these technologies, a bibliometric analysis was conducted. An initial analysis determined the specialization of different countries with respect to each technology. A further detailed analysis was conducted to determine progress in basic research, applied research, and development for each technology.

14.1.1 Current Furnace Market

Figure 14.1 illustrates a breakdown of the Northwest Furnace market, adapted from the 2011 Residential Building Stock Assessment (RBSA) [2]. Based on the Energy Star furnace key product criteria [1], furnaces in the south of the USA require a rating of 90 % AFUE or greater, while in the north of the USA a rating of 95 % or greater. As can be seen from the figure, majority of furnaces in this region are below these specifications. Understanding the lack of adoption of high AFUE furnaces is beyond the scope of this study; however, energy efficient furnace fans are generally included with high AFUE furnaces, unless the fan is retrofitted into an existing system.

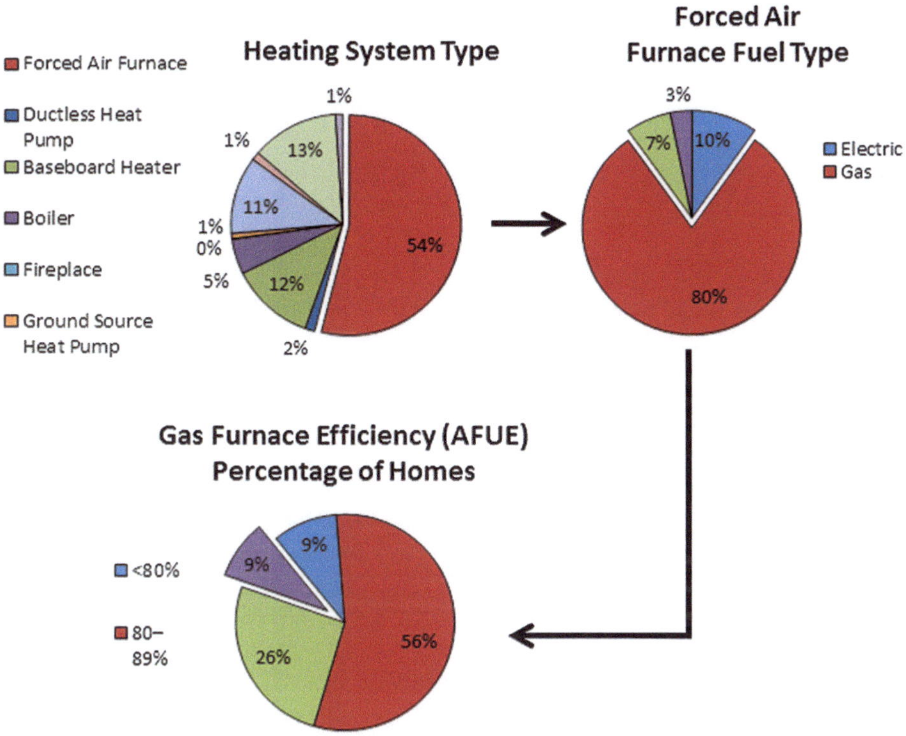

Fig. 14.1 Northwest furnace market

14.1.2 Furnace Fan Motor Technologies

In order to determine what current and future technologies are of interest, previous research by the U.S. Department Of Energy (DOE) was assessed. Figure 14.2 represents a perceptual map of the furnace fan motor technologies that were under review by the U.S. DOE [3]. The main technologies were Permanent Split Capacitor (PSC) Motor, Electronically Commutated Motor (ECM), ECM with Backward Curved Impeller (BCI), and Constant Torque ECM (X-13). The *x*-axis represents the average annual electricity savings in kWh and the y-axis represents the total installation cost in 2011 dollars. The increasing relationship between total installation cost and average annual electricity savings is as expected, due to the general increased costs associated with manufacturing energy efficient devices. It can be noted that majority of the X-13 (Constant Torque ECM) motors have some of the lowest total installation costs, together with a relatively high average annual electricity savings. This technology is therefore of interest due to a shorter payback period.

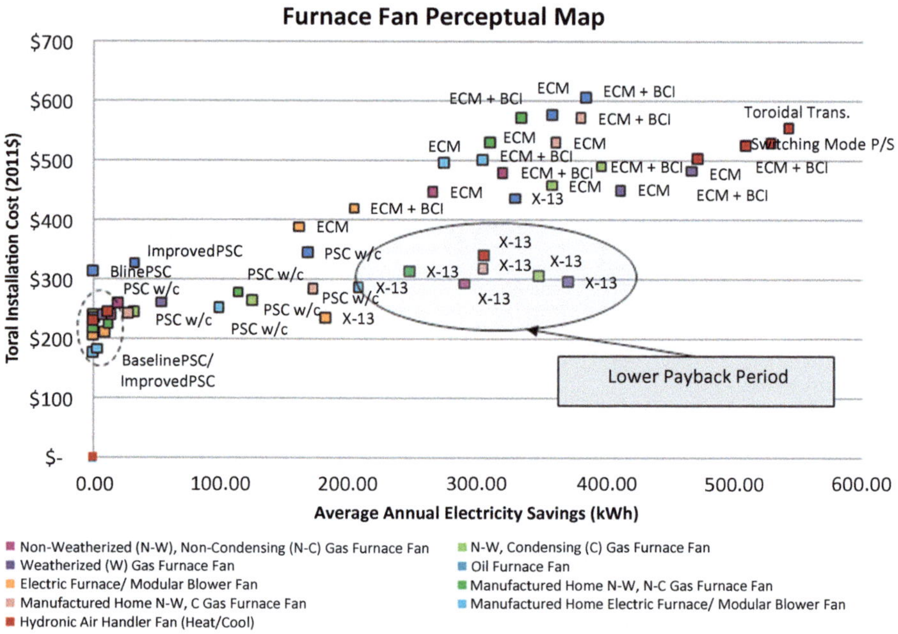

Fig. 14.2 Furnace fan perceptual map (adapted from [3])

14.1.3 Current Furnace Fan Motor Market

Based on the Air-conditioning, Heating, and Refrigeration Institute (AHRI) directory [4] and research conducted by the U.S. DOE on furnace fans [3], it was possible to analyze the current market relating to furnace fan motors. Figures 14.3 and 14.4 both summarize the distribution of the different motor types in current furnaces. It can be noted that ECM motors are all in multi-stage furnaces, and majority are eligible for the **Advanced Main Air Circulating Fan federal tax credit**. Based on the Energy Star furnace key product criteria [1], the furnace fan efficiency should be less than or equal to 2.0 %. As shown in Fig. 14.4, majority of furnaces with ECM motors are below this requirement, which is the opposite for PSC motors.

14.2 Methodology

Figure 14.5 describes the methodology that will be followed in this study. The first step of the methodology is to identify the current technical, organizational, and personal needs for furnace fans, what the current status with respect to these needs are, and the gaps that need to be dealt with to accommodate these needs. The second step is to determine potential solutions to these gaps in order to encourage the

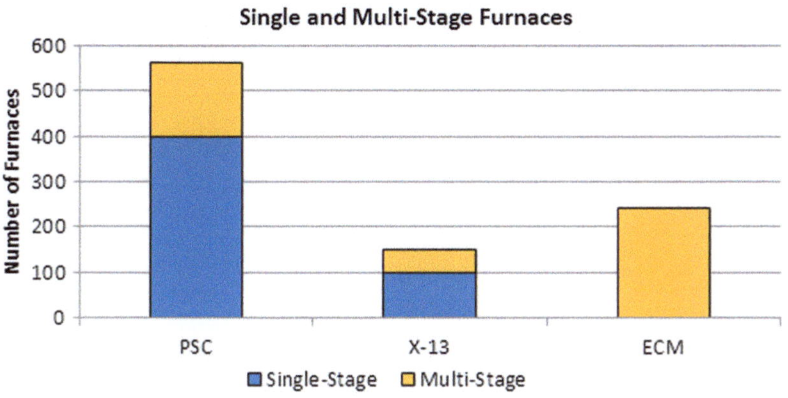

Fig. 14.3 Single and multi-stage furnaces in market

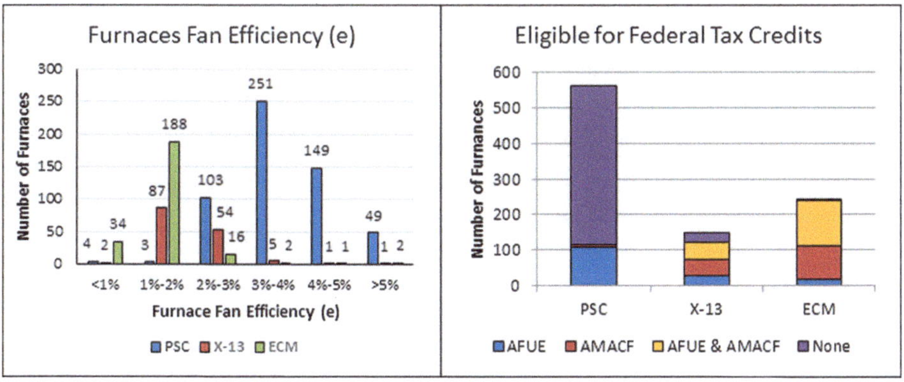

Fig. 14.4 Furnaces fan efficiency (e) and available for tax credits

Fig. 14.5 Research methodology

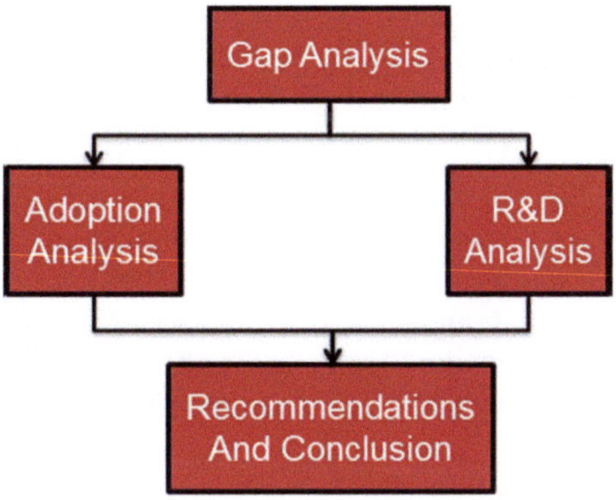

adoption of energy efficient technology alternatives, and to conduct a bibliometric analysis to evaluate the current Research and Development (R&D) status with respect to the previously mentioned technologies. The final step is to combine the analyses of the market perspective and the R&D perspective to recommend future developments with respect to the technology.

14.3 Gap Analysis

The first step of this study is to determine what are the current technical, organizational, and personal gaps relating to furnace fan motor technologies. In order to do so, it is important to first determine what is needed, then what the current status of resolving these needs are, and finally determining the gaps by comparing the current status to the needs. Once the gaps are determined, potential solutions to closing these gaps can be identified.

14.3.1 Technical

Table 14.1 lists the current technical needs, statuses, and gaps, and Table 14.2 lists the potential solutions to these gaps. TGx represents the technical gap x, while TS$x.y$ represents the solution y to gap x. The numbers for the technical solutions are represented in the influence diagram in Fig. 14.6.

14.3.2 Organizational

Table 14.3 lists the current organizational needs, statuses, and gaps. Table 14.4 lists the potential solutions to these gaps. OGx represents the organizational gap x, while OS$x.y$ represents the solution y to gap x. The numbers for the organizational solutions are represented in the influence diagram in Fig. 14.6.

14.3.3 Personal

Table 14.5 lists the current personal needs, statuses, and gaps. Table 14.6 lists the potential solutions to these gaps. PGx represents the personal gap x, while PS$x.y$ represents the solution *for* gap x. The numbers for the personal solutions are represented in the influence diagram in Fig. 14.6.

Table 14.1 Needs, statuses, and gaps (technical)

Needs	Current status	Gaps
Lower initial cost furnace fans with efficiency (e) below 2 %	Motor accounts for approximately 80 % electricity consumption in residential furnaces [5]	TG1—Reduction in manufacturing costs associated with high efficient furnace fans
	Electronically Commutated Motors (ECM) satisfy requirement, however costs are relatively high	
Tradeoffs at High External Static Pressure (ESP)	ECM maintains airflow at the expense of increased power consumption at high ESP [5] [6]	TG2—Evaluation of alternative motor technologies under high ESP conditions
	PSC motor does not increase power consumption at the expense of decreased airflow	TG3—Evaluation of methods for reducing high ESP
		TG4—Improved sizing of furnaces and ducts
NW—Improved duct insulation to meet Energy Star requirements	Northwest Energy Efficiency (NEEA) residential building stock assessment shows 14 % average supply leakage and 19 % average return leakage [2]	TG5—Energy Star requirements of 6 % supply and 3 % return leakages need to be met in NW [2]
Improved efficiency of impeller blades	Existing forward curved blades are generally less efficient than backward curved blades [7]	TG6—Improved performance of impeller blades at acceptable manufacturing costs
	Also blades are currently made from sheet metal [7]	
Finalized test procedure	DOE's furnace fan test method due to be finalized in Dec 2013 [8]	TG7—Fan efficiency (e) does not account for all modes of operation (excludes cooling and standby)
		TG8—SEER test conditions do not reflect real house (higher ESP) installations [9]

Table 14.2 Gaps and solutions (technical)

Gaps	Potential solutions
TG1—Reduction in manufacturing costs associated with high efficient furnace fans	TS1.1—Lower cost Constant-Torque ECM motor
	TS1.2—Reduction in cost of permanent magnets
	TS1.3—Motor technology without permanent magnets (e.g., switched reluctance motor)
TG2—Evaluation of alternative motor technologies under high ESP conditions	TS2.1—Incorporation of ESP into DOE testing procedure
	TS2.2—Evaluate performance of constant torque motors at high ESP
TG3—Evaluation of methods for reducing high ESP	TS3.1—Minimize duct friction [10]
	TS3.2—Select filters with low-pressure drop characteristics [10]
	TS3.3—Reduced constricted ductwork [9]

(continued)

Table 14.2 (continued)

Gaps	Potential solutions
TG4—Improved sizing of furnaces and ducts required	TS4.1—Manual J and Manual D requirements part of 2006 International Residential Code should be enforced instead of suggested [11]
TG5—Energy Star requirements of 6 % supply and 3 % return leakages need to be met in NW	TS5.1—2012 IECC requirements for new construction, additions, and alterations should be both visually tested and pressure tested [12]
	TS5.2—Training and awareness of correct installation procedures and requirements
TG6—Improved performance of impeller blades at acceptable manufacturing costs	TS6.1—Impeller blades optimized for each specific application
	TS6.2—Aerodynamic material used for impeller blades
	TS6.3—Backward curved impeller blades
TG7—Fan efficiency (e) does not account for all modes of operation (excludes cooling and standby)	TS7.1—DOE to update testing procedure to include cooling and standby mode (currently under evaluation) [13]
	TS7.2—Incorporation of different ESP conditions into DOE testing procedure [13]
	TS7.3—DOE not currently addressing standby mode and off mode for Hydronic Air Handlers, DOE plans to incorporate test procedure from furnaces [13]
TG8—SEER test conditions do not reflect real house (higher ESP) installations	TS8.1—Adjust default ESP and fan power values higher to more realistic values [9]
TG9—"Plug and Play" type furnace fan for retrofitting applications	TS9.1—Furnace fans that learn the current application conditions and adapts accordingly to improve efficiency. Simple installation required

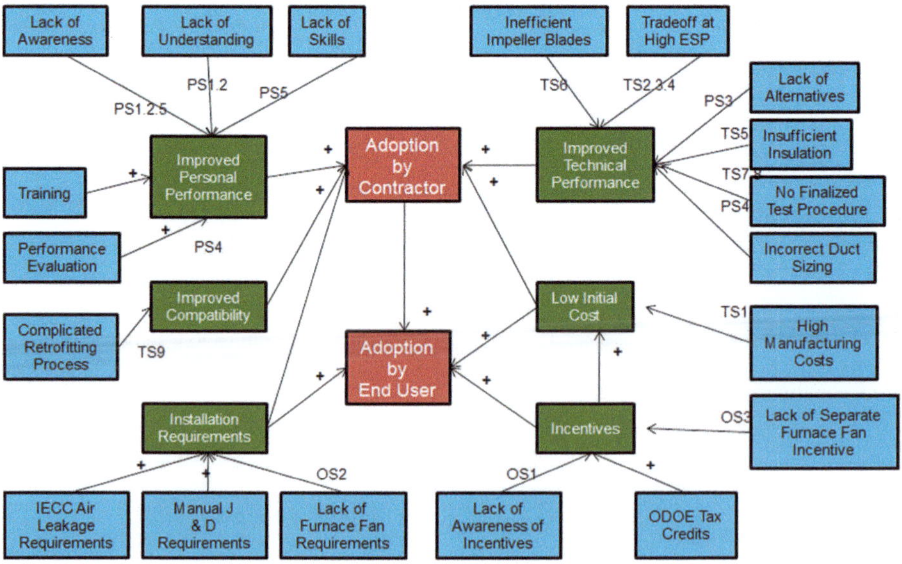

Fig. 14.6 Influence diagram including solutions

Table 14.3 Needs, statuses, and gaps (organizational)

Needs	Current status	Gaps
Increase adoption rate of high efficiency furnace fans	In 2003, PSC accounts for approx. 95 % of furnace fans used in residential furnaces, while ECM accounts for 2.5 % [14]	OG1—Large available market for adoption of high efficient furnace fans
Standards or requirements for high efficient furnace fans	Updated air leakage requirements added to 2012 IECC for residential and commercial construction [12] [14], however no furnace fan requirement	OG2—Include high efficient furnace fan requirements for new installations
Tax credit/incentives for the adoption of high efficiency furnace fans, separate to furnace tax credits and incentives	$50 federal tax credit for "advanced main air circulating fan" in 2011, not 2012 [15]	OG3—Tax credits/incentives are mainly offered for furnaces together with fans, not only for fans
	ODOE—tax credit up to $350 for forced air furnace (≥95 % AFUE) together with fan [16]	

Table 14.4 Gaps and solutions (organizational)

Gaps	Potential solutions
OG1—Large available market for adoption of high efficient furnace fans	OS1.1—Determine appropriate market segments that would adopt this technology [17], and focus on understanding these segments and reasons for adopting
	OS1.2—Increase consumer awareness of the technology and its potential benefits
	OS1.3—Increase consumer awareness of available incentives (NEEA residential building stock assessment showed 47.5 % of households self-funded efficiency improvements [2])
OG2—Include high efficient furnace fan requirements for new installations	OS2.1—Amend furnace fan requirement to next revision of IECC
OG3—Tax credits/incentives are mainly offered for furnaces together with fans, not only for fans	OS3.1—Offer incentives for furnace fans separate from furnaces for retrofitting applications
	OS3.2—Maintain additional incentive for high efficient fans with new installations

Table 14.5 Needs, statuses, and gaps (personal)

Needs	Current status	Gaps
Training for HVAC contractors to correctly size furnaces and ducts required	Manual J and Manual D requirements are widely ignored and rarely enforced [11]	PG1—Increased awareness and understanding of Manual J and Manual D requirements
Training for HVAC contractors to meet air leakage requirements in 2012 International Energy Conservation Code (IECC)	2012 IECC requirements for new construction, additions, and alterations should be both visually tested and pressure tested air [12]	PG2—Increased awareness and understanding of IECC requirements for air leakage and testing procedures
Reduce operating costs and payback period	ECM increases annual kWh savings over PSC. However, further savings can be achieved by reducing the cost of the motor or increasing the efficiency of the motor	PG3—New technologies that may be more efficient or cost-effective than ECM are required
		PG4—Correct sizing of furnaces and ducts for new homes
Training required for product support of ECM motors (including constant torque). Contractors not comfortable with more complex technologies [18]	ECM motors are set up using programming tools for adjusting motor and load parameters	PG5—Increased awareness and skills relating to variable speed and constant torque ECM technology
	Retrofitting requires professional contractors	

Table 14.6 Gaps and SOLUTIONS (PERSONAL)

Gaps	Potential solutions
PG1—Increased awareness and understanding of Manual J and Manual D requirements	PS1.1—Training workshops/courses for use of Manual J and Manual D software for calculations
	PS1.2—Simplify Manual J and Manual D software for calculations
PG2—Increased awareness and understanding of IECC requirements and testing procedures	PS2.1—Training workshops/courses required for correct installation and testing procedures
	PS2.2—Contractor performance evaluation based on installation (3rd party evaluation)
PG3—New technologies that may be more efficient or cost-effective than ECM are required	PS3.1—Evaluate alternative motor technologies (e.g., Switched Reluctance Motor (SRM) [19], Doubly Salient Permanent Magnet (DPPM) Motor [20])
	PS3.2—Evaluate emerging heating technologies (e.g., MicroHeater [21])
PG4—Correct sizing of furnaces and ducts for new homes	PS1.x
	PS4.1—Contractor performance evaluation based on installation (3rd party evaluation)
PG5—Increased awareness and skills relating to variable speed and constant torque ECM technology	PS5.1—Training workshops/courses relating to correct setup (including software) and installation of ECM motors

14.4 Adoption Analysis

Figure 14.6 represents an adoption influence diagram taking into account the needs, gaps, and solutions previously mentioned. The diagram represents the factors that influence the adoption of energy efficient furnace fan motors by the installation contractor and the end user. A positive arrow represents a positive influence while a negative arrow represents a negative influence. The numbers next to the arrows represent the solutions mentioned in the previous table. As an example, training and performance evaluation have positive influences on improved personal performance, while lack of awareness, understanding, and skills have negative influences. In order to remove the negative influences, personal solutions PS1, PS2, and PS5 could be followed. Installation requirements is the only factor that has apposing influences on contractor and end user adoption. It was seen that as more installation requirements are added, the contractor's work would increase which would thereby decrease the will to adopt the technology. The end user on the other hand would prefer the additional requirements since it could potentially improve the quality of the installation.

14.5 Research and Development Analysis

14.5.1 Motor Technologies

In order to further understand the motor technologies, a breakdown of the different single-phase motors was created together with the motors evaluated by the DOE highlighted in bold (Fig. 14.7). For single-phase applications, there are two main

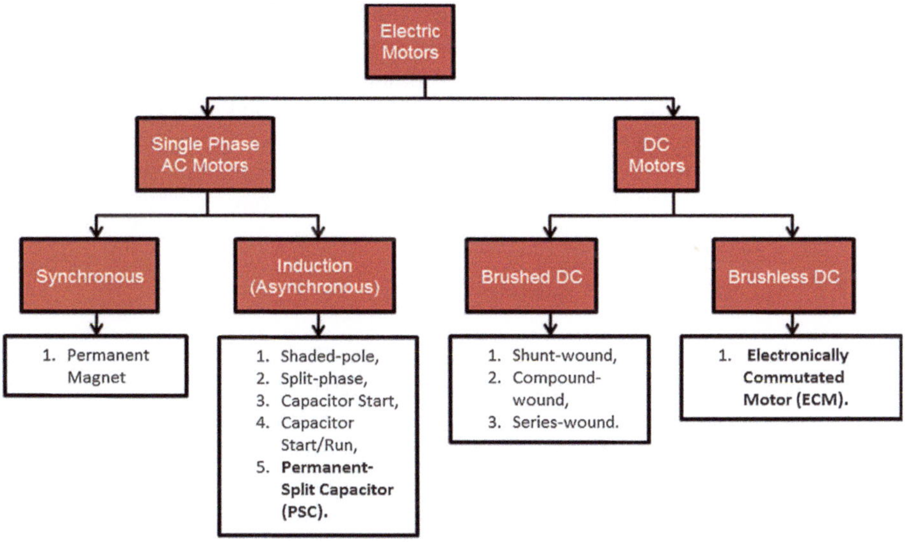

Fig. 14.7 Different motor technologies (adapted from [22])

Table 14.7 R&D stages and source of Information

R&D Stage	Typical source
Basic research	Science Citation Index
Applied research	Engineering Index
Development	US Patents
Application	Newspaper Abstracts Daily
Social impacts	Business and Popular Press

alternatives, either single-phase Alternating Current (AC) motors, or Direct Current (DC) motors. Due to the fact that majority of residential households have single-phase electrical supply, there is little focus on three-phase motors for residential applications. The X-13 motor is also an ECM motor, however with constant torque control in place of the variable speed control. The ECM motor mentioned by DOE is the variable speed alternative.

14.5.2 Bibliometric Analysis

Martino [23] describes the association between the different R&D stages and typical sources of information, as shown in Table 14.7. Due to the focus of this study and the availability of information, the Basic Research and Applied Research stages of R&D were evaluated in both of the following sections, while Development was included in the second section.

14.5.2.1 Search Strategy

The idea behind the search strategy was to combine the terminology used by the DOE together with the terminology identified as part of the evaluation of alternative motor technologies in order to create the search queries. Additional keywords were identified in an iterative manner, by first using the mentioned search queries, then selecting the most relevant results, and finally selecting the appropriate keywords from these results. Keywords were also identified from results that were not relevant, and these were added to the search queries as exclusions.

14.5.2.2 Specialization of Countries

The Revealed Technology Advantage (RTA) represents the strength of an organization, country, etc. in a specific technological field [24]. The RTA for a country in a specific field is calculated by dividing the share of a country's patents for that specific field, by the country's total share of all patents. Avila-Robinson and Miyazaki [25] use the same concept with scientific publications and referred to it as the Revealed Scientific Advantage (RSA). Figure 14.8 illustrates the RSA values for different countries for the three previously mentioned motor technologies. Constant Torque Motor (CTM) will be used from this point on to represent the X-13 motor.

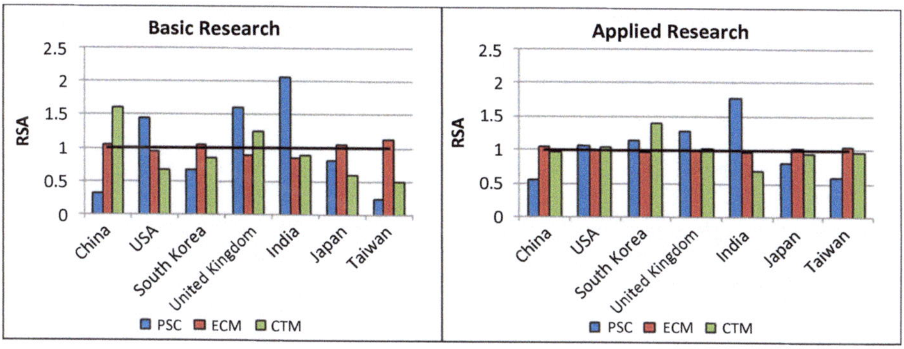

Fig. 14.8 Revealed Scientific Advantage (RSA) and average citations per article

From the RSA formula, it is simple to conclude that the larger the RSA value, the higher the specialization of the country in that particular area. As an example, the USA, United Kingdom, and India have a higher share of publications for PSC motors under basic research than its share of publications for all areas (RSA value greater than one). From these results it can be seen that the USA has a relatively equal distribution of specialization between these three technologies under applied research; however, there seems to be more specialization in PSC motors under basic research.

The problem with only looking at the share of publications to determine specialization is that it does not take the importance or impact of the publications into account. One approach of determining the impact of a publication is by looking at the number of times it was cited. For each country, we could determine the average citations per publication to determine quality; however, this gives excessive weight to highly cited publications and publications that have not yet been cited. Hirsch [26] introduced the h-index to characterize the scientific output of a researcher to resolve these issues. A specific researcher will have an h-index of *n* if *n* of his/her publications have at least *n* citations. The index therefore takes into account both the citations per publication and the number of publications. The same index can also be used to determine the impact and productivity of each country. Figure 14.9 illustrates RSA versus the h-index for each technology and each country. It can be seen that even though some countries have a high specialization in a specific area, they do not have a high impact. It can be concluded from the figure that even though the USA is not specifically specializing in the ECM technology, their research has a higher impact as opposed to the other countries.

14.5.2.3 R&D Focus (USA)

In order to evaluate the current focus of research for each of the motor technologies in the USA, the Compound Annual Growth Rate (CAGR) of publications and patents between 2000 and 2013 was calculated and plotted against the share of total publications. This is illustrated in Fig. 14.10. The technologies are clearly

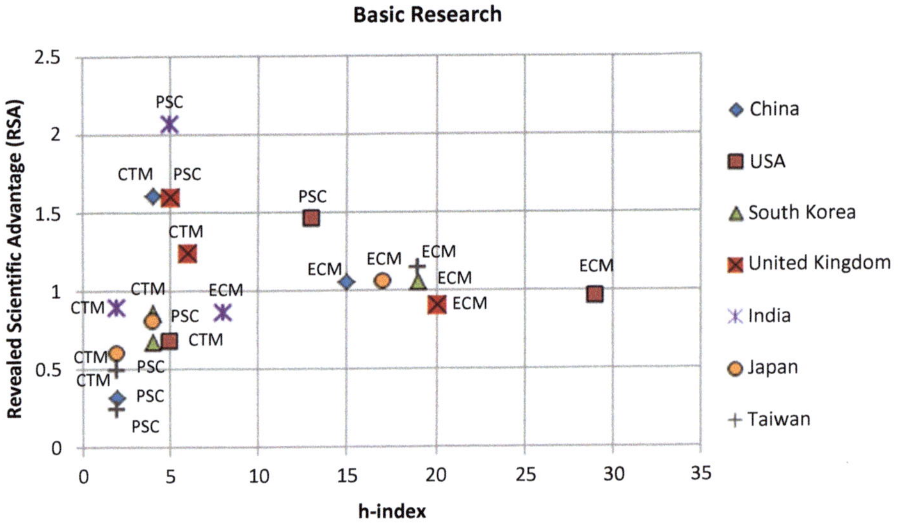

Fig. 14.9 RSA versus average citations per article

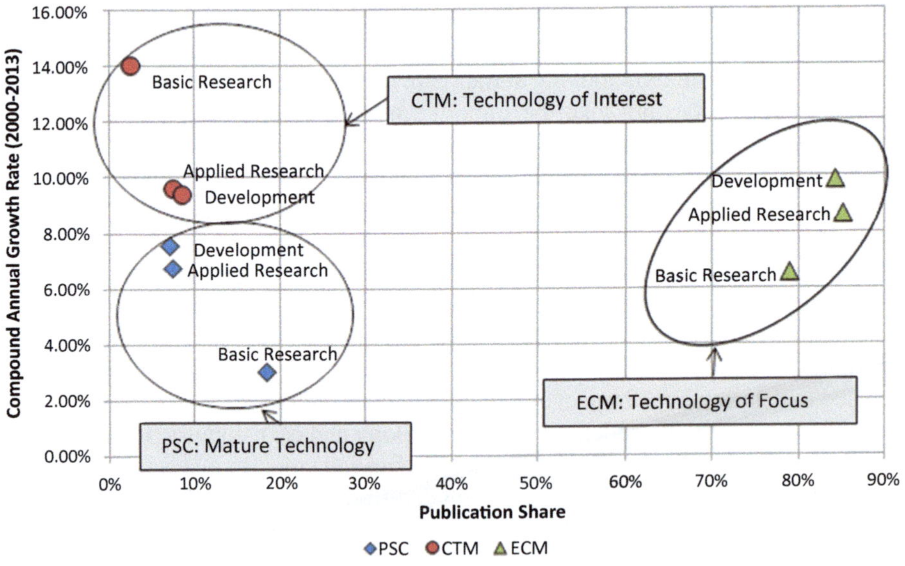

Fig. 14.10 Share of Publications vs. Compound Annual Growth Rate (CAGR) from 2000 to 2013

differentiated from one another by both the share of publications and growth. The CTM technology has a high growth rate but a small publication count and can be seen as the technology of interest. Basic Research for this technology is also growing faster than Applied Research and Development and it can therefore be concluded that this technology is still in the early stages. The PSC technology has a slow growth rate, slightly larger share of the publications, and Applied Research and Development is growing faster than Basic Research. It can therefore be concluded that the technology is in the later stages and due to the slow growth rate, possibly reaching maturity. Finally, the ECM technology has a high growth rate, a large share of the publications, and in the later stages of R&D. Therefore, it can be seen as the main technology of focus.

In order to illustrate the change in CAGR and publication share in the period of 2000–2013, the data was divided into two sets, one set consisting of 2000–2006 data and the other set consisting of 2007 to 2013 data. The direction of the arrow signifies the change from the 2000–2006 period to the 2007–2013 period. It can be seen that majority of the CAGRs reduce from one time period to the other, expect for Basic Research (BR) for the CTM technology and for Basic Research for the PSC technology. Additionally, the share of publications remains relatively consistent from one time period to the other, and it is only the CAGR that has significant changes for some cases. In order to further understand the changes in the technology, it is possible to reduce the time period of each set and also the duration over which the analysis is done (Fig. 14.11).

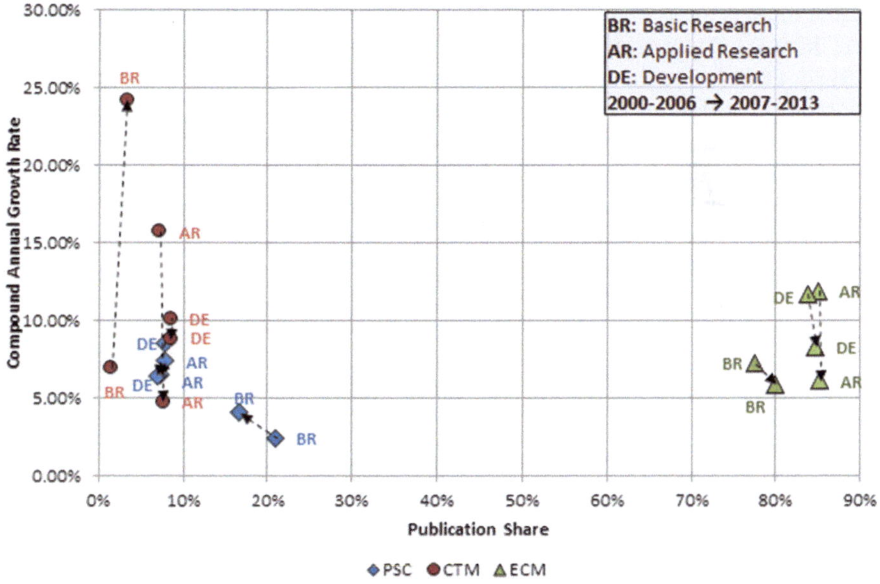

Fig. 14.11 Change in CAGR and Publication Share from 2000–2006 period to 2007–2013 period

14.6 Conclusion

The intention of this chapter is to evaluate the furnace fan motor technology from both the R&D perspective and the market perspective. It is possible to identify what factors are restricting the adoption of the technology and what potential solutions could be followed to increase adoption. Some of the factors are focused on the technology itself while others focused on organizational and personal factors. The bibliometric analysis is used to further analyze the research and development of the technology. From the bibliometric analysis we are able to identify the specialization of each country in each technology and then by selecting a specific country, it is possible to differentiate between the technologies.

The problem however is tying the results of these two analyses together. It was identified in the gaps that one of the restrictions is the payback period of more energy efficient technologies. DOE conducted research on this topic and determined that the lowest payback period was associated with the X-13 motor, or as referred to in this chapter as the CTM. However, it was also noted in the introduction that not all furnaces with X-13 motors meet the 2 % requirement for federal tax credits. Therefore, there is still an improvement required for this technology to successfully meet these requirements. Fortunately as part of the bibliometric analysis, we identified that there is focus in this technology, by noticing the relatively high growth rate in the three R&D stages. It was also determined that the focus on this technology is relatively new due to the low number of publications, thereby signifying that this technology is still in the early stages of R&D.

As identified in the gap analysis, there are many aspects to take into account when evaluating this technology, one of which is the R&D analysis. This chapter gave a basic idea of what aspects of the technology can be evaluated and how they can be tied together. However, the other aspects should be further evaluated to gain a better understanding of the technology.

14.7 Recommendations and Future Work

Even though the gap analysis and adoption analysis gave a clear picture of the current situation with respect to the technology, a more meaningful outcome would be a model that can be used to evaluate the impact of different scenarios. Converting the influence diagram in Fig. 14.6 into some form of a model that can be used for decision-making purposes may benefit organizations trying to encourage the adoption of this technology. This could possibly be achieved by assigning weights/costs to the different solutions presented and a preferred solution chosen by means of an optimization model.

The bibliometric analysis is useful due to the abundance of readily available information that can be easily analyzed. This study only touched on some of the available information to understand the R&D associated with the technologies. By evaluating the information in more depth, it is possible to determine in which areas a technology is used. By identifying which areas an inefficient technology is used and an efficient technology is not used, energy efficient administrators can focus on this area to encourage adoption. This is one means of identifying potential areas of saving energy. Other approaches could be by identifying keywords in publications for each technology and identifying where there is an overlap and where there isn't.

Finally, by determining a way in which to combine the results of the two analyses done in this study into one model, a more in-depth evaluation of the technology could be conducted.

References

1. Energy Star, Furnace key product criteria (2012), http://www.energystar.gov/index.cfm?c=furnaces.pr_crit_furnaces. Accessed 25 Oct 2012
2. D. Baylon, P. Storm, K. Geraghty, B. Davis, Residential building stock assessment: single-family characteristics and energy Use, Northwest Energy Efficiency Alliance. (2012)
3. US Department of Energy: Energy efficiency and renewable Energy, Residential Furnace Fans Preliminary Technical Support Document (2012), http://www1.eere.energy.gov/buildings/appliance_standards/product.aspx/productid/42. Accessed: 20 Oct 2012
4. AHRI, RFR Search (2012), http://www.ahridirectory.org/ahridirectory/pages/rfr/defaultSearch.aspx
5. V. Franco, J. Lutz, A. Lekov, L. Gu, Furnace blower electricity: national and regional savings potential (2008), http://efficiency.lbl.gov/sites/all/files/furnace_blower_electricity_national_and_regional_savings_potential_lbnl_417e.pdf. Accessed: 25 Oct 2012
6. K. Ueno, ECM Efficiency better (and worse) than you think, Home Energy 34–38 (2010)
7. I. S. Walker, Improving air handler efficiency in houses (2004), http://www.inive.org/members_area/medias/pdf/Inive\LBL\LBNL-54760.pdf. Accessed: 25 Oct 2012
8. Appliance Standards Awareness Project, US DOE appliance standards rulemaking schedule (2012), http://www.appliance-stand-ards.org/sites/default/files/doe_schedules/DOE_Schedule_by_Product_17.pdf. Accessed: 30 Oct 2012.
9. H. Sachs, S. Nadel, H. Henderson, D. Shirey, D. Jaynes, Can SEER be saved? in *2006 ACEEE Summer Study on Energy Efficiency in Buildings—Proceedings*, American Council for Energy-Efficient Economy, Washington, DC, 2006
10. State of Wisconsin Department of Administration Division of Energy, Electricity use by new furnaces (2003), http://www.ecw.org/ecwresults/230-1.pdf. Accessed: 30 Oct 2012
11. M. Holladay, Saving energy with manual J and manual D (2010), http://www.greenbuildingadvisor.com/blogs/dept/musings/saving-energy-manual-j-and-manual-d. Accessed: 25 Oct 2012
12. US Department of Energy: Energy Efficiency and Renewable Energy, Building technologies program air leakage Guide (2012), http://www.energycodes.gov/sites/default/files/documents/BECP_Buidling En-ergy Code Resource Guide Air Leakage Guide_Sept2011_v00_lores.pdf. Accessed: 30 Oct 2012
13. US Department of Energy: Energy Efficiency and Renewable Energy, Notice for proposed rulemaking, public meeting on test procedure for residential furnace fans (2012), http://www1.eere.energy.gov/buildings/appliance_standards/pdfs/ff_nopr_pmpres_2012_06_12.pdf. Accessed: 30 Oct 2012
14. US Environmental Protection Agency, Building codes for energy efficiency (2012), http://www.epa.gov/cleanenergy/documents/suca/buildingcodesfactsheet.pdf. Accessed: 29 Oct 2012
15. AHRI, Federal energy efficiency tax credits (2011), http://www.ahrinet.org/+federal+energy+efficiency+tax+credits.aspx. Accessed: 25 Oct 2012

16. Oregon Department of Energy (ODOE) Premium efficiency furnaces, boilers, and air handlers (2012), http://www.oregon.gov/energy/cons/pages/res/tax/hvac-furnaces.aspx. Accessed: 25 Oct 2012
17. S. J. Moss, M. Cubed, Market segment and energy efficiency program design (2008), http://uc-ciee.org/downloads/MarketSegementationWhitePaper.pdf. Accessed: 25 Oct 2012
18. B. Michael, Tips for HVAC sales professional understanding the advantages of X13 motor technology (2009), http://www.cfmdistributors.com/pdfs/technical-info/ts_X13MotorTech.pdf. Accessed: 25 Oct 2012
19. L. E. Teschler, The switch to switched reluctance (2008), http://machinedesign.com/article/the-switch-to-switched-reluctance-1211. Accessed: 30 Oct 2012
20. M.M. Radulescu, C. Martis, I. Husain, Design and performance of a small doubly-salient rotor-permanent-magnet motor. Electr. Pow. Comp. Syst. **30**(8), 823–832 (2002)
21. Flex Your Power, Furnaces (2012), http://www.fypower.org/com/tools/products_results.html?id=100156. Accessed: 30 Oct 2012
22. CEA Technologies, Electric motors: energy efficiency guide (2007), http://www.ceati.com/freepublications/7017B_guide_web.pdf
23. J.P. Martino, A review of selected recent advances in technological forecasting. Technol. Forecast. Soc. Change **70**(8), 719–733 (2003)
24. Y.-S. Chen, K.-C. Chang, The relationship between a firm's patent quality and its market value—The case of US pharmaceutical industry. Technol. Forecast. Soc. Change **77**(1), 20–33 (2010)
25. A. Avila-Robinson, K. Miyazaki, Emerging micro/nanofabrication technologies as drivers of nanotechnological change: paths of knowledge evolution and international patterns of specialization, in *PICMET'12: Technology Management for Emerging Technologies*, 2652–2662 (2012)
26. J.E. Hirsch, An index to quantify an Individual's scientific research output. Proc. Natl. Acad. Sci. U. S. A. **102**(46), 16569–16572 (2005)

Chapter 15
Technology Assessment of Insulation Material for Home Construction

Rachanida Koosawangsri, Zeina Boulos, Nameer Al-Mulla, and Mario Bitzer

Abstract Depleting energy sources is alarming many governments, organizations, and companies to set ambitious goals to reduce their energy use over the next few years. Buildings consume significant portion of energy. One of the most practical strategies to reduce buildings' demand for energy is by avoiding heat losses and implementing energy saving measures. Today's high performance insulation and thermal design can dramatically reduce heat losses. Many technical solutions are already available and applied across all regions, both in new build and renovation. The choice of the most appropriate insulation product has to be decided on a case-by-case basis as it largely depends on the building type and design and climate zone. This paper conducts technology assessment for different type of insulation technology that fits different construction application. Traditional and modern insulation technology has been discussed across this research. R&D recommendations are presented in the conclusion section of this report for improving manufacturing process of new high performance insulation materials to be able to compete in the insulation market.

15.1 Introduction

Use of the excessive source of energy in our planet helped with the expansion of human population and technology advancement. Societies through history have built a broad infrastructure, which needs energy supplies for its proper functioning. Along with this was the developed and inherited practice of aggressive energy consumption in daily activities of industry, commercial, and private life. When energy sources

R. Koosawangsri • M. Bitzer
Portland State University, Portland, OR, USA

Z. Boulos
Portland Community College, Portland, OR, USA

N. Al-Mulla (✉)
Intel Corp, Hillsboro, Oregon, USA
e-mail: nameeralmulla@gmail.com

T.U. Daim et al. (eds.), *Policies and Programs for Sustainable Energy Innovations*, Innovation, Technology, and Knowledge Management,
DOI 10.1007/978-3-319-16033-7_15

begin to deplete at a rapid rate, it became important to realize how societies will manage for adjusting to the limited energy sources left. In that matter, some effort has been made to establish new habits towards more efficient energy practice on organizational and personal levels. Many research and technologies has been utilized to reduce energy consumption in commercial and residential sectors [1]. Space cooling and heating systems in residential building become a necessity over time, thus they account for most part of energy consumption. The percentage of HVAC usage depends on many factors, e.g., the climate zone in which the building exist and the age of the building which determine the level of technologies that could be used in its construction. Insulation in wall, roof, pipes, and frames play an important role in limiting the amount of energy leakage.

15.1.1 Problem Statement

Maintaining buildings' coolness in summer and warmness in winter, while reducing energy cost is a challenge. Almost half of the energy consumed is used on HVAC. Some of that thermal energy is wasted by simply leaking out or in the building. One of the methods to prevent that and therefore increasing the energy efficiency is applying insulation technology in construction. Insulation helps to reduce the energy consumption of heating and cooling systems while achieving better living comfort. Effective insulation protects the environment and reduces greenhouse gas emissions by lowering energy consumption [2]. The industry today is in need for better energy-efficient insulation technologies to meet newer energy star standards and demands. Many type of materials are used to insulate houses and there are even more manufacturing techniques that give these materials various capacities. These capacities are large numbered and it is difficult to compare to achieve the best insulation solution for different application.

15.1.2 Research Objective

According to the Department of Energy, heating and cooling systems use about half of the energy consumed in American homes [3]. Typically, 42 % of the average family's utility bill goes to keeping homes at a comfortable temperature. The energy sources that power these heating and cooling systems emit more than 500 million tons of carbon dioxide and 12 % of the nitrogen oxide emissions [3]. Improving energy efficiency in residential buildings by using insulation is the simplest and most cost-effective way to reduce energy consumption. Using insulation does not require energy to save energy unlike other energy-efficient products [3]. Providing suitable insulation is one of the most important home improvements considering the rise in prices for energy units. This paper will assess different insulation technologies and materials for new house construction to reach higher energy efficiency.

15.1.3 Project Scope

The scope of this research is to assess different insulation materials used in home construction. The assessment will include materials available and commonly used today. In addition, this research will study newer materials that are recently introduced in the market or still in R&D phase.

15.1.3.1 Limitation

The authors based this research on literature review to assess the different insulation materials. There are many literature resources available today on the different types of insulation materials. To complete this assessment the authors selected a sample of insulation materials that best represent most available products. The assessment is narrowed on materials commonly used in new home construction buildings.

The study will focus on the most important criteria when assessing the different insulation materials. For this reason the selected criteria and materials are limited to outer house wall and ceiling applications. Other special use insulation materials were excluded from this research.

The cost data of insulation materials in this research include purchase prices and installation. This does not cover the cost of long-term sustaining of the material if needed. Another limitation in this research regarding scoring of the materials cost is that the authors used a 4-point scale which does not always represent the accurate quantitative comparison. Below is a summary of the research limitations:

- Research based mainly on literature review
- There many building insulation material that could not be included in this research
- Geographic locations and weather differences were not considered in this research
- The criterion studied in this research is a small sample of the most important criterion
- Cost estimates based on initial installation of materials for this research

15.1.4 Structure of the Report

The report considered three perspectives to categorise thermal insulation criteria: technical, environmental, and organizational. A gap analysis is presented to identify areas for development and improvement in next generation of insulation materials. Next, the report presents a technology landscape of existing insulation materials. The authors scored each of the materials against a selected set of criteria, followed by a hierarchical decision model (HDM) to identify the most important criteria for selecting an insulation material. After that, a conclusion of the report is presented with recommendations for future R&D to improve insulation materials based on the studies' final results. In addition, the authors suggested areas for future research and studies in assessment of insulation materials.

15.2 Methodology

The assessment of different insulation materials and technologies are realized with an emphasis on using a multiple perspective approach. A gap analysis model is used to identifying the needs in insulation materials. This gap analysis gives an overview of the different construction insulation technologies and materials. These materials are then categorized in the technology landscape analysis.

To prepare the assessment model we reviewed multiple insulation technology evaluation criteria in literature. Using the multiperspective approach, we were able to organize the criteria in three perspectives. In this study the authors used combined approaches of HDM and scoring to assess different technologies and material of thermal insulation in the construction sector. The researcher were faced with a mix of qualitative, quantitative, and sometimes conflicting factors in scoring the different materials that are taken into consideration in thermal insulation technology. In that facet the team needed to capture insulation expert's opinion to evaluate the importance of the defined criteria of insulation materials and technologies.

15.2.1 Multiple Perspective Approach

Multiple perspective approach is introduced to support the decision making and enables researcher and decision maker to view the problem environment. The multiple perspectives will enhance the ability to make better-informed choices. This model provides a feasible framework that decision makers and researchers can use to better understand and facilitate multiple perspectives in decision making. The team attempt to structure the HDM in light of the multiple perspective approach. Using this approach, we define a broad terms to cover multiple viewpoints. For this particular research the team used three perspectives to categorise thermal insulation criteria: technical, environmental, and organizational.

- Technical perspective: Criteria that relate to technical performance
- Environmental perspective: Criteria that have an impact on the environment
- Organizational perspective: Criteria that make up political motivation, policies and regulations, market special interests, compliance, and security

15.2.2 Gap Analysis

As mentioned earlier that the technology gap analysis will be used to identify the needs for future technology and the differences in current technology situation. The gap analysis should be conducted to achieve the technology objective. By identifying the gap, it will provide the possible direction of the technology to improve in the future. In order to create gap analysis, there are three steps that need to be considered. The first step is specifying the needs of the technology. From this step, the future state or place where the technology should be at will be stated. Second, the current situation for each need will be analyzed to see how the technology can perform in nowadays. Finally, the gaps will be identified by comparing the need for future state and the current state of the technology [4].

15.2.3 Technology Landscape Analysis

A landscape analysis model presents current state for the area of interest and is used to research the area of study. Dimensions and key criteria and features are identified. This landscape will also identify potential alternative options and how they are related. It will also present possible strengths and/or weaknesses for each alternative.

15.2.4 Scoring Model

The scoring model is a method to evaluate and rate several alternatives by having several command variables. The distinctive feature of a scoring model is the ability to put values for assessment criteria that are not measurable or expressible in numbers [5]. It is possible to directly compare qualitative and quantitative attributes and features after translating them all in the same scoring scheme. At the beginning of a scoring process, a scoring scheme and a scoring range has to be defined. The scoring scheme can either consist of numbers, for example, from 1 to 10, or it can also consist of words, for example, starting from bad over poor good to perfect. The challenge in scoring is to categorize the subjective and quantitative features of a product or a technology into the chosen scoring scheme.

Scoring is a heuristic method to compare and rate different independent products, technologies, opportunities, etc. The advantages of scoring are the traceable and comprehensible procedure and the simple and easy way to conduct further comparisons with the scored items. The main disadvantages in this method are the difficulties to score some subjective criteria and the intimateness of a score when it is done. Other problems are the conflict potential in scorings done by more than one expert and the fact that not all criteria are mostly of the same importance. So adding all scores to a final score for one study objective can lead to biased and unbalanced results [5].

15.2.5 Hierarchical Decision Model

HDM is a strategy decision tool to provide direction in strategic planning. The HDM was developed by the Engineering and Technology Management department at Portland State University. The software associated with the HDM creates a special record to collect evaluations from each participant, and displays participants' pairwise comparison for each level of the model. This model simplifies complex decisions, captures judgments of decision makers, and identifies opinion's similarities and differences. In a typical HDM procedure, there are six stages of research. The stages that the decision-making process should go through using HDM are modeling, expert panel selection, data collection, analysis of results, sensitivity analysis, and validation. The benefit of crafting the decision model with the input from the experts is that it enables to express the strategy concept and comparing the long-term and short-term objectives. In addition, it also enables to identify agreements and disagreements among experts. Crafting HDM demonstrates opinion's similarities and differences, and it helps to present them to be discussed or resolved. The structure of HDM consists of mission as the final goal of the decision model, objectives to fulfill the mission, the criteria for each objectives and alternatives as initial target for decision making. This structure helps decision makers choose the best solution among several alternatives across multiple criteria.

15.3 Model Implementation for Insulation Material Assessment

The assessment of insulation technologies and materials for house construction is described in the following chapter. The methodology for the assessment is based on the methods presented in the previous chapter. The core of the model is the parallel conducted scoring and HDM by using the same categories and criteria in both analyses. Thereby we are able to determine the performance of the insulation types on the one hand and on the other hand we will get results for the weight and the importance of each criterion.

15.3.1 Gap Analysis

After the project objective is selected, the direction of the project needs to be identified. The best way to know which direction the insulation technology should be in the future is conducting the gap analysis. In this research, the needs for future insulation technology and the capabilities of current technology will be based on the multiple perspectives; technical, environmental, and organization. In each perspective, the needs will be the same as the criteria in scoring method and HDM that will be described later. However, in each needs, it will contain sub-criteria in order to better understand it. Then, other research papers will help to analyze the current technology capabilities. After that, the gap between the needs and the capabilities will occur. Therefore, the direction of the insulation technology will be identified in order to meet the project objective.

15.3.1.1 Technical Perspectives

In technical perspective, thermal performance will focus on the thickness and thermal conductivity value of the material. For the better performance of the material, the market needs for insulation material to be thinner and lower conductivity. Cost of insulation material and installation will be considered because some material is cheaper than the other, but there are problems with its life-cycle cost because the cheaper material is easily to damage and need to be replaced during its life cycle. Moreover, because different climates need to install different insulation materials, and some materials cannot be used in some areas due to the moisture issue, the new insulation material should be improved to be more adaptable in various climates. The final criterion is ease of construction. In this criterion, the new insulation material should be fast and easy to install and maintain during its life cycle. Moreover, it should decrease the air leakage and thermal bridging after install at construction site. Table 15.1 illustrates the gap of technical perspective for insulation technology.

15.3.1.2 Environmental Perspectives

In environmental perspective, the research will focus on life-time and recycle impacts of the materials. The new insulation material should be environmental friendly and not contain or include hazardous chemicals. Hazardous chemicals can have negative affect on recycling the materials and then they cannot be recycled. An example is the added fire-retardant chemicals on cotton and/or cellulose insulation. Table 15.2 dictates the gap between the needs and capabilities for environmental perspective.

15.3.1.3 Organization Perspectives

The last perspective is organization. The criteria that this research paper focused are availability of the material and building code requirement from government. Nowadays the incentives from government are not enough for the residents to install or upgrade to new type of insulation material. Moreover, some materials are not widely available in the market because of lacking of support. Also, the health and safety issue and fire and humidity resistance should be included in the regulation for new insulation materials. The gap analysis for organization perspective is shown in Table 15.3.

15.3.2 Technology Landscape Analysis

The type of insulation materials differ in many criteria including thermal performance, the ability to apply in the construction site, form and shape, fire and temperature resistance, cost, durability, and many more. The authors use the technology

Table 15.1 Gap analysis, technical prospective

Needs	Capabilities	Gaps
Thermal performance		
– Thinner insulation materials with same performance	– The thickness of insulation will depend on *R*-value and range of thermal conductivity [6]	– Evaluation of insulation material that has better *R*-value without increasing thickness
– Lower thermal conductivity without increasing the thickness	– For traditional material, polyurethane has lower thermal conductivity from 20 to 30 mW/(mK) [7]	– Improved in thermal conductivity material
	– For state-of-art material, it has thermal conductivity less than 4 mW/(mK) such as VIP and aerogel [7]	
Cost-effective insulation methods	Total costs per square foot: \$1.18–\$1.58 for wall [8], \$1.26–\$1.7 for attic [9]	– Reduction in material and installation costs
		– Reduction in the life-cycle costs of insulation
Durability		
– Applicable insulation material for all climate zones	Different climate zones use different insulation materials [10]	Evaluation of liable material (properties change with temperature and humidity)
Ease of construction		
– Fast and easy to install	– Depends on insulation type: foam, batts, rigid panels, etc.	– Evaluation of new technology that easy to install and no need for professionals
– No air leakage and thermal bridging	– Mostly need of professionals [6]	– Evaluation of installation methods or insulation materials that automatically avoid air leakage or thermal bridging
– Easy to maintain during life cycle-easy to replace	– Proper installation is important for performance of insulation [11]	– Improved the maintenance method
	– Most insulations are included in the walls or roof that is hard to replace and maintain [6]	

Table 15.2 Gap analysis, environmental prospective

Needs	Capabilities	Gaps
Life-time impact	– Some of material use CFCs, HCFCs, and CO_2 when installed that is hazardous gas [12]	– Materials that are environment friendly when disposed and installed
	– Specific treatment need to use to dispose the material ex. Extruded polystyrene [12]	
Recycle impact	– Fiberglass is semi-hazardous and difficult to recycle [12]	– Easer recycling material and procedure
	– Recycle materials are usually coated with fire—retardants that cannot be recycled [13]	

Table 15.3 Gap analysis, organization perspective

Needs	Capabilities	Gaps
Availability		
– Increased government incentive	– Limited government incentives such as Fed: tax credit of up to $500 or 10 % [14]	– Improved government incentives
– Increased adoption rate of high efficiency material	– The state-of-art materials are not commonly used in the market	– Large available market for higher efficiency material
Building code requirements		
– Increase occupant health, comfort and safety	– Regulation on dust for respirable dust: 5 mg/m^3 and total nuisance dust: 10 mg/m^3 and fiber: 1 F/mL [15]	– Included the regulation of health and safety in the new material
– Fire-resistance	– Some of material produce toxic in case of fire such as extruded polystyrene and polyurethane foam [12]	– Included A1 class fire-resistance requirement
– Effect as vapor barrier	– Fire class between A1–B2 depend on material [15]	– Included humidity resistance requirement
	– Resistance of vapor diffusion varies from the factor of 200 [15]	
	– Cellulose is easily to damage when contact with moisture [16]	

landscape to categorize the several insulation types. The insulation materials included in this research are divided into two groups: the traditional and state-of-the-art materials [7]. Materials in the traditional include commonly used insulation for construction in today market. Materials in the state-of-the-art group are the newer insulation technologies that are still not used in construction but are with promising features for use in this field. As shown in Fig. 15.1, the traditional materials are further divided into four technology groups by application type. Group 1 includes the foam materials, Polystyrene, Polyisocyanurate (ISO), PUR, and Phenolic. As the name implies, the materials in this group are liquid that will convert to foam when applied on site and then convert to solid. Group 2 includes batts material, Fiberglass, Rockwool, Cotton, and Polyethylene. The materials in this group are sold and delivered to site in large rolls. Group 3 includes rigid board, EPS, XPS, ISO, Fiberglass, PUR, and Perlite. This type of material is delivered in large board form and then cut to size on the construction site. Finally, group 4 is the loose fill materials, Fiberglass, Rockwool, Cellulose, and Perlite. These are in the form of lose material. There are four promising state-of-the-art material chosen for this research; Aerogel, Vacuum Insulation Panels (VIP), Gas Filled Panels (GFP), and Phase Change Materials (PCM) [7].

The following sections will further describe the different insulation technologies and materials like they are grouped in the technology landscape.

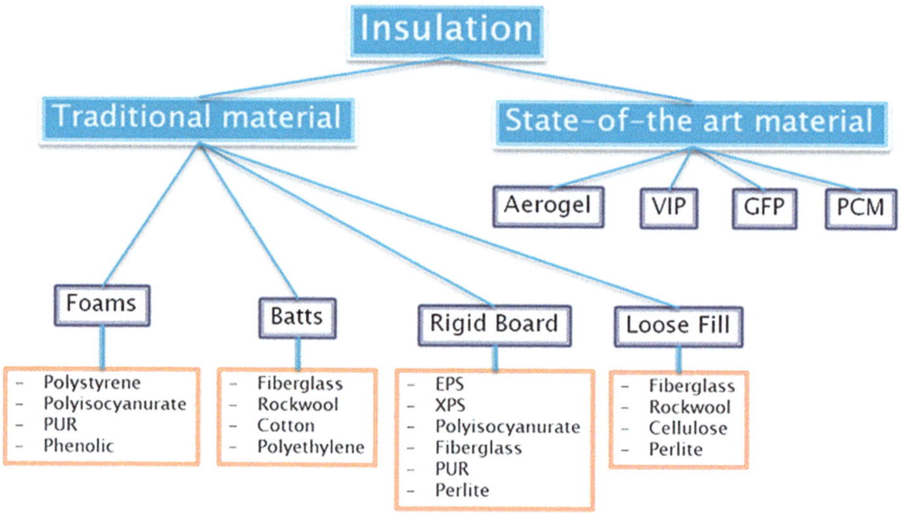

Fig. 15.1 Insulation materials landscape

15.3.2.1 Traditional Materials

15.3.2.1.1 Foam Insulation Material

Spray foam is made by mixing two or more liquid chemicals. The mixing and reacting materials respond quickly and expanding to create foam that insulates air seals and provides a moisture barrier. Spray foam insulation is known to resist heat transfer well, and it offers a highly effective solution in reducing unwanted air infiltration through cracks, seams, and joints [17]. There are different types of spray foam, which are basically either high pressure foam and/or low pressure foam. Different types of foam could be installed in existing or new constructions. Different types of foam are suitable for different applications. The liquids are delivered in different drums or containers to the construction location to be mixed and reacted; the result will be expanded foam. This process required a professional worker [17].

15.3.2.1.2 Batts Insulation Material

Batts, roll or blanket insulation is one form of insulation material that is the most common used and available, and also relatively inexpensive [13]. It can be manufactured from various materials such as fiberglass and Rockwool. This type of insulation is delivered in very large roll with the width that suit with standard spacing of wall studs [6]. However, the insulation is easier to install when compared with other types. The home owners can install it themselves, by cutting or trimming the batts, or hire a professional installer [13]. The facing of the batts, such as foil-kraft paper, vinyl and flame-retardant facing will be used to act as a barrier from air, moisture, and fire. It can be installed to unfinished wall, floor, and ceiling. The big concern when using this type of insulation is that its thermal performance tends to decrease easily. The *R*-value will decrease when pressing or cutting the batts and also thermal bridging will occur between cut edges [18]. Thus, this insulation type need to be installed carefully when there are joist spacing and other obstruction such as wires, electric boxes, and pipes in the wall. In this research paper, the materials that will be considered for scoring methodology are fiberglass, Rockwool, cotton, and Polyethylene.

15.3.2.1.3 Rigid Board Insulation Material

Rigid insulation panels and boards are made out of fibrous materials or foams [12]. The big difference to the already described foam or batts insulation is the rigid shape of the panels. They are all premanufactured and come in boards or panels to the construction site. So they are pretty simple to install by only putting or gluing them on the walls that should be insulated [19]. Many board insulations are faced with reflective foils to increase the thermal performance or with a water-resistant layer to use it as a vapor barrier [6]. Rigid board insulation is often used in prefabricated structures, as insulative wall sheathing and it is widely used for foundation insulation. Materials to produce rigid board insulation are mostly overlapping with foam insulation materials. The most established materials that are reviewed in the following scoring model are Expanded Polystyrene (EPS), Extruded Polystyrene (XPS), Polyisocyanurate (ISO), fiberglass, Polyurethane (PUR), and Perlite.

15.3.2.1.4 Loose Fill Insulation Material

Loose fill materials chosen for this study are fiberglass, Rockwool, cellulose, and Perlite. These are available in the form of loose material and spray installed on site. Some of the loose fill types can be sprayed either dry or wet depending on the area it is going to be installed at. Usually it is installed dry in roofs. The wet feature is useful when installing the material in vertical walls.

Fiberglass is high in thermal insulations properties. Rockwool is made from natural materials and is allergy free safe on the environment. Cellulose products are made from old newspapers that are shredded and sold as loose fill insulation materials. It is safe as is however is fast to catch fire considering it is made from paper. Fire-resistant chemicals are mixed with the shredded papers in the manufacturing process to eliminate the risk of catching fire during the life span of the installation. Perlite is natural rock material that is low in thermal conductivity. The Rock is crushed to smaller form and used inside cement and brick walls to enhance the insulation of the building.

15.3.2.2 State-of-Art Materials

15.3.2.2.1 Aerogel

Aerogel is one of the top promising technologies in thermal insulation area [22]. Aerogel is made up of a gel that has had its liquid component replaced with air; in fact the material is 99 % air. It's quite thin, breathable, and fireproof, doesn't absorb water, and is very strong considering its weigh. Aerogel has a very low thermal conductivity of (14 mW/mK) which results in an *R*-value that is twice better than what other typical insulation provide [7]. Aerogel is fairly expensive compared to traditional insulation materials and out of the price range for the average homeowner. It is a great material for insulation where size matters [22]. Only two companies currently have a commercial product available—Aspen Aerogel and Thermoblock. Aerogel has been used before in insulation for a number of NASA projects including the Mars Rover and space suites [23]. There are many applications for aerogel beside insulation like super insulating blankets which is made with aerogel. They are also the world's lightest solid materials, and mechanically robust aerogels [24].

15.3.2.2.2 Vacuum Insolation Panels

VIPs are one of the more efficient materials compared to other alternatives because of its low thermal conductivity (3–4 mW/mK) [7]. In the structure of this insulation, open porous material will be enveloped by thick metal sheets or materialize polymeric layers to act as a barrier for environment and handling protection. The materials that are used as a porous core need to have suitable pore size in order to maintain the vacuum. The core materials can be foam, powder, fiber, and fiber/powder composites such as PUR, EPS, silica, and expanded perlite [25]. Then, gaseous heat transfer is suppressed to the core materials. Despite the very low thermal conductivity that helps in reducing the thickness, the installation process and application type need to be considered because it is the most critical aspect of this material. Even though VIP has very low thermal conductivity, it tends to increase easier than other state-of-art material over time. Moreover, handling and maintaining this material needs to be very cautious because the vacuum can be lost, and thermal conductivity will be increased. Another drawback of this material is that it cannot be cut to adapt and adjust the shape at construction site because cutting will result in losing the vacuum [7].

15.3.2.2.3 Gas-Filled Panels

The technology and functioning of Gas-filled Panel is similar to the Vacuum Insulated Panels. The core of the panel is created by a baffle structure which is filled with a gas with a lower thermal conductivity than air [7], such as Argon (Ar), Krypton (Kr), or Xenon (Xe). The whole structure of GFPs is very fragile so all process from handling to applying it to walls must be carried out with high caution. GFPs are still in a research state and first prototypes are not able to reach the theoretically calculated low thermal performance which was expected [26]. They show similar advantages and disadvantages than VIPs but are not able to reach the thermal performance and they are more expensive because of the gas that is included in the GFPs [26].

15.3.2.2.4 Phase Change Material

PCMs are able to absorb heat and release it as temperature changes [27]. This absorbing and releasing of temperature is not a unique characteristic of these materials. What makes it unique is the sensitivity of the material to the surrounding temperature changes and the ability to store the heat. PCM change material state from solid to liquid when it absorbs the heat. The PCM will maintain the liquid status until temperature drops causing the heat to be released. As the heat is released the PCM will return to its sold state.

Because of the way PCM store the heat, PCM is very promising for heat storage technologies. Currently PCM is used in construction as insulation improvement in building materials in concrete mixes or bricks. PCM can also be mixed in Wool or fiber batts insulations to improve on their thermal insulation performance. Because PCM is used as an additive or part of a concrete mix and not used in a standalone format, this added difficulty in evaluating this product against the rest of the insulation materials studied in this report.

15.3.3 Scoring Model

The first step in the assessment of construction insulation technologies and materials was the definition of the assessment criteria. These criteria will be the same for the scoring and the HDM. This will enable the study to both evaluate the actual available technologies and identify further research areas for the state-of-the-art technologies to become competitive or superior to the most used insulation materials so far. The definition of the criteria is based on an extensive literature review about needs and features of construction insulation by applying a multiple perspective approach described earlier. The criteria are described on the following section.

Technical Criteria

- Thermal performance: This is mainly based on the thermal conductivity expressed in the unit W/mK. It can be interpreted as the ability of a material to transfer and transport thermal energy through the material. The smaller the thermal conductivity of the material the better is its insulation performance. For insulation materials, the reciprocal value of the thermal conductivity, the *R*-value, is commonly used [28].

- Cost: This criterion only includes the one-time purchase and installation cost. There is no evaluation of life-cycle cost. The scoring is mainly based on the cost per *R*-value of a material and a comparison of prices for the same insulation thickness.
- Durability: It is to evaluate the time impact on insulation material, e.g., the behavior of the *R*-value over time, water and moisture effects, thermal expansion and contraction, settling over time, etc.
- Ease of construction: This criterion is to evaluate the impact of insulation material/technology on workmanship requirements, ease and speed of construction, ease of operation, maintenance and replacement.

Environmental Criteria

- Life-time impact: The negative environmental impacts caused during the production and usage is considered in this criterion. This includes the used ingredients of the insulation material as well as potential toxic and hazardous products which could outgas during the life time.
- Recycle impact: This criterion covers the easiness and possibilities to recycle or dispose the insulation materials.

Organizational Criteria

- Availability: We will evaluate the market availability in this criterion. Assessment criteria are the easiness to purchase the material, the number of companies which manufacture it, the distribution channels, etc.
- Building code requirements: The fulfillment of the legal requirements and government specifications will be evaluated with this criterion.

This assessment paper uses a scoring model to assess and compare all the introduced and described construction insulation technologies and materials. The scoring model uses values from 1 to 4 with 1 being the worst and 4 being the best score. We are aware of the fact that scoring with only four values can distort the results of quantitative criteria like the thermal performance or cost but we intentionally used the same scoring range for every criterion to simplify the interpretations of the scoring results. The scoring is based on a literature review about the capabilities of the different materials, the advantages and disadvantages of the technologies in general and each material itself as well as the potential issues related with each material.

The results of the scoring are presented in the following paragraphs. The scores for each criterion will be showed in a separate chart. The scores of the traditional materials are in blue color and the state-of-the-art technologies are colored in green. The first evaluation criterion is the thermal performance. We looked at the range of all evaluated thermal conductivities for every material and calculated the scores based on these values. Some insulation materials can be produced and manufactured in different ways or special treatments or add-ons can slightly influence the thermal conductivity of insulation materials. That is why we calculated the average of all values found for each specific material. A perlite rigid board has the worst average thermal conductivity of 0.05 W/mK [12], whereas VIP has the best with

Table 15.4 Scoring scale calculation

Score	Thermal conductivity [W/mK]
4	0.004–0.0155
3	0.0155–0.027
2	0.027–0.0385
1	0.0385–0.05

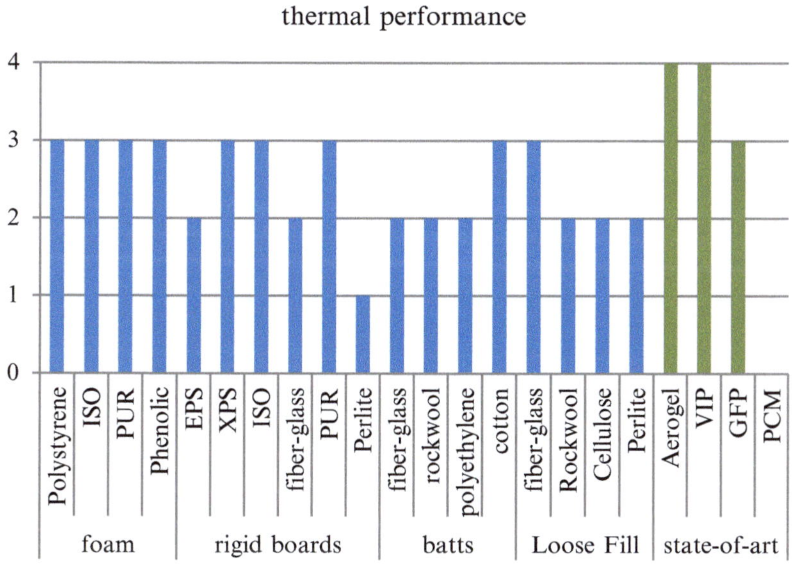

Fig. 15.2 Scoring results of the thermal performance criteria

0.004 W/mK [7]. Again, the smaller the thermal conductivity, the better the insulation effects. The difference between the worst and the best value is 0.046 W/mK which leads in the score system to a single score size of 0.0115 W/mK.

$$\frac{0.046}{4} = 0.0115$$

Based on this calculation, each material can be scored in the following scheme (Table 15.4):

The scoring shows, Fig. 15.2, that aerogel and VIP are the only two products with a score of 4 in matters of thermal performance. All foam materials, XPS boards and ISO boards, as well as cotton batts and fiber-glass loose fill have a score of 3. It is interesting that fiber-glass batts have a different score than fiber-glass as loose fill. The open cell structure of fiber-glass used in loose fill insulation results in a lower thermal conductivity [12]. So even in terms of thermal performance, the same material used as a different insulation technology can influence the scoring. This is also the case in other criteria as we can see later. PCM insulation is not scored

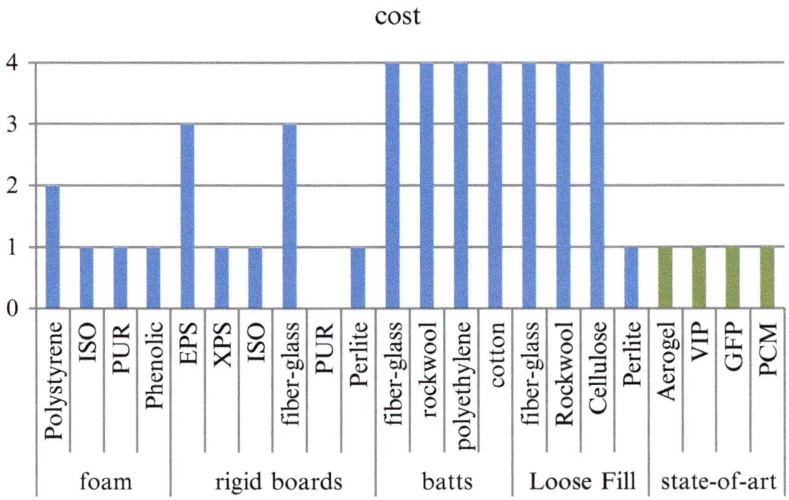

Fig. 15.3 Scoring results of the cost criteria

because there is no single insulation only consisting of PCM and the material characteristics are strongly changing due to the ability of changing phases within the material. If PCM is used as a thermal insulation it is always combined with other insulation material and the thermal conductivity is mainly dependent on the other insulation [29].

The scoring results of the cost criterion, Fig. 15.3, are based on price comparison between the different materials as well as the information found in research paper. The following chart shows the result of the cost scoring.

Except of perlite loose fill, all loose fill and batt insulation are the cheapest insulation materials. Rigid boards and foam are more expensive and all state-of-the-art insulation technologies are more expensive than the traditional ones and they are all scored with the worst score of 1 [12]. Even in terms of a comparison of the costs per *R*-value, where the advanced materials could benefit from because there is less material needed to achieve the same *R*-value, the state-of-the-art technologies are still the most expensive ones [12]. So including their better thermal performance in the cost evaluation does not result in higher scores for the new technologies.

The durability defines the long-term behavior of the materials. Aerogel and Perlite are showing, Fig. 15.4, no aging effects and are scored with 4 points [12]. The thermal performance of cotton, fiber-glass, and Rockwool can be reduced through moisture or increasing compression because of settlement [6]. All the other traditional materials are affected by a decreasing *R*-value over time [12]. VIP and GFP show a slow loss of vacuum and gas, respectively. They can be easily damaged either at the construction site or later by drilling holes or putting nails in the walls [7]. That is why VIP and GFP are scored with worst.

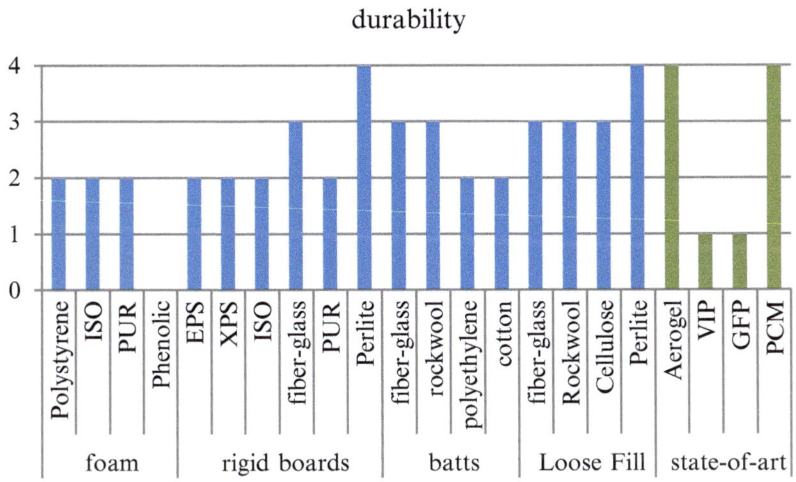

Fig. 15.4 Scoring results of the durability criteria

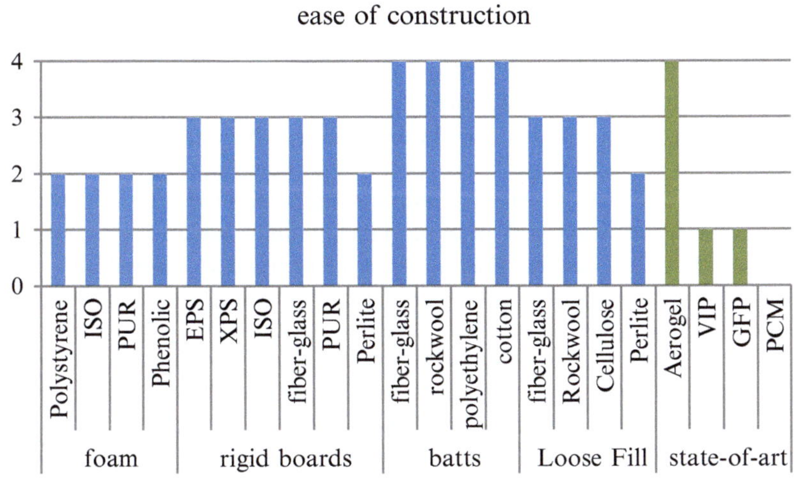

Fig. 15.5 Scoring results of the ease of construction criteria

The next evaluation criterion is the easiness of construction, Fig. 15.5. The insulation technologies show greater differences compared to the other technologies than the materials in its respective technology. The easiness of construction is more depending on the shape and the structure than on the material itself. It is, for example, a big difference if the insulation is brought to the construction site as liquids and the foam is expanded directly in the wall or if the premanufactured foam board needs to be put on the wall. This is why all the foam materials and batts have the same score and all rigid boards and loose fill are scored similar as well.

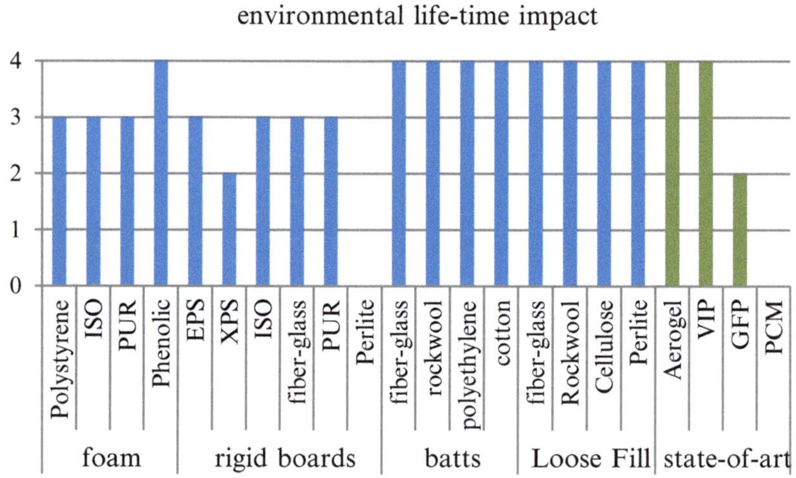

Fig. 15.6 Scoring results of the environmental life-time impact criteria

All insulation materials that are formed as batts are easy and convenient to install [15]. They all qualify for DIY and they need no curing or drying time so the resulted installation time is also short [30]. All Rigid boards and loose fill, except Perlite, are scored with 3 points. Rigid boards are easily cuttable and adjustable on the construction site and there is no special equipment needed to install it [12]. Loose fill materials are very easy for attic insulation and these loose materials are well suited for places where it is difficult to install other insulations [6]. Foam insulation fit in all cavities because they are sprayed or filled as a liquid into the walls and the foam expands afterwards. However, this advantage is equalized by the need of special equipment and professional installers. It also needs drying time and the quality and the thickness of the insulation can hardly be controlled, so all foam insulation get 2 points. Aerogel for wall insulation is produced in a similar shape and form than batts and it is installed in the same way [31]. It has also the widest range of different thermal insulation applications and is therefore scored with 4 points. VIP and GFP are absolutely not adjustable at the construction site and the large premanufactured panels need professional installation workers [32]. That is why we scored them with the worst score for ease of construction.

The next two scoring criteria are evaluating the environmental perspective of the insulation technologies and materials. As explained above this perspective is divided in the environmental impacts during the life time, Fig. 15.6, and after the usage of it by looking at the recycle, Fig. 15.7, and dispose issues. All materials that are scored with 4 points are showing no special negative environmental impacts both for manufacturing them and during the usage as an insulation material. The polystyrene materials include a toxic brominated flame retarder and either CO_2 or HCFC is used as an expanding agent [12]. GFPs are filled with toxic noble gases and are therefore scored with only 2 points [7].

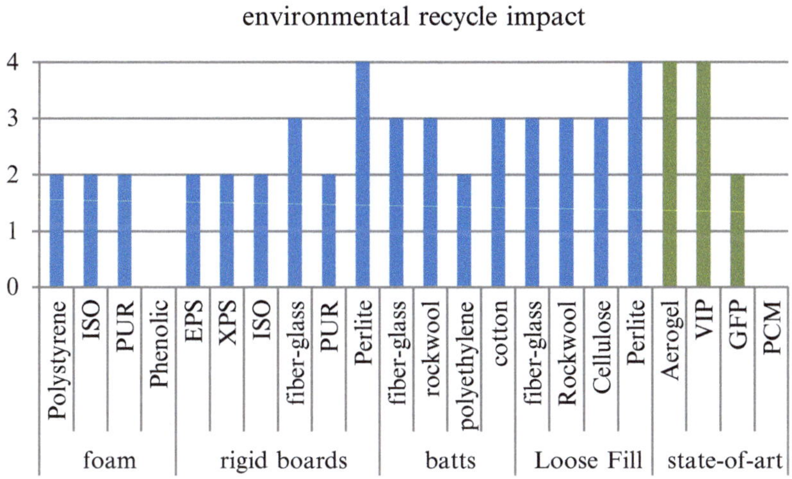

Fig. 15.7 Scoring results of the environmental recycle impact criteria

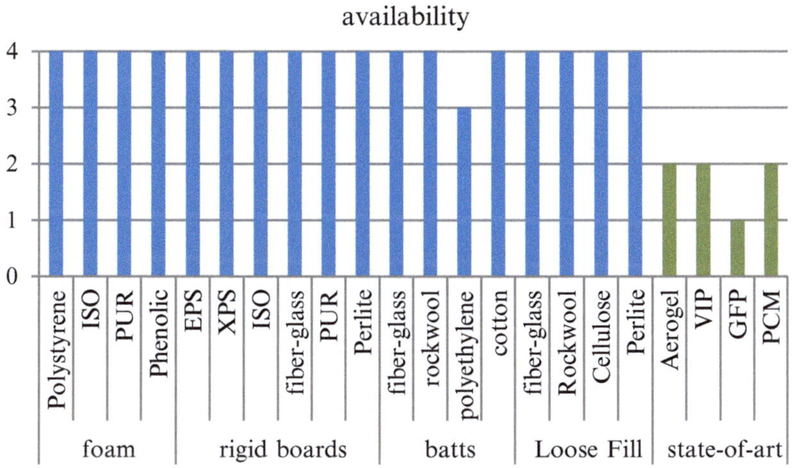

Fig. 15.8 Scoring results of the availability criteria

The negative recycle and dispose impacts are based on a similar scoring like the criterion above. Having 4 points means no special negative impact and easily disposable or even recyclable. This is only the case for Perlite, Aerogel, and VIP [7]. Fiber-glass, Rockwool, cotton, and cellulose insulations can be normally disposed without big impact on the environment [33]. The fire retarders are causing special treatment before disposing and this is why all flammable materials that include fire retarders are scored with only 2 points. However, there is no material which is absolutely not disposable or recyclable, so no material is scored with 1 point.

The last two criteria are evaluating the insulation materials more from an organizational perspective. The availability, Fig. 15.8, looks at the easiness of purchasing the materials and the different sources to order and purchase the insulation.

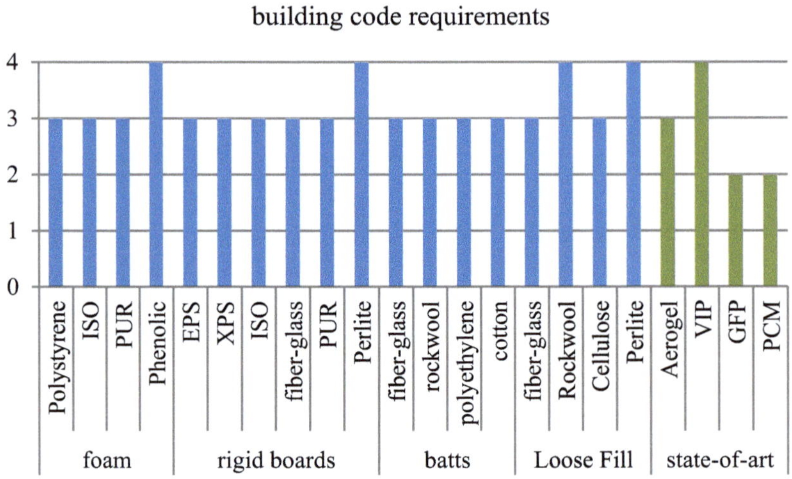

Fig. 15.9 Scoring results of the building code requirements criteria

Except of polyethylene batts, which require special order, all traditional materials are scored with 4 points because there are many companies which manufacture or trade with these materials and most of them can be purchased in building centers. This is pretty different with the state-of-the-art materials. There are only few companies which are offering these products and often times it also needs special construction companies that are able to use these technologies [34]. With GFP materials we were not able to find one company that offers GFP and it seems that this technology has not left the research state yet.

The last assessment criterion is the building code requirements, Fig. 15.9. The scoring in here is based on the fulfillment of the legal instructions and obligations. Phenolic, Perlite, Rockwool, and VIP are scored highest because these materials are all causing no skin and/or odor irritations as well as they are all free of health hazardous ingredients and they are also the most fire-resistant materials [12]. All other traditional materials are less fire resistant but are still good in all terms of health concerns and irritations. There are no legal restrictions known for Aerogel and VIP as well so they got the best score in here, too [7].

The scoring of the materials is complete and the next chapter will introduce our used HDM and show the results of the different experts.

15.3.4 Hierarchical Decision Model

HDM tool has been used to collect expert input on prioritizing the insulation technologies in construction sector. The decision is to find the insulation technology alternatives that have the highest potential contribution to overall objectives and goals of energy saving. In this section we will emphasize on what the HDM

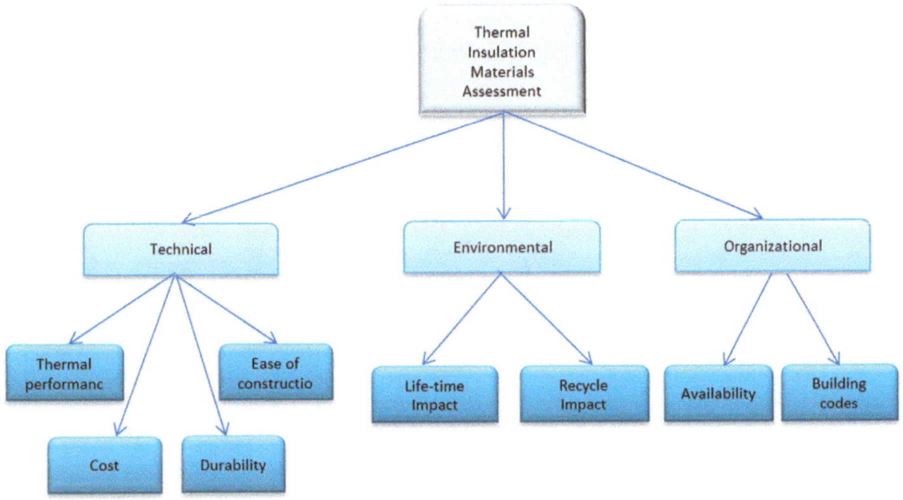

Fig. 15.10 Hierarchical decision model

outcome communicated to the project objective and criticize the inconsistencies and disagreements of the expert judgments by highlighting their effect on the final result. In this section all the data are collected and analyzed toward the final assessment. The evaluation of thermal insulation technologies included wide range of variables which has been identified by the researcher based on literature review. The HDM was developed and validated in multiple iterations before it was before it was finalized for submission to the identified experts. The team has contacted 16 experts in the field of construction insulation by e-mail. The e-mail included the HDM link to access the model, a brief explanation of our study purpose and expectations. Three experts replied by quantifying the HDM model based on their knowledge and experience in the field. The different criteria at all levels of the HDM clearly described within the model to define each of these terms and clarify the sub-aspect entailed in each term. One of the major strengths of the HDM is the use of pairwise comparisons to highlight accurate ratio scale priorities, as opposed to using traditional approaches of assigning weights which is difficult to justify. In this scenario a HDM has been delivered to evaluate the insulation quality criteria which are divided into three different aspects (technical, environmental, and organizational). The model was developed to examine the contribution of various insulation technology criteria to the mission of selecting the most important insulation features. The following is a graphical presentation of our HDM model, Fig. 15.10.

Typical HDM methods use the *mean* values to aggregate the opinions of the experts. The results will weight criteria with respect to each criterion. Other indictors to be considered in HDM analysis are the inconsistency and the disagreement rates. Inconsistency in the judgment is defined as the mean of the population standard deviation of each expert in each decision element. A conservative limit for acceptable inconsistency is 0.01. Disagreement is the rate of difference in viewpoint among experts. The disagreement in the group's subjective judgments is calculated

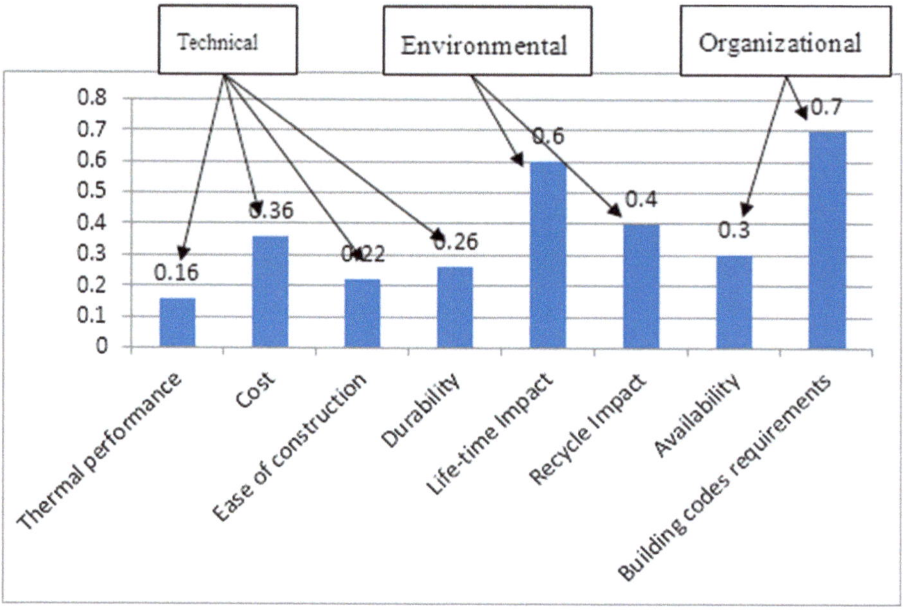

Fig. 15.11 Data captured from Expert 1

using the value assigned to each decision element and the mean of responses. A conservative limit for acceptable disagreement is 0.1.

For the purpose of this model each expert has quantified a certain level of the hierarchical model that fits the best to their expertise area. Expert 1 and Expert 3 have quantified the third level of the model, while Expert 2 has quantified the second level. The following is a graphical presentation of individual experts' judgment.

Expert 1: Figure 15.11, this expert evaluated at the third level. The pairwise comparison data for the technical criteria shows that cost got the highest at 0.36 followed by durability at 0.26. Ease of construction and thermal performance got 0.22 and 0.16, respectively. In the environmental criteria, life-time impact received 0.60 and recycle impact received 0.40. Last, in the organizational criteria, availability received 0.30 and building code requirement received 0.70.

Expert 2: Like Expert 1, this expert, Fig. 15.12, judgment contributed to the second level of the model where the researcher attempt to identify the three perspectives (technical, environmental, organizational) to which the different insulation technology criteria are categorized. It's noticeable from the figure that this expert has raised the importance of technical aspect above the other two perspectives. In his opinion environmental perspective participates the least in making decision when choosing insulation. Expert 2 result has a level of 0.02 inconsistencies.

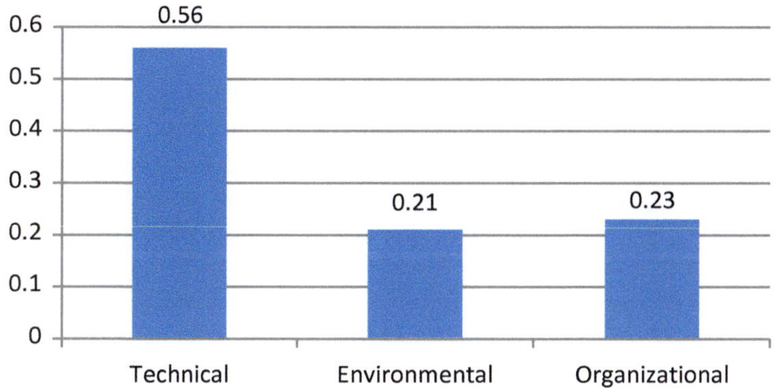

Fig. 15.12 Data captured from Expert 2

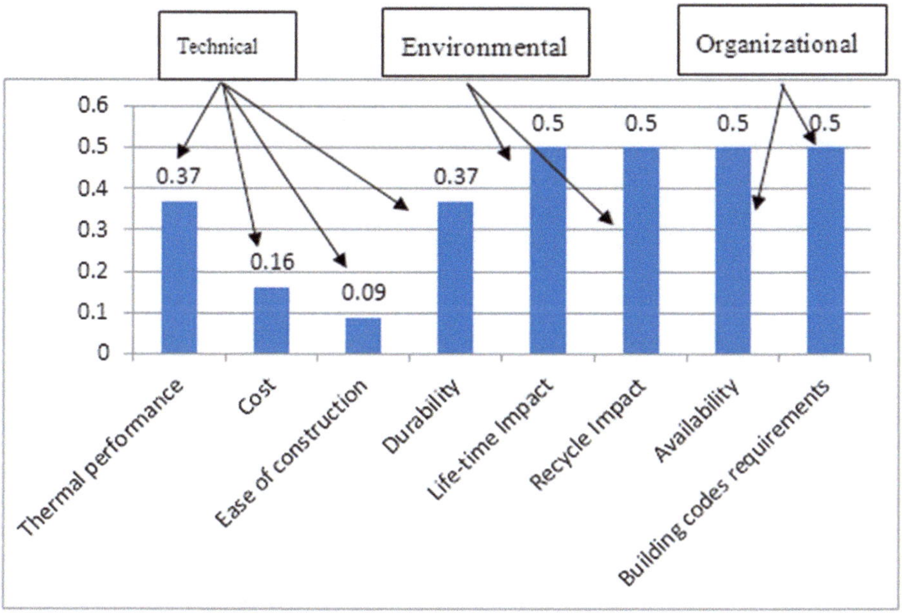

Fig. 15.13 Data captured from Expert 3

Expert 3: Figure 15.13, like Expert 1, this expert evaluated at the third level. The pairwise comparison data for the technical criteria shows that thermal performance and durability received 0.37 equally. Cost and Ease of construction got 0.16 and 0.09, respectively. In the environmental criteria, life-time impact and recycle impact received 0.50 each. Similarly, in the organizational criteria, availability and building code requirement received 0.50 each.

15.4 Analysis and Discussion

The presentation of the used methodology to assess thermal insulation technologies were mainly focused on the introduced criteria. The scoring results were showed separately for each criteria and the HDM helped to find the weight and the importance of the criteria. The following analysis of the scoring results will focus more on the insulation technologies and materials and it will combine the scoring results with the HDM data. The performance analysis of the state-of-the-art technologies is only focused on aerogel and VIP. VIP and GFP are pretty similar technologies and the scoring showed that VIP is superior to GFP. So we decided to exclude GFP of the further analysis. This is also supported by literature resources who argument in the same direction that future research should focus more on VIP than on GFP [7]. The technology of PCM is not included because PCM is not possible stand-alone insulation material. It has to be included in other insulation or construction material to be used as house insulation. So it is not completely matched to all the other technologies and materials and a comparison would lead to biased results.

To analyze the strengths and weaknesses of the state-of-the-art technologies they are compared with the traditional materials. There are criteria where Aerogel and VIP are scored better and there are others were the majority of the traditional materials perform better but most of the criteria show indifferent scores with no clear trend against or in favor of the use of state-of-the-art technologies. The following chart, Fig. 15.14, displays the scoring results for Aerogel and VIP as well as XPS and fiber-glass loose fill as two wide used insulation methods.

The scoring table shows that Aerogel and VIP are superior in thermal performance and they both are scored best in the environmental criteria. The weaknesses of the two state-of-the-art technologies are the availability and especially the cost crite-

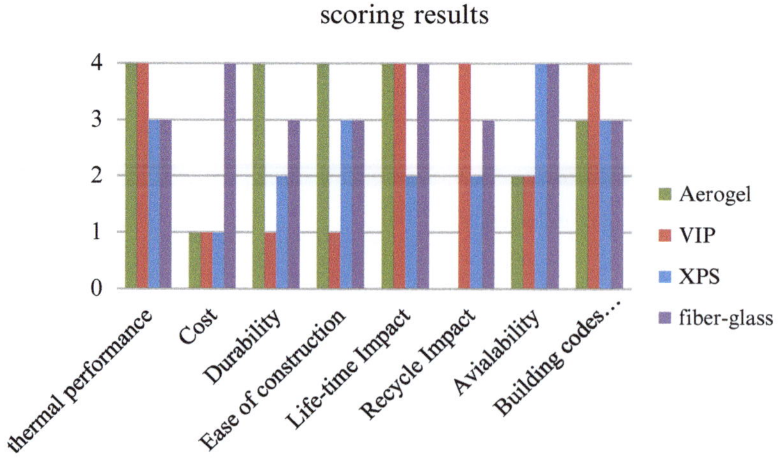

Fig. 15.14 Scoring results; comparing top two State-of-the-art with two wide used traditional materials

rion. The following HDM result analysis highlights the most important criteria to re-evaluate Aerogel and VIP with regard to the most important categories. This will help to define areas on which future research about Aerogel and VIP have to focus in order to be comparative or generally superior to the traditional types.

After collecting the inputs from the three experts, the team had to examine the validity of the result by observing the inconsistency and the disagreement level to make sure they are within the acceptable level. Although Expert 2 and Expert 3 have 0.02, 0.03 inconsistency in their judgment, the overall inconsistency level was within the limit for HDM restrictions, so there was no need to re-contact any of the expert for a revision. Disagreement of 0.00 was calculated among the three experts opinion. Two of the experts were highly agreed with each other in term of their preferable among model variables. While the third expert has slightly differentiated his input in term of what he think is the important consideration for selecting insulation technology. Although, the disagreement was within the acceptable level for HDM, the team tried to inspect the reason of the different evaluation among experts. The conclusion was that experts grounded their judgment by their own background. However combining the three experts input reflect more powerful result that beyond the individual preference. The composite input of the three experts is presented in the following chart, Fig. 15.15.

Durability of the insulation technology and material has been emphasized the most in the composite result of the three expert inputs. While both cost and thermal performance came at the second place. The three criteria mentioned are technical aspect of insulation technology. However the criteria that came at the third place in the composite result was the building code requirement which is categorized under

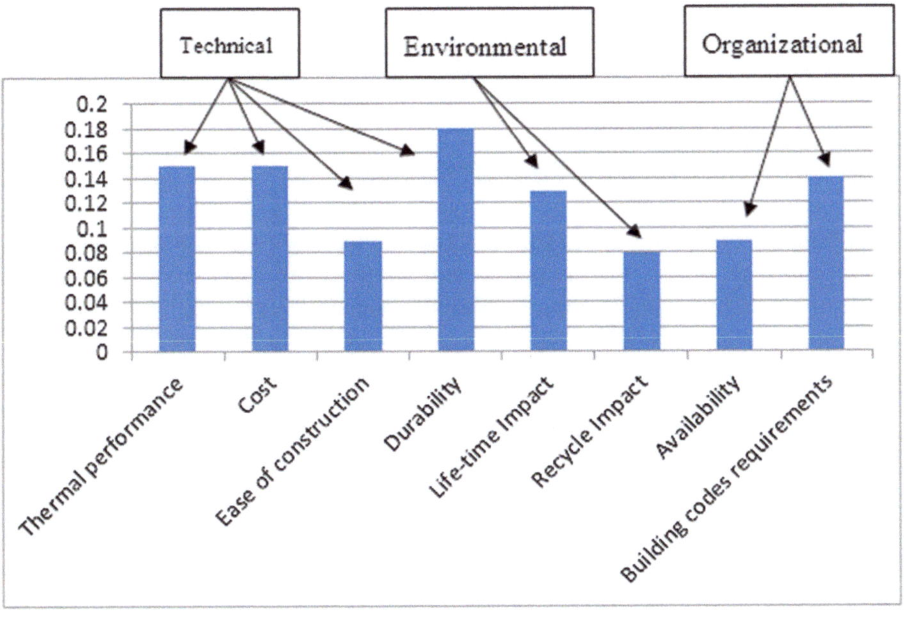

Fig. 15.15 HDM results (mean values)

Table 15.5 Scoring results of top 5 criteria per expert judgment from HDM results

Top five criteria	Aerogel	VIP	XPS	Fiber-glass (loose fill)
Thermal performance	4	4	3	3
Cost	1	1	1	4
Durability	4	1	2	3
Life-time impact	4	4	2	4
Building code requirement	3	4	3	3

the organizational perspective. Although one of the experts evaluated the criteria under environmental aspect as the most important among all others, none of the environmental criteria took place within the top three.

At this point of the study, the researchers were able to identify the available insulation technology and material and to present them in the Technology land-Scape model. As was presented in the landscape analysis section, the materials were categorized in two sections, traditional and state of art. Where traditional include all existing insulation for construction in today market, while the state of art included the insulation technology that still in developing phase under R&D lab. Using literature review, scoring model and expert opinion, the team were able to recognize the most two promising insulation technology from the state of art category as mention above. Considering HDM outcome, Table 15.5: Scoring results of top five criteria per expert judgment from HDM results, Aerogel and VIP occur to score highly at four out of the top five criteria according to expert judgment. However, the two of them scored very low from cost perspective, one of the top five important criteria. The following table shows the scores which Aerogel and VIP accomplish for thermal performance, cost, durability, life-time impact and building code requirement.

After analyzing the results of the scoring and the HDM, the next section will include research conclusions and recommendations for further R&D.

15.5 Conclusion and Recommendation

After the project objective was selected, the methodology for assessing insulation technologies and materials were performed. Starting with a gap analysis, it is used to identify the direction for future advanced insulation technologies. After that a landscape analysis was conducted to categorize the insulation materials which are already in the market and in a promising research phase for future applications. These two major groups of insulation materials were identified, traditional and state-of-the-art technologies. Further improvements for the state-of-the-art technologies at the conclusion of this research are recommended. These improvements will better place the state-of-the-art products in the market to compete with traditional ones in the home construction field. Therefore, two methods were used in parallel to assess the capabilities of all thermal insulation materials. One is scoring the insulation technologies based on the literature review. Then, it is followed with creating HDM based on expert judgments in each evaluation criteria that are the same as in scoring method.

From the result, the authors conclude that Aerogel will be the most promising insulation material for future use with the highest score in four out of five important criteria; durability, life-time impact, recycle impact and building code requirement. However, the analysis shows that cost is high for the state-of-the-art materials to compete with the traditional materials. Aerogel scored very low in this area even though this cost criterion is identified as important by the expert judgments. Similarly, VIP which scored better in environmental area, but the cost is the weakness of this technology. Both materials receive better score in thermal performance when compared with traditional technologies. However, the high cost of Aerogel and VIP are preventing them from further diffusion in the construction insulation market. Because of the cost factor, traditional technologies still have the advantage to be used in construction today.

Recommendations after assessing and analyzing construction insulation materials are for research and development to focus on reducing cost for the state-of-the-art materials. The main focus should be to improve manufacturing processes to lower the production costs especially for Aerogel and VIP. Both Aerogel and VIP proved to perform high as an insulation material. Improving the manufacturing process will help to reduce the production cost leading to better marketability and to become superior construction insulation materials. Further process innovations needed for these to technologies to establish cost-effective mass production procedures.

The large panel size of VIP and GFP as well as the challenge and experience needed to install them at the construction site are big disadvantages compared to other available insulation materials. This disadvantage requires R&D to improve ease of construction for VIP and GFP. As the study shown, these two products are hard to customize for installation on site because of potential gas or air leakage. Further R&D can focus on flexibility of these products for use in the construction field by the installer.

In addition if government and regulations gave higher incentives for the higher performance insulation materials, the state-of-art materials will have a greater advantage over the traditional materials. This will help these technologies to be widely used and be more available to the market.

The analysis in this report is useful for R&D in the insulation industry. The HDM results highlighted the critical criteria chosen by experts when selecting an insulation material. The landscape analysis and scoring results in this report can be used for homeowners and construction companies to choose their preferred insulation material for their specific need.

15.6 Future Research

The authors suggest areas for future research and studies in assessment of insulation materials.

- Further research to asses insulation materials using other methodologies
- Further research for additional government incentives
- Further research to include life-cycle cost calculations
- Comparison of market diffusion of other construction technologies and materials

Appendix

Scoring Table of Foam Materials

	Category	Traditional			
	Type	Foam			
	Materials	Polystyrene	Polyisocyanurate	PUR	Phenolic
Thermal performance	Min thermal conductivity	0.033	0.02	0.02	0.02
	Max thermal conductivity		0.028	0.038	0.025
	Average	0.033	0.024	0.029	0.0225
	Score	3	3	3	3
Cost	Cost per *R*-value [13]		High	High	
	Price per sq. ft for 1 in. thick				
	Score		1	1	
Durability	[13]	*R*-value decrease with time	*R*-value decrease with time	*R*-value decrease with time	
	Score	2	2	2	
Ease of construction	Advantages	Fills in all cavities; possible solution for existing buildings without wall insulation	Fills in all cavities	Fills in all cavities	Fills in all cavities
	Disadvantages	Needs drying time; needs professional installer: cannot control thickness; hard to control quality	Needs drying time; needs professional installer: cannot control thickness; hard to control quality	Needs drying time; needs professional installer: cannot control thickness; hard to control quality	Needs drying time; needs professional installer: cannot control thickness; hard to control quality
	Score	2	2	2	2

(continued)

(continued)

	Category	Traditional			
	Type	Foam			
	Materials	Polystyrene	Polyisocyanurate	PUR	Phenolic
Life-time impact		Includes brominated flame retardant HBCD (toxic) (hexabromo-cyclododecane) (included in all polystyrene insulations)			
	Score	3	3	3	4
Recycle impact		[27]	[27]	[27]	
	Score	2	2	2	
Availability					
	Score	4	4	4	4
Building codes requirements	Fire resistance [13]				
	Fire resistance				
	Health Hazardous				
	Odor/skin irritation				
	Score	3	3	3	4

Scoring Table of Rigid Boards Materials

	Category	Traditional					
	Type	Rigid boards					
	Materials	EPS	XPS	Polyisocyanurate	Fiber-glass rigid board	PUR boards	Perlite
Thermal performance	Min thermal conductivity	0.029	0.025	0.023	0.032	0.02	0.04
	Max thermal conductivity	0.045	0.037		0.063	0.03	0.06
	Average	0.037	0.031	0.023	0.0475	0.025	0.05
	Score	2	3	3	2	3	1
Cost	Cost per *R*-value [13]	Lowest for rigid board types	High	High	Medium		High
	Price per sq. ft for 1 in. thick	0.19	0.42	0.7			
	Score	3	1	1	3		1
Durability	[13]	*R*-value decrease with time	*R*-value decrease with time	*R*-value decrease with time	Better durability than fiber-glass batts	*R*-value decrease with time	High
	Score	2	2	2	3	2	4
Ease of construction	Advantages	Easily cuttable and adjustable on construction site	Easily cuttable and adjustable on construction site	Cuttable, but more difficult than polystyrene	Easily cuttable and adjustable on construction site	Easily cuttable and adjustable on construction site	
	Disadvantages	Fragile	Fragile				
	Score	3	3	3	3	3	2

Life-time impact		Uses pentane gas as the expanding agent, toxic; includes brominated flame retardant HBCD (toxic) (hexabromocyclododecane) (included in all polystyrene insulations)	Uses HCFC or CFC gases as the expanding agent, toxic fumes	Uses CO_2 or CFC gases as the expanding agent, toxic fumes	Quite safe, may be some out-gassing of resins used as binders	Serious health concerns and hazards in case of a fire	
		Environmental rating: A+ (best)	Environmental rating: with HFC: E (worst)				
	Score	3	2	3	3	3	0
Recycle impact		[27]	[27]	[27]		[27]	
	Score	2	2	2	3	2	4
Availability							
	Score	4	4	4	4	4	4
Building codes requirements	Fire resistance [13]						
	Fire resistance						
	Health Hazardous						
	Odor/skin irritation						
	Score	3	3	3	3	3	4

Scoring Table of Batts Materials

	Category	Traditional			
	Type	Batts			
	Materials	Fiber-glass	Rockwool	Polyethylene	Cotton
Thermal performance	Min thermal conductivity	0.033	0.037	0.041	0.029
	Max thermal conductivity	0.04			
	Average	0.0365	0.037	0.041	0.029
	Score	2	2	2	3
Cost	Cost per *R*-value [13]	Low	Low	Low	Low
	Price per sq. ft for 1 in. thick	0.055–0.085			0.0625–0.09775
	Score	4	4	4	4
Durability	[13]	Compression reduces *R*-value	Compression reduces *R*-value	*R*-value decrease with time	*R*-value can change over time: can be significantly lower due to typically deficient installation
	Score	3	3	2	2
Ease of construction	Advantages	Fitted between studs, joists or rafters [15], No settling, No dry time require [4]; easy to replace	Fitted between studs, joists or rafters [15]		No settling, No drying time require, DIY but need motorized cutting tool [4]
	Disadvantages	Protection glasses and gloves required for cutting	Protection glasses and gloves required for cutting	Difficult to handle and cut with standard tools	Must be properly fitted to completely fill the wall without being compressed by pipes or wires
	Score	4	4	4	4
Life-time Impact		4.5	4.5	Made from recycled plastic milk bottles	0.5
	Score	4	4	4	4

(continued)

(continued)

	Category	Traditional			
	Type	Batts			
	Materials	Fiber-glass	Rockwool	Polyethylene	Cotton
Recycle impact		Recycle content [24]	Recycle content [24]	Hard to dispose because of fire retarders	Be recycled or composed
	Score	3	3	2	3
Availability				Special order required	
	Score	4	4	3	4
Building codes requirements	Fire resistance [13]	Good	Excellent	Poor	
	Fire resistance	–4–260 °C [13] melting at 1,300 °F (704 °C) [30]	–240–800 °C [13], melt at 2,150 °F (1,177 °C) [30]	–40–90 °C [13], doesn't burn readily, melt when expose to flame	Flammable, must be treat with fire retardant [4]
	Health hazardous	Formaldehyde binders		Organic (off-gassing, toxic smoke) [13], treat with fire retardant	nontoxic (the same low-toxicity and biodegradable flame retardant and insect/rodent repellent used in cellulose insulation and infant clothing) [28]
	Odor/skin irritation	Inorganic, Irritating dust during installation [13]	Inorganic, Irritating dust during installation [13]	Non-irritating to work with	Can install it without using respiratory or skin exposure protection
	Score	3	3	4	3

Scoring Table of Loose Fill Materials

	Category	Traditional			
	Type	Loose fill			
	Materials	Fiberglass (open cell structure)	Rockwool (open cell structure)	Cellulose	Perlite
Thermal performance	Min thermal conductivity	0.03	0.04	0.046	0.04
	Max thermal conductivity	0.038		0.054	0.06
	Average	0.034	0.04	0.05	0.05
	Score	3	2	2	2
Cost	Cost per *R*-value [13]	Low	Low	Low	High
	Price per sq. ft for 1 in. thick	–0.48		–0.85	
	Score	4	4	4	1
Durability	[13]	Compression and moisture degrade *R*-value	Compression and moisture degrade *R*-value	Compression and moisture degrade *R*-value	Good
	Score	3	3	3	4
Ease of construction	Advantages	Easy for the attic; well suited for places where it is difficult to install other types; generally fast to install	Easy for the attic; well suited for places where it is difficult to install other types; generally fast to install	Easy for the attic; well suited for places where it is difficult to install other types; generally fast to install	Can be used and included in concrete
	Disadvantages	Settles after time if used in vertical applications; true *R*-value Depends on quality of workmanship, amount of installation material; need special equipment and professional worker	Settles after time if used in vertical applications; true *R*-value depends on quality of Workmanship, amount of installation material; need special equipment and professional worker	True *R*-value depends on quality of workmanship, amount of Installation material and moisture content; needs drying time if sprayed wet	Limited use mostly between bricks
	Score	3	3	3	2

(continued)

(continued)

	Category	Traditional			
	Type	Loose fill			
	Materials	Fiberglass (open cell structure)	Rockwool (open cell structure)	Cellulose	Perlite
Life-time impact				0.25	
	Score	4	4	4	4
Recycle impact					Made out of rock-disposable
	Score	3	3	3	4
Availability					
	Score	4	4	4	4
Building codes requirements	Fire resistance [13]	Very good	Excellent	Very good	Excellent
	Fire resistance				
	Health Hazardous			Add fire resisting chemical	
	Odor/skin irritation			Produces lower dust during installation	
	Score	3	4	3	4

Scoring Table of State-of-the-Art Materials

	Category	State-of-the-art			
	Type				
	Materials	Aerogel	VIP	GFP	PCM
Thermal performance	Min thermal conductivity	0.013	0.004	0.01	
	Max thermal conductivity			0.046	
	Average	0.013	0.004	0.028	#DIV/0!
	Score	4	4	3	
Cost	Cost per *R*-value [13]				
	Price per sq. ft for 1 in. thick				
	Score	1	1	1	1

(continued)

(continued)

	Category	State-of-the-art			
	Type				
	Materials	Aerogel	VIP	GFP	PCM
Durability	[13]	Offers constant design performance, no aging effects	Loss of vacuum over time; easily damageable by daily activity	Potential gas loss; easily damageable by daily activity	High because included in the wall material
	Score	4	1	1	4
Ease of construction	Advantages	Wide range of building application; fast to install in new buildings: ease of maintenance			If embedded in the construction wall material (bricks, concrete, sheetrock) no extra insulation is needed); can be added to other insulation to improve their thermal Performance
	Disadvantages		Need professional installation workers, not adjustable on construction site	Need professional installation workers, not adjustable on construction site	No stand-alone insulation
	Score	4	1	1	-
Life-time impact		Free of toxic ingredients		Depends on used gas	
	Score	4	4	2	
Recycle impact		Free of toxic ingredients	No toxic materials in it	No toxic materials in it	Embedded I other materials- hard to recycle; disposal?
	Score	4	4	4	
Availability					
	Score	2	2	1	2
Building codes requirements	Fire resistance [13]				
	Fire resistance				
	Health Hazardous				
	Odor/skin irritation				
	Score	3	4	2	1

References

1. Center for Climate and Energy Solutions, Building overview, PEW Center (2009)
2. Department of Industry, Insulation (2013), http://www.livinggreener.gov.au/energy/heating-cooling/insulation#why_is_insulation_important. Accessed 1 Dec 2013
3. U.S. Department of Energy, Today in Energy, Heating and cooling no longer majority of U.S. home energy use (2013), http://www.eia.gov/todayinenergy/detail.cfm?id=10271. Accessed 1 Dec 2013
4. Mind tools, Gap analysis; Identifying what needs to be done in a project (2013), http://www.mindtools.com/pages/article/gap-analysis.htm. Accessed 26 Oct 2013
5. Springer Gabler, Cost-benefit analysis (2013), http://wirtschaftslexikon.gabler.de/Archiv/4761/nutzwertanalyse-v11.html. Accessed 3 Dec 2013
6. Oak Ridge National Laboratory, *Insulation Fact Sheeet* (The Building Technology Program Office of Energy Efficiency and Renewable Energy, Washington, DC, 2008)
7. B.P. Jelle, Traditional, state-of-the-art and future thermal building insulation materials and solutions-properties, requirements and possibilities. Energy Build. **43**, 2549–2563 (2011)
8. Homewyse, Cost to install wall insulation-2013 cost calculator (2013), http://www.homewyse.com/services/cost_to_install_wall_insulation.html. Accessed 25 Oct 2013
9. Homewyse, Cost to insulate an attic-2013 cost calculator (2013), http://www.homewyse.com/services/cost_to_insulate_attic.html. Accessed 25 Oct 2013
10. U.S. Department of Energy, Tips: insulation (2012), http://energy.gov/energysaver/articles/tips-insulation. Accessed 26 Oct 2013
11. U.S. Environmental Protection Agency, Properly installed insulation, Energy Star, pp. 1–2.
12. M.S. Al-Homoud, Performance characteristics and practical applications of common building thermal insulation materials. Build. Environ. **40**, 353–366 (2005)
13. U.S. Department of Energy, Types of insulation (2012), http://energy.gov/energysaver/articles/types-insulation. Accessed 26 Oct 2013
14. North America Insulation Manufactures Association, IRS energy efficiency tax credit for homeowners (2013), http://www.naima.org/insulation-resources/tax-rebates-incentives-residential/irs-energy-efficiency-tax-credit-for-homeowners.html. Accessed 27 Oct 2013
15. A. Papadopoulos, State of the art in thermal insulation materials and aims for future developments. Energy Build. **37**, 77–86 (2005)
16. North America Insulation Manufactures Association, *Building Insulation; A Performance Comparison for Today's Environmental Home Builder* (NAIMA, Alexandria, 2009)
17. American Chemistry Council, Inc., How does spray polyurethane foam (SPF) insulation work? (2013), http://www.spraypolyurethane.org/Main-Menu-Category/Consumers/SPF-Insulation-Basics. Accessed 28 Nov 2013
18. "Insulation," Insulation, pp. 101–108.
19. Huntsman, Polyurethanes Insulation for construction(Huntsman, 2007).
20. Lowe's, Blown insulation R value calculator (2013), http://www.lowes.com/cd_Blown+In+R+Value+Insulation+Calculator_477243886_. Accessed 21 Nov 2013
21. Perlite P15 Medium aluminum aggregate. Building Design, http://www.buildingdesignindex.co.uk/browse/category/1-40/type/11/Building-materials/Companies. Accessed 21 Nov 2013
22. B. Meinhold, Space-age aerogel insulation now cheaper and available in sheets (2010), http://inhabitat.com/exciting-advances-in-insulation-with-aerogel/. Accessed 20 Nov 2013
23. H. Quinlan, How aerogels work (2013), http://www.howstuffworks.com/aerogel.htm. Accessed 20 Nov 2013
24. B. Dodson, Flexible, high-strength polymer aerogels deliver "super-insulation" properties (2012), http://www.gizmag.com/polymer-aerogel-stronger-flexible-nasa/23955/. Accessed 24 Nov 2013
25. M. Alam, H. Singh, M.C. Limbachiya, Vacuum insulation panels (VIPs) for building construction industry—a review of the contemporary developments and future directions. Appl. Energy **88**, 3592–3602 (2011)

26. R. Baetensa, B.P. Jelle, A. Gustavsen, S. Grynning, Gas-filled panels for building applications: A state-of-the-art review. Energy Build. **42**, 1969–1975 (2010)
27. Insulla.com, Products; How do phase change materials work? (2011), http://www.insulla.com/thermal_insulation_tiles_pcm_work.php. Accessed 21 Nov 2013
28. Energy Saving Trust, *Insulation materials chart thermal properties and environmental ratings* (Energy Saving Trust, London, 2010)
29. C. Castellón, M. Medrano, J. Roca, M. Nogués, A. Castell, L.F. Cabeza, *Use of Microencapsulated Phase Change Materials in Building Applications* (ASHRAE, 2007), pp. 1–6
30. A. Papadopoulos, E. Giama, Environmental performance evaluation of thermal insulation materials and its impact on the building. Build. Environ. **42**, 2178–2187 (2007)
31. R. Baetensa, B.P. Jelle, A. Gustavsend, Aerogel insulation for building applications: A state-of-the-art review. Energy Build. **43**, 761–769 (2011)
32. B. T. Griffith, D. Arasteh, D. Türler, Gas-filled panels: an update on applications in the building thermal envelope (1995), pp. 1–14
33. H. Ulrich, A. Odinak, B. TUCKER, A. Sayigh, Recycling of polyurethane and polyisocyanurate foam. Polym. Eng. Sci. **18**(11), 844–848 (1978)
34. Aspen Aerogel, Ultra Thin Spaceloft Insulation (2013), pp. 1–6

MIX
Papier aus verantwortungsvollen Quellen
Paper from responsible sources
FSC® C105338

If you have any concerns about our products,
you can contact us on
ProductSafety@springernature.com

In case Publisher is established outside the EU,
the EU authorized representative is:
Springer Nature Customer Service Center GmbH
Europaplatz 3, 69115 Heidelberg, Germany

Printed by Libri Plureos GmbH
in Hamburg, Germany